de Gruyter Studies in Mathematics 2

Michel Métivier

Semimartingales

a Course on Stochastic Processes

W DE G

Walter de Gruyter · Berlin · New York · 1982

Author:

Michel Métivier
Professor at the Ecole Polytechnique
Centre de Mathematiques Appliquées
Palaiseau Cedex

Library of Congress Cataloging in Publication Data

Métivier, Michel, 1931–
Semimartingales : a course on stochastic processes.
(De Gruyter studies in mathematics ; 2)
Bibliography: p.
Includes indexes.
1. Stochastic processes. I. Title. II. Series.
QA274.M47 1982 519.2'87 82–17986
ISBN 3-11-008674-3

CIP-Kurztitelaufnahme der Deutschen Bibliothek

Métivier, Michel:
Semimartingales : a course on stochast. processes
Michel Métivier. – Berlin ; New York : de
Gruyter, 1982.
(De Gruyter studies in mathematics ; 2)
ISBN 3-11-008674-3
NE: GT

 Typesetting and Printing: Fa. Tutte, Salzweg. – Binding: Lüderitz & Bauer Buchgewerbe GmbH, Berlin. – Cover design: Rudolf Hübler, Berlin. **Printed in Germany.**

Preface

This book has its origin in courses given by the author in Erlangen in 1976, in lectures given in Berkeley during the summer 1979 and in a course in München in the second semester of 1980.

Until recently, many important results in the general theory of stochastic processes, in particular those developed by the "Strasbourgschool", were considered by many probalists as devices only for specialists in the field. It turns out, however, that the growing interest for non-Markovian processes and point processes, for example, because of their importance in modelling complex systems, makes it more and more important for "non-specialists" to be acquainted with concepts such as martingales, semimartingales, predictable projection, stochastic integrals with respect to semimartingales, etc.

By chance, the mathematical thinking in the ten past years has produced not only new and sophisticated results but makes it possible to present in a quite concise way a corpus of basic notions and tools, which may be regarded as essential for what is, after all, the goal of many: the description of stochastic systems, the ability to study their behaviour and the possibility of writing formulas and computational algorithms to evaluate and identify them (without mentioning their optimization!).

Over the years, the description of stochastic processes was based on the consideration of moments and in particular covariance. A more modern trend is to give a "dynamical" description based on the consideration of the evolution law of the processes. This is perfectly appropriate to the study of Markov processes. In this case the "dynamical structure" of the process leads to equations providing users with formulas and equations to describe and compute its evolution. But more generally one may give a "dynamical description" of a process, Markovian or not, by considering its relation with an increasing family of σ-algebras $(\mathscr{F}_t)_{t \in \mathbb{R}^+}$ of events, where $\mathscr{F}_t$ expresses the information theoretically available until time t. The notion of generator of a Markov process has, in the case of non-Markovian processes, a kind of substitute, which may be expressed in terms of a "Dual predictable projection". In this general setting, the notions of martingales, semimartingales, stopping times and predictability play a fundamental role. Stochastic equations are also appropriate tools for describing general stochastic systems and the stochastic calculus cannot be developed without the same notions of martingales, semimartingales, predictability and stopping times.

The purpose of this book is precisely to present these fundamental concepts in their full force in a rather concise way and to show, through exercises and paragraphs devoted to applications, what they are useful for.

The first part of the book which contains chaps. 1 to 4, is devoted to the exposition of fundamental notions and results for processes. Chapter 1, after giving examples of processes, deals mainly with the two basic concepts of stopping times and predictability. Chapter 2 gives a quite extensive treatment of martingales and quasimartingales as to their regularity and convergence properties. Applications to the study of some stochastic algorithms give some indication of the power of this tool in the study of asymptotic properties. Chapter 3 studies the properties of processes admitting a σ-additive Doleans measure. This point of view provides a general and synthetic understanding of many properties and particularly those tied with Dellacherie's predictable projection and dual predictable projection. The Doob-Meyer decomposition theorem finds here its more natural and easy exposition. We are also able to give the version of this theorem for Banach-valued processes. The introduction of pointprocesses as measure-valued processes at the end of this chapter, is an illustration of the interest of such an extension.

Chapter 4 concludes this part of the book with the study of the Hilbert space of square integrable martingales. This is in some sense the "modern dynamic analogue" of the classical theory of second order processes (in the non-stationary situation). Semimartingales are then introduced as the sum of a process which is "locally" a square integrable martingale and a process with paths of finite variation. A recent characterization of these processes through a quite useful domination property closes the chapter. The way is thus opened to stochastic calculus in chapters 5 and 6.

The second part is actually devoted to stochastic calculus.

Chapter 5 constructs the stochastic integral with respect to a semimartingale. This integral, which is a mapping from processes into processes, is defined as a continuous extension of the integral of elementary predictable processes for seminorms associated with the control processes of the semimartingale. The main properties of the stochastic integral are immediate from this definition. As in the previous chapters, we consider separately the real case and the vector case.

The transformation formula, which is the core of stochastic calculus, is then proved in its various forms.

Chapter 6 presents the most classical and elementary applications of the transformation formula. It introduces typical "martingale problems" and questions of absolute continuity.

The stochastic integral with respect to point processes (more generally, random measures) is studied in chapter 7. The consideration of the point process of jumps of a semimartingale leads naturally to the concepts of local characteristics of a semimartingale. Martingale problems are introduced in their general form at the end of this chapter.

Stochastic equations are the subject of chapter 8. The equations considered are generalizations of the Ito-Skorokhod equations, where the driving terms are general semimartingales and "white random measures". Strong solutions are proved to exist and to be unique under a quite general Lipschitz-type hypothesis. Non-explosion criteria are given. A pathwise regularity for strong solutions depending on a parameter is proved. This chapter and the book close on a short introduction to the problem of weak solutions of a stochastic equation in simple cases.

It is the intention of this book to provide the reader with many examples as close as possible to situations of practical interest and help him through a bibliography covering many extensions and technical developments. To the first goal correspond the numerous exercises and, to the second, a few bibliographical notes at the end of the chapters.

Acknowledgements:

The author wishes to express his gratitude to Professors H. Bauer, P. Gabriel and W. Schuder who accepted this book in the new collection edited by de Gruyter. He feels much indebted to many people who helped very much during the preparation of the manuscript: specially Professors E. Wong and H. Kellerer who, in Berkeley and München respectively, offered opportunities of delivering advanced courses on the subject. Thanks are due to many colleagues who expressed valuable criticism on successive drafts and encouragements. It is my pleasure to mention in a very special way Professors K. L. Chung, J. Jacod, J. Pellaumail, C. Dellacherie, P. A. Meyer, G. Letta, G. Da Pratto, J. Groh and S. Orey. A special mention is deserved by Mrs J. Bailleul who did most of the successive typings of the manuscript. My thanks are finally due to the staff at de Gruyter for displaying a lot of patience in our final cooperation.

Paris, July 1982 Michel Métivier

Contents

Part I: Martingales - Quasimartingales - Semimartingales

Part 1:

Martingales – Quasimartingales – Semimartingales

Chapter 1

Basic notions on stochastic processes

The aim of this chapter is to give the minimal account of stochastic processes necessary to deal with quasimartingales and their applications.

The notions presented here are developed in the books by Meyer and Dellacherie [Mey 1] [Del 1] (DeM 1]. In some sense, what follows is our own self-contained abbreviated account and is extremely close to our presentation in [Met 3].

1. Stochastic basis – Stochastic processes

Stochastic basis

A *stochastic basis* is a quadruple $(\Omega, (\mathscr{F}_t)_{t \in T}, \mathfrak{A}, P)$ where Ω is a set, T a subset of $\mathbb{R}^+$, $\mathfrak{A}$ a σ-algebra of subsets of Ω, $(\mathscr{F}_t)_{t \in T}$ an increasing family of sub-σ-algebras of $\mathfrak{A}$ (increasing means: $s \leqslant t \in T \Rightarrow \mathscr{F}_s \subset \mathscr{F}_t$), and P a probability on $(\Omega, \mathfrak{A})$.

When $\mathfrak{A}$ is not specified, it is assumed that $\mathfrak{A}$ equals the σ-algebra generated by $\bigcup_{t \in T} \mathscr{F}_t$, denoted by $\bigvee_{t \in T} \mathscr{F}_t$. When $T = \mathbb{R}^+$, the σ-algebra $\bigvee_{t \in \mathbb{R}^+} \mathscr{F}_t$ is denoted by $\mathscr{F}_\infty$.

The family $(\mathscr{F}_t)_{t \in T}$ is often called a *filtration* of Ω. When the filtration is clear, we will abbreviate and write $(\mathscr{F})$ or $\mathscr{F}.$ instead of $(\mathscr{F}_t)_{t \in T}$. The "physical" meaning of $\mathscr{F}_t$ is the following. $\mathscr{F}_t$ ist the σ-algebra of events occuring up to time t – the "past events up to t".

With the family $(\mathscr{F}_t)_{t \in \mathbb{R}^+}$ are associated the following families $(\mathscr{F}_{t^+})_{t \in \mathbb{R}^+}$ and $(\mathscr{F}_{t^-})_{t \in \mathbb{R}^+}$ of σ-algebras.

$$\mathscr{F}_{t^+} := \bigcap_{s > t} \mathscr{F}_s ;$$

$$\mathscr{F}_{t^-} := \bigvee_{s < t} \mathscr{F}_s \text{ (the } \sigma\text{-algebra which is generated by } \bigcup_{s<t} \mathscr{F}_s).$$

For $t = 0$, we set $\mathscr{F}_{0^-} = \mathscr{F}_0$.

1.1 Definition. The family $(\mathscr{F}_t)_{t \in \mathbb{R}^+}$ is said to be *right-continuous* if for every t, $\mathscr{F}_t = \mathscr{F}_{t^+}$.

The basis $(\Omega, \mathfrak{A}, (\mathscr{F}_t)_{t \in \mathbb{R}^+}, P)$ is said to fulfill the "*usual hypotheses*" when the family

$(\mathscr{F}_t)_{t\in\mathbb{R}^+}$ is right-continuous and every set F which belongs to the P-completion of the σ-algebra $\mathscr{F}_\infty$ with $P(F)=0$ belongs to every $\mathscr{F}_t$, $t\in\mathbb{R}^+$.

Stochastic processes

In applications, the random evolution of a system is described by saying that the "state of the system" is a "function of time and randomness'. This leads to the following definition.

1.2 Defintion. Let $(\mathfrak{x},\mathscr{B})$ be a measurable space. A *stochastic process* on the probability space $(\Omega,\mathfrak{A},P)$ with state-space $(\mathfrak{x},\mathscr{B})$ and index set T is a mapping from $(T\times\Omega)$ into $\mathfrak{x}$ such that, for every $t\in T$, $\omega\curvearrowright X(t,\omega)$ is $\mathfrak{A}$-measurable.

The mappings $X(.,\omega): t\curvearrowright X(t,\omega)$, $\omega\in\Omega$, are called the *paths* of X.

Notation: The random variable $\omega\curvearrowright X(t,\omega)$ is often denoted by X_t, and the process itself by $(X_t: t\in T)$.

The mapping $\omega\curvearrowright X(.,\omega)$ is the *random function* of the process.

We give examples of processes in the following paragraph.

1.3 Definition. A stochastic process X with state-space a vector space $\mathfrak{x}$ is said to be *evanescent* or P-*null* if $P(\{\omega:\exists t\in T, X(t,\omega)\neq 0\})=0$.

A subset A of $T\times\Omega$ is said to be *evanescent* if 1_A is an evanescent process.

Two processes X and Y with the same state-space and index set T are said to be *indistinguishable* or *P-equal* if the set $\{X\neq Y\}\subset T\times\Omega$ is evanescent.

Measurability and regularity properties of processes

1.4 Defintion. Let $(\Omega,\mathfrak{A}(\mathscr{F}_t)_{t\in T^+},P)$ be a stochastic basis and X a process on $(\Omega,\mathfrak{A})$ with index set $T\subset\mathbb{R}^+$ and state space $(\mathfrak{x},\mathscr{B})$. The process X is said to be

- *adapted* (to the filtration $(\mathscr{F})$) if, for every $t\in T$, the mapping $\omega\curvearrowright X(t,\omega)$ is $\mathscr{F}_t$-measurable (one sometimes says *non-anticipating*);
- *measurable* if the mapping X is measurable from $(T\times\Omega,\ \mathscr{B}_T\otimes\mathfrak{A})$ into $(\mathfrak{x},\mathscr{B})$, where $\mathscr{B}_T$ denotes the Borel σ-algebra of T:
- *progressively measurable* if, for every $t\in T$, the mapping X restricted to $[0,t]\cap T\times\Omega$ is $\mathscr{B}_{[0,t]}\times\mathscr{F}_t$-measurable, where $\mathscr{B}_{[0,t]}$ denotes the Borel σ-algebra of subsets of $[0,t]\cap T$.

A subset A of $\mathbb{R}^+\times\Omega$ is called *adapted* (resp. *measurable, progressively measurable*) if the function 1_A is an adapted (resp. measurable, progressively measurable) process.

It is immediately seen from the definition that the adapted (resp. measurable, progressively measurable) sets form a σ-algebra $\mathscr{M}_0$ (resp. $\mathscr{M}$, $\mathscr{M}_1$), and that a process X is adapted (resp. measurable, progressively measurable) if it is $\mathcal{M}_0$-measurable (resp. $\mathcal{M}$-measurable, $\mathcal{M}_1$-measurable).

It is clear that $\mathcal{M}_0 \supset \mathcal{M}_1$.

1.5 Definition. Let $\mathfrak{x}$ be a topological space. A process X with state-space $\mathfrak{x}$ and index set $\mathbb{R}^+$ will be said *regular* when X is adapted and all its paths have left and right limits for every $t \in \mathbb{R}^+$.

A process X is called *right-continuous* (resp. *left-continuous*) when its paths are right-continuous (resp. left-continuous) functions.

We will use the abbreviation R. R. C. (resp. R. L. C.) to mean *regular right-continuous* (resp. *regular left-continuous*). We have the following.

1.6 Theorem. *Every adapted right-continuous (therefore, every R. R. C.) process, and every adapted left-continuous (therefore, every R. L. C.) process are progressively measurable.*

Proof: We consider the process X on $[0, s] \times \Omega$. For every $n \in \mathbb{N}$, we define

$$X_1^n(t, \omega) := \sum_{k=0}^{k=2^n-1} 1_{\left[\frac{ks}{2^n}, \frac{(k+1)s}{2^n}\right[}(t) X_{ks/2^n}(\omega)$$

$$X_2^n(t, \omega) := 1_{\left[0, \frac{s}{2^n}\right]}(t) X_0(\omega) + \sum_{k>0} 1_{\left]\frac{ks}{2^n}, \frac{(k+1)s}{2^n}\right]}(t) \cdot X_{(k+1)s/2^n}(\omega).$$

Since X is adapted, the processes X_1^n and X_2^n are clearly $\mathcal{B}_{[0,s]} \otimes \mathcal{F}_s$-measurable on $[0, s] \times \Omega$. If X is right-continuous (resp. left-continuous), the sequence $(X_2^n)_{n \in \mathbb{N}}$ (resp. $(X_1^n)_{n \in \mathbb{N}}$) converges on $[0, s] \times \Omega$ towards X Therefore, X is progressively measurable. □

2. Examples and construction of stochastic processes

In this paragraph, we recall the definitions of two basic processes, namely, the Brownian and the Poisson processes. We will also describe the way in which most stochastic processes appear in a wide range of applications.

2.1 Brownian motion with respect to a given Stochastic Basis

Let X be a process with state space $\mathfrak{x} := \mathbb{R}^n$. We call it an n-dimensional Brownian motion with respect to the given stochastic basis if it has the following properties.

(i) X has *increments independent of the past:* For every $s < t$, the random variable (increment) $X_t - X_s$ is independent of the σ-algebra $\mathcal{F}_s$.

(ii) For every $s < t$, $X_t - X_s$ is a Gaussian random variable with mean zero and variance matrix $(t - s)C$, where C is a given matrix.

The Brownian motion is said to start at x if $X_0 = x$ a. s.

The existence of such a mathematical object is easily proved as follows. We take $\Omega := \mathfrak{x}^{\mathbb{R}^+}$ and define X_t as the canonical projection of $\mathfrak{x}^{\mathbb{R}^+}$ onto the "$t-th$ coordinate"; i. e., $X_t(\omega) := \omega(t)$, where ω is any function from $\mathbb{R}^+$ into $\mathfrak{x}$.

We write $\mathscr{F}_t$ for the σ-algebra generated by $X_u : u \leq t$. The hypothesis about the increments shows that P is entirely defined on $\mathscr{F}_\infty$ by the iterated integrals

$$(2.1.1) \qquad \int \varphi(X_0, X_{t_1} \dots X_{t_n}) dP = \int (v_{t_1} * \varepsilon_x) dy_1 \int (v_{t_2 - t_1} * \varepsilon_{y_1}) dy_2 \dots \int \varphi(x, y_1 \dots, y_n)(v_{t_n - t_{n-1}} * \varepsilon_{y_{n-1}}) dy_n\,, \quad ^{1)}$$

where $t_1 < \dots < t_n$ ranges over all finite subsets of $]0, \infty[$, φ is any bounded Borel function on $\mathfrak{x}^{n+1}$ and v_h is the Gaussian probability on $\mathfrak{x}$ with mean zero and covariance h. C.

It is a theorem of measure theory (Kolmogorov's theorem: see [Nev 1] chap. 3, [Met 6] chap. 6) that a probability P is uniquely determined on $(\Omega, \mathscr{F}_\infty)$ by formula (2.1.1).

2.2 The canonical Brownian motion

If it is clear from the construction that the paths of the above defined Brownian motion $(\Omega, (\mathscr{F}_t), (X_t), P_x)$ starting from x have no regularity property. It can, however, be proved (see [KrP] [Nev 1], or exercises E. 1 and E. 2) that the outer measure $P_x^*(\Omega_c)$ of the set Ω_c of continuous paths ω is one.

Therefore, one defines a continuous Brownian motion by taking the "restriction" μ_x of P_x^* to $(\Omega_c, \mathscr{F}_\infty \cap \Omega_c)$ and defining $X_t(\omega) := \omega(t)$.

This Brownian motion is called the *canonical Brownian motion* with starting point x.

The corresponding canonical stochastic basis is $(\Omega_c, (\mathscr{C}_t)_{t \in \mathbb{R}^+}, \mathscr{C}_\infty, \mu_x)$, where $\mathscr{C}_t = \mathscr{F}_t \cap \Omega_c$.

When C is the identity matrix, the probability μ_0 is often referred to, as the *Wiener measure*.

2.3 Poisson process

A Poisson process with parameter $\lambda > 0$ is a process N with the set $\mathbb{N}$ of integers as state space and with the following properties.

(i) N has *increments independent of the past*.

(ii) For every $s < t$, $N_t - N_s$ is a Poisson random variable with parameter $(t-s)\lambda$ $\left(\text{i. e. } P\{N_t - N_s = k\} = e^{-\lambda(t-s)} \dfrac{\lambda^k (t-s)^k}{k!}\right)$.

The Poisson process is said to start at k if $X_0 = k$ a. s.

The existence of a process with such properties is shown exactly as in the case of

---

[1]) ε_x denotes the Dirac measure at point x and $*$ is the convolution operation.

Brownian motion, replacing the Gaussian probabilities ν_h by Poisson laws with parameters h, λ.

It follows trivially from the definition that the paths of the Poisson process are almost certainly increasing. A more detailed study (see Exercise 3) shows that a.s. each path has jumps of magnitude 1 and has, therefore, a finite number of them on every finite interval.

As in the case of Brownian motion, there exists a regular representation of the Poisson process. In this "canonical" representation, the stochastic basis $(\Omega, \mathscr{F}_t)_{t \in \mathbb{R}^+}, P)$ is as follows: Ω is the set of right-continuous functions from $\mathbb{R}^+$ into $\mathbb{N}$, with left limits at every time t; $\mathscr{F}_t$ is the σ-algebra generated by the mappings x_s: $\omega \curvearrowright X_s(\omega) := \omega(s)$, $s \leqslant t$; and P is uniquely determined by the conditions (i), (ii) and $X_0 = k$.

2.4 Construction of stochastic processes

As is clear from the above examples, constructing a stochastic process with state space $\mathfrak{x}$ can be described as defining a probability P on a suitable set Ω of mappings from $\mathbb{R}^+$ into $\mathfrak{x}$ endowed with the canonical σ-algebra (i.e., the σ-algebra generated by the coordinate functions X_t: $X_t(\omega) := \omega(t)$).

The canonical stochastic basis is then $(\Omega, \mathscr{F}_t)_{t \in \mathbb{R}^+}, \mathscr{F}_\infty, P)$, where $\mathscr{F}_t$ is generated by $\{X_s : s \leqslant t\}$.

The characterization of P (if it exists) may be obtained from the knowledge of finite-dimensional probabilities (i.e. the joint probability of $(X_0, X_{t_1} \dots X_{tn})$ for every $t_1 < \dots < t_n$), as it was the case in examples (2.1) and (2.3) above.

There are many cases where P is characterized in another way. A typical case is the following one. Once a process is given in some canonical basis, for example $(\Omega_c, (\mathscr{C}_t), \mathscr{C}_\infty, \mu)$ as above with the Wiener measure μ_0, a new process can be defined through any transformation of the measure into another probability. A very particular case consists, for instance, in considering a probability P having some given density with respect to μ_0.

2.5 Remark: "*Processes with independent increments*" and "*Processes with increments independent of the past*".

Let Y be an $\mathbb{R}^d$-dimensional process on a stochastic basis $(\Omega, \mathscr{F}_t)_{t \in \mathbb{R}^+}, P)$. We denote by $\mathscr{F}_t^Y$ the σ-algebra which is generated by $\{Y_s : s \leqslant t\}$. We thus obtain a filtration which may be different from $(\mathscr{F}_t)_{t \in \mathbb{R}^+}$. If Y is adapted, we have $\mathscr{F}_t^F \subset \mathscr{F}_t$ for every t, and if Y has increments independent of the past for the filtration $\mathscr{F}$, it has the same property with respect to the filtration $\mathscr{F}^F$. In this latter case, this property can be expressed as follows: for every family $0 = t_0 < t_2 < t_1 \dots < t_n$ of times, the random variables $Y_{t_0}, Y_{t_2} - Y_{t_1}, \dots, Y_{t_n} - Y_{t_{n-1}}$ are independent. This property is often refered to as the "*indepentent increments property*".

When we speak of Brownian-process or Poisson-process, we should be careful to note the filtration with respect to which it is meant. In the case of canonical processes, it is with respect to their own filtration.

A simple illustration of this fact is the following to be found in a paper of Ito [Ito 1]: Let β be a process, which is a Brownian motion with respect to its own filtration $\mathscr{F}^{\beta}$. Let Y be the constant process $Y := \beta_1$, and $\mathscr{F}$ the filtration generated by β and Y. Then β is no longer a Brownian motion with respect to this filtration. For more details, see exercise E. 1 at the end of Chapter II.

2.6 An example of a process in Physics: transport-process

A particle is submitted to random "independent" shocks which produce, when they occur, instantaneous changes in its direction, without changing its speed (the norm of its velocity). One assumes that the particle has a uniform movement between the shocks and that the direction of the velocity immediately after each shock is "independent of the direction just before the shock". A stochastic process which models this phenomenon can be described in the following way: Let (τ_n) be a sequence of independent identically distributed positive random variables defined on a probability space, say $(\mathbb{R}^{+\mathbb{N}}, \mathbb{R}_{\mathbb{R}^+}^{\otimes \mathbb{N}}, \mu^{\otimes \mathbb{N}})$ where μ denotes the common probability law of the $\tau_n' s$. These random variables are intended to model the intervals of time between the successive shock. Let $(\vec{u}_n$ be a sequence of identically distributed random unit vectors in $\mathbb{R}^3$. If S is the unit sphere in $\mathbb{R}^3$ and ℓ the uniform distribution on S, the sequence of variables $(\tau_n, u_n)_{n \in \mathbb{N}}$, which completely defines the movement of the particle, when it starts at point x at time zero and initial velocity $\vec{u}_0$ can be considered as defined on the probability space:

$$(\Omega, \mathfrak{A}, P) := (\mathbb{R}^{+\mathbb{N}} \times S^{\mathbb{N}}, \mathscr{B}_{\mathbb{R}^+}^{\otimes \mathbb{N}} \otimes \mathscr{B}_S^{\otimes \mathbb{N}}, \mu^{\otimes \mathbb{N}} \otimes \ell^{\otimes \mathbb{N}}).$$

Denoting by $T_n := \tau_1 + \ldots + \tau_n$ the time of the nth shock and by U_n the speed of the particle at time T_n the position $X(t, \omega)$ of the particle at time t is given by

$$X(t, \omega) = x + \int_0^t \sum_{n \geqslant 0} 1_{[T_n\, T_{n+1}[}(s) \vec{u}_n(s)\, ds,$$

(we set $T_0 = 0$).

If $\mathscr{F}_t$ denotes the σ-algebra generated by the events $\{T_n \leqslant t, u_n \in B\}$ $n \in \mathbb{N}$, $B \in \mathscr{B}_S$, the process x is clearly adapted with continuous paths.

2.7 Markov processes

We consider a state-space $\mathscr{X}$, which is a locally compact subspace of $\mathbb{R}^d$ with $\mathscr{B}$ its Borel σ-algebra. We denote by C_0 the space of continuous functions on $\mathscr{X}$, tending to zero at infinity. A Feller *semi-group of probability transitions* on $\mathscr{X}$ is a family $(T_t)_{t \geqslant 0}$ of linear mappings from C_0 into C_0 with the following properties.

(PT 1) $T_t f \geqslant 0$ for every $f \geqslant 0$ $f \in C_0$.

(PT 2) $T_t 1 = 1$ (1 denoting the constant function equal to 1).

(S.G) For every $s \geqslant 0$ and $t \geqslant 0$ $T_{t+s} = T_t \circ T_s$ (semi-group property).

These properties (PT 1) and (PT 2) imply that for every $x \in \mathscr{X}$ and $t \in \mathbb{R}^+$ there exists a unique probability measure $\Pi_t(x, dy)$ on $\mathscr{B}$ such that $T_t f(x) = \int_{\mathscr{X}} f(y) \Pi_t(x, dy)$ for every $f \in C_0$.

A process $(X_t)_{t \in \mathbb{R}^+}$ with state space $\mathscr{X}$, defined on the stochastic basis $(\Omega, \mathfrak{A}, (\mathscr{F}_t)_{t \in \mathbb{R}^+}, P)$ and adapted to the filtration $(\mathscr{F})$, is said to be a *Markov process with transitions* (T_t) (or (Π_t)) if for every $s < t$ and $f \in C_0$

(M) $$E(f(X_t) \mid \mathscr{F}_s) = T_{t-s} f(X_s).$$

In other words, for every $s < t$ the *conditional law* of X_t given $\mathscr{F}_s$ is $\Pi_{t-s}(X_s, dy)$.

Let us denote by ν_0 the law of the random variable X_0. It is clear from the above that, for every φ continuous on $\mathscr{X}^{n+1}$ and converging to zero at infinity and every finite sequence $0 := t_0 < t_1 < \ldots < t_n$ of times

(2.7.1) $$E[\varphi(X_0, X_{t_1}, \ldots X_{t_n})] = \int \nu_0(dx_0) \int \Pi_{t_1}(x_0, dx_1) \ldots \int \Pi_{t_n - t_{n-1}}(x_{n-1}, dx_n) \varphi(x_0, x_1, \ldots, x_n)$$

A classical monotone class-argument shows that this formula also holds for every Borel bounded φ. If we are given a semi-group $(T_t)_{t \geq 0}$ of probability transitions and a point $x \in \mathscr{X}$ we can always contruct a stochastic basis and on this basis a Markov-process $(X_t)_{t \in \mathbb{R}^+}$ with transitions (T_t) and such that $X_0 = x$ a.s. We can indeed always take $\Omega := \mathscr{X}^{\mathbb{R}^+}$, for $(X_t)_{t \geq 0}$ the projection process and for $(\mathscr{F}_t)$ the associated filtration as in 2.1. By virtue of Kolmogorov's theorem quoted above, a probability P_x on $(\Omega, \mathfrak{A})$, where $\mathfrak{A}$ is the σ-algebra generated by $\{X_t : t \in \mathbb{R}^+\}$, is uniquely defined by the iterated integrals

(2.7.2) $$\int \varphi(X_0, X_{t_1}, \ldots X_{t_n}) dP_x = \int \varepsilon_x(dx_0) \int \Pi_{t_1}(x_0, dx_1) \ldots \int \Pi_{t_n - t_{n-1}}(x_{n-1}, dx_n) \varphi(x_0, x_1 \ldots, x_n)$$

for every bounded Borel φ on $\mathscr{X}^{n+1}$.

It can easily be seen that Brownian motion and Poisson processes are examples of Markov processes (see exercise E. 14 at the end of the chapter).

Markov processes with Feller probability transitions and continuous paths are often called *Feller processes*. The example of the Poisson process starting at $x \in \mathbb{R}$ shows that a Markov process with Feller probability transition may have jumps and is not necessarily a Feller process.

3. Well-measurable or optional and predictable processes

We suppose a stochastic basis $(\Omega, \mathfrak{A}, (\mathscr{F}_t)_{t \in \mathbb{R}^+}, P)$ is given.

3.1 Definition. A subset of $\mathbb{R}^+ \times \Omega$ is called *well-measurable* (or *optional*) if it belongs to the σ-algebra $\mathscr{G}$ which is generated by the real R.R.C. processes. A process X

with state space $(\mathfrak{x}, \mathscr{B})$ is called well-measurable or optional when it is $\mathscr{G}$-measurable.

It is an immediate consequence of Theorem 1.6 that $\mathscr{G} \subset \mathscr{M}_1$.

3.2 Definition. The σ-algebra of subsets of $\mathbb{R}^+ \times \Omega$, which is generated by the adapted continuous real processes, is called the σ-algebra of *predictable* sets and will be denoted by $\mathscr{P}$. A process which is $\mathscr{P}$-measurable is called predictable.

3.3 Theorem. (1) *The following inclusion holds.* $\mathscr{P} \subset \mathscr{G}$.

(2) *The σ-algebra $\mathscr{P}$ is generated by the adapted left continuous processes and by the following families of sets as well.*

$$\mathscr{R} := \{]s, t] \times F : s \leqslant t,\ F \in \mathscr{F}_s\} \bigcup \{\{0\} \times F,\ F \in \mathscr{F}_0\},$$
$$\mathscr{R}_1 := \{]s, t] \times F : s \leqslant t,\ F \in \mathscr{F}_{s-}\} \bigcup \{\{0\} \times F,\ F \in \mathscr{F}_0\},$$
$$\mathscr{R}_2 := \{[s, t[\times F : s \leqslant t,\ F \in \mathscr{F}_{s-}\} \text{ (we set } \mathscr{F}_{0-} := \mathscr{F}_0).$$

Proof: $\mathscr{P}$ being generated by the class of continuous processes, the inclusion $\mathscr{P} \subset \mathscr{G}$ is trivial, as is the inclusion of $\mathscr{P}$ in the σ-algebra $\mathscr{P}'$, which is generated by the adapted left-continuous processes.

For every $R \in \mathscr{R}$, the indicator function 1_R is a R. L. C. process and therefore we have $\sigma(\mathscr{R}) \subset \mathscr{P}'$. To prove the converse inclusion, it is enough to show that every bounded adapted left-continuous process X can be approximated by processes of the form

$$\sum_{i \in I} \lambda_i 1_{]s_i, t_i] \times F_i},\ ]s_i, t_i] \times F_i \in \mathscr{R},\ I \text{ denumerable.}$$

But, for every $n \in \mathbb{N}$ and $k \in \mathbb{N}$, $X\left(\frac{k}{2^n}\right)$ is $\mathscr{F}_{(k/2^n)}$-measurable, and one can choose $\lambda^i_{n,k} \in \mathbb{R}^+$ and $F^i_{n,k} \in \mathscr{F}_{(k/2^n)}$ such that

$$\sup_{\omega \in \Omega} \left| X\left(\frac{k}{2^n}, \omega\right) - \sum_{i \in \mathbb{N}} \lambda^i_{n,k} 1_{F^i_{n,k}}(\omega) \right| \leqslant \frac{1}{2^n}.$$

Therefore, the sequence $(X^n : n \in \mathbb{N})$ of processes where

$$X^n(t, \omega) := \sum_k \sum_i \lambda^i_{n,k} 1_{]\frac{k}{2}, \frac{k+1}{2}] \times F^i_{n,k}}(t, \omega)$$

converges pointwise towards X on $\mathbb{R}^+ \times \Omega$. This proves $\mathscr{P}' \subset \sigma(\mathscr{R})$ and $\mathscr{P} \subset \sigma(\mathscr{R})$.

To prove the equality of $\mathscr{P}$, $\sigma(\mathscr{R})$ and $\mathscr{P}'$, it is now sufficient to show that $\mathscr{P} \supset \sigma(\mathscr{R})$. We consider $]s, t] \times F \in \mathscr{R}$. There exists a sequence (φ_n) of continuous real functions φ_n on $\mathbb{R}^+$ such that

$$1_{]s, t]} = \lim_n \varphi_n$$

and

$$\varphi_n(u) = 0 \quad \text{for every} \quad u \in]0, s].$$

The processes $1_F \cdot \varphi_n$, $n \in \mathbb{N}$, are adapted and continuous. Since

$$1_{]s,t] \times F} = \lim 1_F \cdot \varphi_n ,$$

the process $1_{]s,t] \times F}$ is $\mathscr{P}$-measurable and, therefore, $\mathscr{P} \supset \sigma(\mathscr{R})$.

It is now only left to prove that $\sigma(\mathscr{R}) = \sigma(\mathscr{R}_1) = \sigma(\mathscr{R}_2)$. But it is clear that $\mathscr{R}_1$ (resp. $\mathscr{R}_2$) generates the same σ-algebra as

$$\mathscr{R}_1' := \{]s,t] \times F : s \leqslant t,\, F \in \bigcup_{r<s} \mathscr{F}_r\} \bigcup \{\{0\} \times F,\, F \in \mathscr{F}_0\}$$

$$\text{(resp.} \qquad \mathscr{R}_2' := \{[s,t[\times F : s \leqslant t,\, F \in \bigcup_{r<s} \mathscr{F}_r\} \bigcup \{\{0\} \times F,\, F \in \mathscr{F}_0\}).$$

Since

$$1_{]s,t] \times F} = \lim_n 1_{[s+(1/n),\, t+(1/n)[\times F} ,$$

the inclusion $\sigma(\mathscr{R}_1) \subset \sigma(\mathscr{R}_2$ is immediate. Conversely, we have

$$1_{[s,t[\times F} = \lim_n 1_{]s-(1/n),\, t-(1/n)] \times F} .$$

As a consequence, we obtain

$$\sigma(\mathscr{R}_1) \subset \sigma(\mathscr{R}_2) = \sigma(\mathscr{R}_2') \subset \sigma(\mathscr{R}_1') = \sigma(\mathscr{R}_1) . \tag{3.3.1}$$

Since $\sigma(\mathscr{R}_1) \subset \sigma(\mathscr{R}$ follows immediately from the definitions and since $\sigma(\mathscr{R}) \subset \sigma(\mathscr{R}_1)$ follows from

$$1_{]s,t] \times F} = \lim_n 1_{]s+(1/n),\, t+(1/n)] \times F} ,$$

the proof of the theorem is now complete. □

Notation: We shall denote by $\mathscr{A}$ the Boolean ring which is generated by $\mathscr{R}$. It is immediately seen that the elements of $\mathscr{A}$ are the finite unions of sets from $\mathscr{R}$.

3.4 Definition. The elements of $\mathscr{R}$ are called *predictable rectangles.*

3.5 Corollary. *The σ-algebra $\mathscr{P}$ is the same for the three filtrations $(\mathscr{F}_t)$, $(\mathscr{F}_{t^-})$ and $(\mathscr{F}_{t^+})$.*

4. Stopping times

Stopping times in general

For every mapping $T: \Omega \to \bar{\mathbb{R}}^+$, we define the following subsets of $\mathbb{R}^+ \times \Omega$.

$$[0,T] := \{(t,\omega) : 0 \leqslant t \leqslant T(\omega),\, t \in \mathbb{R}^+,\, \omega \in \Omega\} ,$$
$$[0,T[:= \{(t,\omega) : 0 \leqslant t < T(\omega),\, t \in \mathbb{R}^+,\, \omega \in \Omega\} .$$

4.1 Definition and Example. The mapping $T: \Omega \to \bar{\mathbb{R}}^+$ is called a *stopping time* (with respect to the given filtration if, for every $t \in \mathbb{R}$, the subset $\{T \leqslant t\}$ of Ω belongs to $\mathscr{F}_t$.

The filtration $(\mathscr{F}_t)$ in example 2.6 has been defined in such a way that the times of shocks T_n are stopping times. For other examples, see 4.12 and 4.15.

4.2 Proposition. *A numerical function $T: \Omega \to \bar{\mathbb{R}}^+$ is a stopping time iff the process $1_{[0,T[}$ is well-measurable.*

Proof: The process $1_{[0,T[}$ is right-continuous and is moreover regular iff it is adapted. This last property is equivalent to: $\forall t \in \mathbb{R}^+$, $\{T \leqslant t\} \in \mathscr{F}_t$. Therefore, the proposition is clear. □

4.3 Proposition. *Let S and T be two stopping times. Then, $S \vee T$ and $S \wedge T$ are stopping times.*

If (S_n) is a sequence of stopping times, then $\sup_n S_n$ is a stopping time.

Proof: This follows immediately from Proposition 4.2 and the relations

$$\begin{aligned} 1_{[0,S \vee T[} &= \sup(1_{[0,T[}, 1_{[0,S[}), \\ 1_{[0,S \wedge T[} &= \inf(1_{[0,T[}, 1_{[0,S[}), \\ 1_{[0,\sup_n S_n[} &= \sup_n 1_{[0,S_n[}. \;\square \end{aligned}$$

4.4 Proposition. *Every stopping time relatively to the family $(\mathscr{F}_t)_{t \in \mathbb{R}^+}$ of σ-algebras is a stopping time with respext to $(\mathscr{F}_{t^+})_{t \in \mathbb{R}^+}$.*

T is a stopping time with respect to $(\mathscr{F}_{t^+})_{t \in \mathbb{R}^+}$ iff $1_{[0,T]}$ is predictable or, equivalently, iff $\{T < t\} \in \mathscr{F}_t$ for every t.

Proof: The first statement of the proposition follows immediately from the definition and the inclusion $\mathscr{F}_t \subset \mathscr{F}_{t^+}$.

Let us remark now that, according to the definition of $(\mathscr{F}_{t^+})$, the following properties are equivalent.

(a) $$\{T \leqslant t\} = \bigcap_n \left\{T < t + \frac{1}{n}\right\} \in \mathscr{F}_{t^+} \text{ for every } t.$$

(b) $$\{T < t\} \in \mathscr{F}_t \text{ for every } t.$$

The property (b) is equivalent to the adaptation property for left continuous processes $1_{[0,T]}$ and, therefore, according to Theorem 3.3, to the predictability of $1_{[0,T]}$. □

Corollary. Let (S_n) be a sequence of stopping times. Then, $S := \inf_n S_n$ is a stopping time for $(\mathscr{F}_{t^+})$.

Proof: The process $1_{[0,S]} = \inf 1_{[0,S_n]}$ is in fact predictable, and one applies 4.4. □

4.5 Definition. Let S and T be two stopping times with $S \leqslant T$. We define the following *stochastic intervals* $]S,T]$ (resp. $[S,T]$, $[S,T[$, $]S,T]$) as follows.

$$
\begin{aligned}
]S,T] &:= \{(t,\omega): S(\omega) < t \leqslant T(\omega), t \in \mathbb{R}^+, \ \omega \in \Omega\}, \\
[S,T] &:= \{(t,\omega): S(\omega) \leqslant t \leqslant T(\omega), t \in \mathbb{R}^+, \ \omega \in \Omega\}, \\
[S,T[&:= \{(t,\omega): S(\omega) \leqslant t < T(\omega), t \in \mathbb{R}^+, \ \omega \in \Omega\}, \\
]S,T[&:= \{(t,\omega): S(\omega) < t < T(\omega), t \in \mathbb{R}^+, \ \omega \in \Omega\}.
\end{aligned}
$$

We use also the following notation.

$$[T] := \{(t,\omega): t = T(\omega) < \infty, t \in \mathbb{R}^+, \omega \in \Omega\}.$$

$[T]$ is known as the graph of the stopping time, but it must be stressed that the stochastic intervals $]S,T]$, $]S,T[$, etc... and $[T]$ *are by definition subsets of* $\mathbb{R}^+ \times \Omega$, while T and S are allowed to take their values in $\bar{\mathbb{R}}^+$.

4.6 Proposition. *The σ-algebra $\mathscr{P}$ of predictable sets is generated by the family of stochastic intervals of the following form: $]S,T]$ and $\{0\} \times F$, $F \in \mathscr{F}_0$.*

$\mathscr{P}$ is also generated by the family of closed stochastic intervals $\{[0,T]: T$ stopping time$\}$.

Proof: According to Proposition 3.3, the R.L.C. processes $1_{]S,T]}$, $1_{[0,T]}$, $1_{\{0\} \times F}$ are predictable. Conversely, since, for every $F \in \mathscr{F}_s$ and $s < t$, the random variable $S := s \cdot 1_F + t \cdot 1_{\complement F}$ is a stopping time, the process $1_{]s,t] \times F}$ can be written $1_{]S,t]}$, where $]S,t]$ is a stochastic interval. Following Proposition 3.3, the family of processes

$$\{1_{]S,T]}: S \leqslant T, S \text{ and } T \text{ stopping times}\} \bigcup \{1_{\{0\} \times F}: F \in \mathscr{F}_0\}$$

generates $\mathscr{P}$. This proves the first part of Proposition 4.6.

Since $]S,T] = [0,T] - [0,S]$ and $\{0\} \times F = \bigcap_n [0,S_n]$, with $S_n := \frac{1}{n} \cdot 1_F$, a stopping time for $F \in \mathscr{F}_0$, the second part of the proposition follows immediately from the first. □

We will prove later (Proposition 5.7) that $\mathscr{G}$ is generated by the stochastic intervals $[0,T[$ when the "usual hypotheses" are fulfilled.

4.7 Proposition. *The graph $[T]$ of every stopping time T is an optional subset of* $\mathbb{R}^+ \times \Omega$

Proof: It is readily checked that for every $n \in \mathbb{N}$, the random variable

$$T_n := \sum_{k \geqslant 0} \frac{k+1}{2^n} 1_{\left[\frac{k}{2^n} \leqslant T < \frac{k+1}{2^n}\right]}$$

is a stopping time. Since

$$[T] = \bigcap_n [T, T_n[= \bigcap_n ([0, T_n[- [0, T[),$$

the optionality of $[T]$ follows from Proposition 4.2. □

4.8 Remark and example. The question arises from Proposition 4.7 whether $[T]$ is not only optional but also predictable. We give the following example, supposedly due to Dellacherie, which shows that it is not always the case.

We set $\Omega = [0, 1]$. For every $t < 1, \mathscr{F}_t$ stands for the σ-algebra of subsets of Ω, which is generated by $]t, 1]$ and the Borel subsets of $[0, t]$. For every $t \geqslant 1$, $\mathscr{F}_t$ is the Borel σ-algebra of Ω.

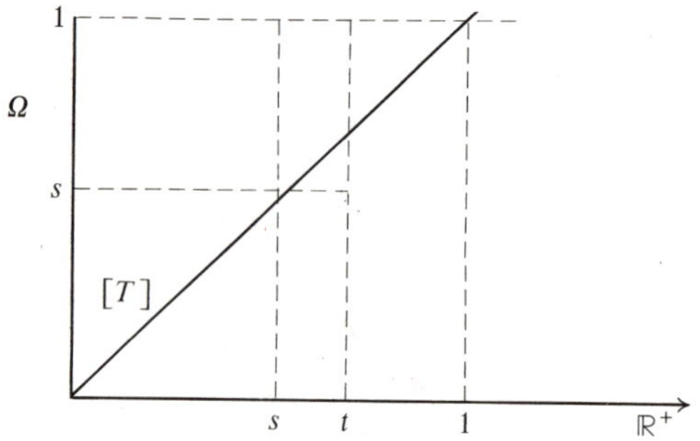

We define

$$T(\omega) := \omega.$$

It is easily seen that T is a stopping

Because of the definition of $\mathscr{F}_s$ for $s < 1$, the following equality holds for every $F \in \mathscr{F}_s$ and $s < t \leqslant 1$:

$$(]s, t] \times F) \cap [0, T] = \begin{cases} \emptyset & \text{if } F \cap]s, 1] = \emptyset, \\ (]s, t] \times \Omega) \cap [0, T] & \text{if } F \cap]s, 1] =]s, 1]. \end{cases}$$

This says that the trace of $\mathscr{R}$ on $[0, T]$ is the same as the trace on $[0, T]$ of the family $\{]s, t] \times \Omega : s < t \leqslant 1\} \bigcup \{\{0\} \times \Omega\}$.

Hence, it follows immediately that the predictable subsets of $[0, T]$ are the intersections $[0, T] \cap (B \times \Omega)$, where B is any Borel subset of $[0, 1]$. This shows that $[T]$ cannot be predictable.

Predictable stopping times

It should be noted that there exist stopping times with a predictable graph. Constant stopping times are the simplest examples.

This example and Remark 4.8 give full meaning to the following definition.

4.9 Definition. A stopping time T is called *predictable* when $[T]$ is a predictable subset of $\mathbb{R}^+ \times \Omega$.

As an example of predictable stopping times which are not constant, let us consider the random variable $T+\alpha$, where T is any stopping time and α a strictly positive number. It is readily checked that $T+\alpha$ is a stopping time and from

$$]T+\alpha\left(1-\frac{1}{n}\right), T+\alpha] \in \mathscr{P},$$

follows

$$[T+\alpha] = \bigcap_{n\in\mathbb{N}}]T+\alpha\left(1-\frac{1}{n}\right), T+\alpha] \in \mathscr{P},$$

and therefore, the predictability of $T+\alpha$.

At this point it should be emphasized that a stopping time with finitely many values is not necessarily predictable (see exercice 5).

4.10 Proposition. *A stopping time T is predictable iff $]0,T[$ is predictable. This is also equivalent to the predictability of $[0,T[$.*

Proof: Since $]0,T]$ is predictable for every stopping time T (Proposition 4.4), $]0,T[=]0,T]-[T]$ is also predictable if and only if $[T]$ is predictable. Since $\{T>0\}\in\mathscr{F}_0$, we have $\{0\}\times\{T>0\}\in\mathscr{R}$. Therefore, the stochastic intervals $[0,T[$ and $]0,T[$ are predictable or not at the same time. □

4.11 Proposition. *Let S and T be two predictable stopping times. Then, $S\wedge T$ and $S\vee T$ are also predictable stopping times.*

For every sequence (T_n) of predictable stopping times, $\sup_{n\in\mathbb{N}} T_n$ is a predictable stopping time.

For every stopping time T, there exists a decreasing sequence (T_n) of predictable stopping times such that

$$T = \inf_{n\in\mathbb{N}} T_n .$$

Proof: The two first statements of this proposition are a straightforward consequence of the following equalities and of Proposition 4.10

$$[0,S\wedge T[= [0,S[\cap[0,T[$$
$$[0,S\vee T[= [0,S[\cap[0,T[$$
$$[0,\sup_{n\in\mathbb{N}} T_n[= \bigcup_{n\in\mathbb{N}} [0,T_n[.$$

As already noticed, the random variables $T_n := T+\frac{1}{n}$ are predictable stopping times. This proves the third statement. □

4.12 Corollary. *The σ-algebra $\mathscr{P}$ is generated by the family*

$$\{[0,T[: T \text{ predictable stopping time}\}$$

of stochastic intervals. The following family is also a system of generators

$$\{]S,T[\,, S \text{ and } T \text{ predictable stopping times, } S \leqslant T\}.$$

Beginning of a set

4.13 Definition. Let A be a subset of $\mathbb{R}^+ \times \Omega$. We call the *beginning of A* the $\mathbb{R}^+$-valued random function D_A defined by

$$D_A(\omega) := \begin{cases} \inf\{t:(\omega,t)\in A, t\in\mathbb{R}^+\} & \text{if } \{t:(\omega,t)\in A, t\in\mathbb{R}^+\} \neq \emptyset; \\ +\infty & \text{if } \{t:(\omega,t)\in A, t\in\mathbb{R}^+\} = \emptyset \end{cases}$$

When the "usual hypotheses" (see Definition 1.1) are fulfilled, the following proposition appears as a trivial consequence of Proposition 4.15 below and its Corollary 4.16. However, we think it is interesting to mention the following elementary property of beginnings, which holds without any special assumption about the filtration $(\mathscr{F}_t)_{t\in\mathbb{R}^+}$.

4.14 Proposition (1) *Let X be a real, adapted, continuous process indexed by $\mathbb{R}^+$ and $A := \{X = 0\}$. Then, the beginning of A is a predictable stopping time.*

(2) *Let X be an R. R. C. process indexed by $\mathbb{R}^+$ with values in $\mathbb{R}^d$ and G be an open set. The gebinning of the set $A := \{(t,\omega);\ X(t,\omega) \notin G\}$ is a stopping time for the filtration $(\mathscr{F}_{t^+})$.*

Proof: (1) The continuity of X implies

$$\{D_A > t\} = \bigcup_{n\in\mathbb{N}} \bigcap_{\substack{r<t \\ r\in\mathbb{Q}\cup\{t\}}} \left\{|X_r| \geqslant \frac{1}{n}\right\}.$$

Therefore $\{D_A > t\} \in \mathscr{F}_t$ and D_A is a stopping time. Since $[D_A] = [0, D_A] \cap \{X = 0\}$, the predictability of D_A follows from 3.2 and Proposition 4.4.

(2) From the right-continuity of X it follows that

$$\{D_A \geqslant t\} = \bigcap_{\substack{r\in\mathbb{Q} \\ r<t}} \{X_r \in G\}.$$

Therefore, $\{D_A \geqslant t\} \in \mathscr{F}_t$ and the assertion (2) follows from Proposition 4.4. □

4.15 Proposition. *If the "usual hypotheses" are fulfilled by the filtration $(\mathscr{F}_t)_{t\in\mathbb{R}^+}$, the beginning of every progressively measurable subset of $\mathbb{R}^+ \times \Omega$ is a stopping time.*

Proof: We make use of the following result in measure theory (see P. A. Meyer [Mey. 1] chap. 3–9). Let $(\mathscr{X}, \mathscr{B}_{\mathscr{X}})$ be a locally compact topological space with denumerable basis endowed with the σ-algebra $\mathscr{B}_{\mathscr{X}}$ of Borel sets. Let $(\Omega, \mathscr{F}, P)$ be a complete probability space. Then, for every $A \in \mathscr{F} \otimes \mathscr{B}_{\mathscr{X}}$, the projection $\pi(A)$ of A into Ω belongs to $\mathscr{F}$.

Since the σ-algebras $\mathscr{F}_t$ are assumed to be complete, we apply this latter property to the sets

$$A_t := A \cap ([0, t[\times \Omega) \in \mathscr{B}_t \otimes \mathscr{F}_t,$$

where A is progressively measurable, $t \in \mathbb{R}^+$ and $\mathscr{B}_t$ is the σ-algebra of Borel subsets of $[0, t]$. But, actually, $\{D_A < t\} = \pi(A_t)$, and therefore $\{D_A < t\} \in \mathscr{F}_t$. This last property shows that D_A is a stopping time with respect to $(\mathscr{F}_{t+})$ (see Proposition 4.4). The "usual hypotheses" tell us that $\mathscr{F}_{t+} = \mathscr{F}_t$. This proves the proposition. □

4.16 Corollary. *We assume the "usual hypotheses". Let A be a predictable set. Then the beginning D_A is predictable if $[D_A] \cup A$ is predictable.*

Proof: We know that $[0, D_A]$ is predictable. This follows from Propositions 4.15 and 4.4 The Corollary is therefore a trivial consequence of

$$[D_A] = ([D_A] \cup A) \cap [0, D_A]. \quad \square$$

5. The σ-algebras $\mathscr{F}_T$ and $\mathscr{F}_{T-}$

5.1 Definition. To every stopping time T we now associate two σ-algebras $\mathscr{F}_T$ and $\mathscr{F}_{T-}$. It is easy to check that the family $\mathscr{F}_T$ of sets $F \in \mathscr{F}_\infty$ with the extra property that

$$F \cap \{T \leqslant t\} \in \mathscr{F}_t \quad \text{for every} \quad t \in \mathbb{R}^+,$$

is a σ-algebra. This σ-algebra will be called the *σ-algebra of the past events up to T*

The σ-algebra $\mathscr{F}_{T-}$ is defined as being generated by the following family of events, which belong to $\mathscr{F}_\infty$.

$$\{F \cap \{t < T\} : F \in \mathscr{F}_t, t \in \mathbb{R}^+\} \cup \mathscr{F}_0$$

This σ-algebra is known as the *σ-algebra of the past events up to T.*

It is easy to show that $\mathscr{F}_{T-}$ is also generated by the family

$$\{F \cap \{t \leqslant T\} : F \in \mathscr{F}_{t-}, t \in \mathbb{R}^+\}$$

(we always agree to set $\mathscr{F}_{0-} = \mathscr{F}_0$).

Since

$$\forall F \in \mathscr{F}_t, \forall s, t \in \mathbb{R}, F \cap \{t < T\} \cap \{T \leqslant s\} = F \cap \{t < T \leqslant s\} \in \mathscr{F}_s,$$

it follows trivially that

$$\mathscr{F}_{T-} \subset \mathscr{F}_T.$$

It is an immediate consequence of the definition that T is $\mathscr{F}_{T-}$-measurable.

5.2 Remark. (a) For every constant stopping time $T: T(\omega) := t \in \mathbb{R}^+$, $\mathscr{F}_T = \mathscr{F}_t$ and $\mathscr{F}_{T-} = \mathscr{F}_{t-}$ hold.

(b) The filtration $(\mathscr{F}_t)_{t \in \mathbb{R}^+}$ may very well be continuous (i.e. $\mathscr{F}_t = \mathscr{F}_{t+} = \mathscr{F}_{t-}$), without implying that $\mathscr{F}_{T-} = \mathscr{F}_T$ for every stopping time T (see exercise 6).

5.3 Proposition. *Let S and T be two stopping times with $S \leqslant T$. Then, the following inclusions hold.*

$$\mathscr{F}_S \subset \mathscr{F}_T \text{ and } \mathscr{F}_{S-} \subset \mathscr{F}_{T-}.$$

Let (T_n) be a monotone sequence of stopping times and $T := \lim_{n\to\infty} T_n$.

a) *If (T_n) is decreasing and if $\mathscr{F}_t = \mathscr{F}_{t+}$ for every $t \in \mathbb{R}^+$, then*

$$\mathscr{F}_T = \bigcap_{n \in \mathbb{N}} \mathscr{F}_{T_n}.$$

b) *If (T_n) is increasing, then*

$$\mathscr{F}_{T-} = \bigvee_{n \in \mathbb{N}} \mathscr{F}_{T_n^-}.$$

Proof: (1) Let S and T be two stopping times with $S \leqslant T$. From the equality $F \cap \{T \leqslant t\} = F \cap \{S \leqslant t\} \cap \{T \leqslant t\}$, it immediately follows that $\mathscr{F}_S \subset \mathscr{F}_T$. Since $F \cap \{t < S\} = F \cap \{t < S\} \cap \{t < T\}$, we immediately have the following implications

$$F \in \mathscr{F}_t \Rightarrow F \cap \{t < S\} \in \mathscr{F}_t \Rightarrow F \cap \{t < S\} \cap \{t < T\} \in \mathscr{F}_T.$$

Hence, $F_{S-} \subset \mathscr{F}_{T-}$.

(2) (a) In view of the first statement of the proposition, it is enough to prove

$$\bigcap_{n \in \mathbb{N}} \mathscr{F}_{T_n} \subset \mathscr{F}_T.$$

Let $F \in \bigcap_{n \in \mathbb{N}} \mathscr{F}_{T_n}$. From the definition of $\mathscr{F}_{T_n}$, we obtain

$$\forall t \in \mathbb{R}^+, \mathrm{F} \cap \{T_n \leqslant t\} \in \mathscr{F}_t,$$

and therefore

$$\forall t \in \mathbb{R}^+, F \cap \{T_n < t\} = \bigcup_{p \in \mathbb{N}} F \cap \left\{T_n \leqslant t - \frac{1}{p}\right\} \in \mathscr{F}_t.$$

Hence,

$$\forall t \in \mathbb{R}^+, F \cap \{T < t\} = \bigcup_{n \in \mathbb{N}} F \cap \{T_n < t\} \in \mathscr{F}_t.$$

The assumption $\mathscr{F}_t = \mathscr{F}_{t+}$ for every t then implies

$$\forall t \in \mathbb{R}^+, F \cap \{T \leqslant t\} = \bigcap_{n \in \mathbb{N}} F \cap \left\{T < t + \frac{1}{n}\right\} \in \mathscr{F}_t.$$

This proves (a).

(2) (b) Because of the first statement of the proposition, we have only to show that every event $F \cap \{t < T\}$, where $F \in \mathscr{F}_t$ and $t \in \mathbb{R}^+$, belongs to $\bigvee_{n \in \mathbb{N}} \mathscr{F}_{T_n^-}$. But this follows immediately from the relation

$$F \cap \{t < T\} = \bigcup_{n \in \mathbb{N}} F \cap \{t < T_n\}. \quad \square$$

5.4 Proposition. *Let T be a stopping time. For every $H \in F_\infty$ we set*

$$T_H(\omega) := \begin{cases} T(\omega) & \text{if} \quad \omega \in H, \\ +\infty & \text{if} \quad \omega \notin H. \end{cases}$$

If $H \in \mathscr{F}_T$, then T_H is a stopping time.

Proof: We may indeed write

$$\{T_H \leqslant t\} = H \cap \{T \leqslant t\} \in \mathscr{F}_t. \quad \square$$

Let X be a process with values in the vector space E. For every stopping time T, we denote by $1_{\{T < \infty\}} \cdot X_T$ the random variable defined by

$$1_{\{T < \infty\}} \cdot X_T(\omega) := \begin{cases} X(T(\omega), \omega) & \text{if } T(\omega) < \infty\,; \\ 0 & \text{if } T(\omega) = \infty\,. \end{cases}$$

5.5 Theorem. *For every progressively measurable process X with values in $(E, \mathscr{B})$ and every stopping time T, the mapping $1_{\{T < \infty\}} \cdot X_T$ is $\mathscr{F}_T$-measurable.*

If the process X is predictable, then $1_{\{T < \infty\}} \cdot X_T$ is $\mathscr{F}_{T-}$-measurable.

Proof: Let us consider, for every $t \in \mathbb{R}^+$, the following mapping ψ_1 from $\{T < t\}$ into $\mathbb{R}^+ \times \Omega : \omega \curvearrowright (T(\omega), \omega)$, and the mapping ψ_2 from $\mathbb{R}^+ \times \Omega$ into $E : (u, \omega) \curvearrowright X(u, \omega)$. Since X is progressively measurable, $\psi_1 \circ \psi_2$ is a measurable mapping from $(\{T \leqslant t\}, \mathscr{F}_t \cap \{T \leqslant t\})$ into $(E, \mathscr{B})$. Therefore, the following holds.

$$\forall B \in \mathscr{B}, \{\omega : X(T(\omega), \omega) \in B\} \cap \{T \leqslant t\} \in \mathscr{F}_t,$$

which implies the $\mathscr{F}_T$-measurability of $1_{\{T < \infty\}} \cdot X_T$.

To prove the second statement of the theorem, we now consider the processes X of the following form

$$X = 1_{]\sigma, \tau]},$$

where σ and τ are stopping times with $\sigma \leqslant \tau$, or processes Y of the form

$$Y = 1_{\{\{0\} \times A\}},$$

where $A \in \mathscr{F}_0$.

From Proposition 4.6, we know that the family of processes X and Y generates the σ-algebra $\mathscr{P}$. The family of sets $]\sigma, \tau]$ and $\{0\} \times A$ is clearly stable for the operations of intersection and contains $\mathbb{R}^+ \times \Omega$. Considering the family $\mathscr{H}$ of predictable processes for which the second statement of the theorem is true, a classical monotone class-argument shows that it is sufficient to prove the property for processes X and Y.

But, for every stopping time T, we may write

$$1_{\{T<\infty\}} \cdot Y_T = 1_A \cdot 1_{\{T=0\}} .$$

and

$$\begin{aligned} 1_{\{T<\infty\}} \cdot X_T &= 1_{\{\sigma<T\leqslant\tau\}} \cdot 1_{\{T<\infty\}} \\ &= (1_{\{T\leqslant\tau\}} - 1_{\{T\leqslant\sigma\}}) \cdot 1_{\{T<\infty\}} . \end{aligned}$$

The function $1_A \cdot 1_{\{T=0\}}$ is clearly $\mathscr{F}_{T-}$-measurable and from the relation

(5.5.1) $$\{\tau<T\} = \bigcup_{r\in\mathbb{Q}} \{\tau\leqslant r<T\} = \bigcup_{r\in\mathbb{Q}} \{\tau\leqslant r\}\cap\{r<T\}\in\mathscr{F}_{T-},$$

and from the $\mathscr{F}_{T-}$-measurability of T, follows the $\mathscr{F}_{T-}$-measurability of $\{\sigma<T\leqslant\tau\}\cap\{T<\infty\}$. This proves the second statement of the theorem. □

5.6 Proposition. *Let S and T be two stopping times. Then*

$$\{S\leqslant T\}\in\mathscr{F}_T \text{ and } \{S<T\}\in\mathscr{F}_{T-} .$$

If S is assumed to be predictable, then

$$\{S\leqslant T\}\in\mathscr{F}_{T-} .$$

Proof: The statement $\{S<T\}\in\mathscr{F}_{T-}$ follows immediately from (5.5.1), merely by replacing τ by S.

If we apply Proposition 5.5 to the process $X := 1_{[S,\infty[}$ and if we notice that

$$1_{\{T<\infty\}} \cdot X_T = 1_{\{S\leqslant T\}} \quad {}_{\{T<\infty\}},$$

we immediately obtain the rest of the Proposition. □

5.7 Theorem. *If the "usual hypotheses" are fulfilled, the σ-algebra $\mathscr{G}$ of optional sets is generated by the following family of stochastic intervals.*

$$\{[S,T[\,: S \text{ and } T \text{ stopping times with } S\leqslant T\}.$$

Proof: We denote by $\mathscr{G}'$ the σ-algebra generated by this family of stochastic intervals. The inclusion $\mathscr{G}'\subset\mathscr{G}$ follows from Proposition 4.2.

Let us now consider processes X of the following form.

$$X = 1_H \cdot 1_{[S,T[},$$

where S and T are two stopping times with $S\leqslant T$ and $H\in\mathscr{F}_S$. Since

$$X = 1_{[S_H, T_H[},$$

Proposition 5.4 shows that X is $\mathscr{G}'$-measurable.

Hence, we easily deduce that every process X of the form

$$X = X_S \cdot 1_{[S, T[},$$

where X_S is $\mathscr{F}_S$-measurable, is $\mathscr{G}'$-measurable.

To complete the proof, we now show that every $R.R.C.$ real process X is the pointwise limit of a sequence $(X^{\varepsilon_p})_{p \in \mathbb{N}}$ of processes which are themselves denumerable linear combinations of processes of the above form.

Let X be an $R.R.C.$ real process. For every $\varepsilon > 0$, we define inductively the following sequence (T_n^ε) of numerical random variables by

$$T_1^\varepsilon(\omega) := 0,$$
$$T_{n+1}^\varepsilon(\omega) := \inf\{t : t > T_n^\varepsilon(\omega), |X_t(\omega) - X_{T_n^\varepsilon}(\omega)| \geqslant \varepsilon\},$$

with always the same convention that $\inf \phi := +\infty$. Let us suppose that T is a stopping time. Then, the set

$$\{(t, \omega) : t > T_n^\varepsilon(\omega), |X_t(\omega) - X_{T_n^\varepsilon}(\omega)| \geqslant \varepsilon\}$$

which can be written

$$\{Y \geqslant \varepsilon\}$$

with

$$Y := 1_{]T_n^\varepsilon, \infty[} \cdot (X - 1_{\{T_n^\varepsilon < \infty\}} \cdot X_{T_n^\varepsilon})$$

is $\mathscr{G}$-measurable, according to the beginning of the proof. Since $T_{n+1}^\varepsilon(\omega)$ is the beginning of this set, it is a stopping time from Proposition 4.15. By recurrence on n, we find that all the T_n^ε's are stopping times.

From the right-continuity of X it follows that

(5.7.1) $$|X_{T_{n+1}^\varepsilon}(\omega) - X_{T_n^\varepsilon}(\omega)| \geqslant \varepsilon \text{ for every } \omega \in \{T_{n+1}^\varepsilon < \infty\}$$

and

(5.7.2) $$|X_t(\omega) - X_{T_n^\varepsilon}(\omega)| < \varepsilon \text{ for every } t, \omega) \in [T_n^\varepsilon, T_{n+1}^\varepsilon[\,.$$

Let us define

$$a(\omega) := \sup_{n \in \mathbb{N}} T_{n+1}^\varepsilon(\omega) \in \bar{\mathbb{R}}^+.$$

In view of (5.7.1), the following limit

$$\lim_{s \uparrow a(\omega)} X_t(\omega)$$

does not exist on the set $\{a < \infty\}$. The regularity of X then implies $a(\omega) = +\infty$ for every ω. This allows us to define the process X^ε by

$$X^\varepsilon(t, \omega) := \sum_{n \in \mathbb{N}} X_{T_n^\varepsilon} 1_{[T_n^\varepsilon, T_{n+1}^\varepsilon[}$$

Because of (5.7.2), the processes X^{ε} converges uniformly towards X when $\varepsilon \downarrow 0$. Letting $\varepsilon_p = 1/p$, we obtain the desired sequence of approximating optional processes for X.

5.8 Remark. It is possible to prove (see [Del. 1] p. 81, and Exercise E–13) that every adapted right continuous process is optional.

6. Admissible measures

6.1 Definition. Let μ be a measure defined on a sub-σ-algebra $\mathscr{A}$ of $\mathscr{B}_{\mathbb{R}^+} \otimes \mathscr{F}_\infty$. Such a measure will be called *admissible* if, for every $A \in \mathscr{A}$, which is evanescent, the following holds

$$\mu(A) = 0 .$$

6.2 Examples. (1) Let V be an adapted process, the paths of which are real, increasing, right-continuous function. For every $A \in \mathscr{B} \otimes \mathscr{F}_\infty$, we set

(6.2.1) $$\mu(A) := \int [\int 1_A(u, \omega)\, dV(u)]\, P(d\omega) .$$

μ is clearly an admissible measure on $(\mathbb{R}^+ \times \Omega, \mathscr{B} \otimes \mathscr{F}_\infty)$.

(2) An important special case of the previous example is the following. $V := 1_{[T, \infty[}$, where T is a stopping time. Then, for every bounded optional process X, we obtain

(6.2.2) $$\int X\, d\mu = E(X_T \cdot 1_{\{T < \infty\}}) .$$

In the sequel, this kind of admissible measures will often be encountered. *We shall denote by μ_T the admissible measure associated with the stopping time T.*

(3) Let M be a process for which $E|M_t| < \infty$ for all t. If, for every $]s, t] \times F \in \mathscr{R}$, we set

(6.2.3) $$(]s, t] \times F) := E(1_F \cdot (M_t - M_s)),$$

we obtain a set function defined on $\mathscr{R}$, associated with the process M.

Let us remark that the restriction to $\mathscr{R}$ of the admissible measure defined by (6.2.1) is a set-function of this type with M replaced by V. A general problem, which will be solved later (Cf. §12), consists in giving extra conditions on M, in such a way that the set-function defined on $\mathscr{R}$ by (6.2.3) can be extended into a measure on $\mathscr{P}$. It is clear from the definition that such an extension is an admissible measure.

6.3 Definitions. In order to advance our study of admissible measures, we introduce several classes of predictable subsets of $\mathbb{R}^+ \times \Omega$.

First, we call *bounded* every set A in $\mathbb{R}^+ \times \Omega$, such that, for some $t > 0$, we have

$$A \subset [0, t] \times \Omega .$$

We denote by $\mathscr{C}$ the class of sets A of the form

$$A := \{X = 0\},$$

where X is an adapted, real, continuous process.

We write $\mathscr{H}$ for the class of bounded elements of $\mathscr{C}$.

Following the usual notation, for any class $\mathscr{D}$ of subsets of a set X, we will write $\mathscr{D}_\delta$ (resp. $\mathscr{D}_\sigma$, $\mathscr{D}_d$, $\mathscr{D}_s$) for the class of subsets which can be written $\bigcap_{n \in \mathbb{N}} D_n$ (resp. $\bigcup_{n \in \mathbb{N}} D_n$, resp. $\bigcap_{n \in I} D_n$, resp. $\bigcup_{n \in I} D_n$), where $(D_n)_{n \in \mathbb{N}}$ (resp. $(D_n)_{n \in I}$) is a denumerable (resp. finite) sequence of sets in $\mathscr{D}$.

Moreover, we introduce the following class.

$$\mathscr{O} := \{\complement C : C \in \mathscr{C}\}.$$

6.4 Proposition. *The following relations hold*

(a) $\mathscr{H}_s = \mathscr{H}_\delta = \mathscr{H}$; $\mathscr{C}_s = \mathscr{C}_\delta = \mathscr{C}$;

(b) $\mathscr{O}_\sigma = \mathscr{O}_d = \mathscr{O}$;

(c) $\mathscr{C}_\sigma = \mathscr{H}_\sigma \supset \mathscr{O}$;

(d) $\mathscr{O}_\delta \supset \mathscr{C}$.

Proof: The relations

$$\mathscr{H}_s = \mathscr{H}_d = \mathscr{H} \quad \text{and} \quad \mathscr{C}_s = \mathscr{C}_d = \mathscr{C}$$

follow immediately from the fact that, for every pair (X, Y) of adapted, real, continuous processes, the processes $X \vee Y$ and $X \wedge Y$ are equally adapted and continuous.

It follows trivially that

$$\mathscr{O}_s = \mathscr{O}_d = \mathscr{O}.$$

To prove $\mathscr{C}_\delta = \mathscr{C}$, we consider a sequence (K_n) extracted from $\mathscr{C}$. Let X_n be such that $\{X_n = 0\} = K_n$. Setting $X_n \wedge 1$ instead of X_n if necessary, we may assume that the processes X_n are real, adapted, continuous and bounded. But the process

$$\mathrm{X} := \sum_{n \in \mathbb{N}} \frac{1}{2^n} X_n$$

enjoys the same properties. Since

$$\bigcap_{n \in \mathbb{N}} K_n = \{X = 0\},$$

we have proved that $\bigcap_{n \in \mathbb{N}} K_n \in \mathscr{C}$.

If the K_n's are bounded, $\bigcap_{n \in \mathbb{N}} K_n$ is clearly bounded. Therefore, $\mathscr{H}_\delta = \mathscr{H}$.

The definition of $\mathcal{O}$ and the property $\mathcal{C}_\delta = \mathcal{C}$ together imply immediately $\mathcal{O}_\sigma = \mathcal{O}$. In the same way $\mathcal{C}_s = \mathcal{C}$ implies $\mathcal{O}_d = \mathcal{O}$.

Let us notice that, for every $t \in \mathbb{R}^+$, the set $[0, t] \times \Omega$ belongs to $\mathcal{H}$. There exists indeed a continuous function φ on $\mathbb{R}^+$, which is zero on $[0, t]$ and strictly positive on $]t, \infty[$. The associated deterministic process $X(t, \omega) := \varphi(t)$ has $[0, t] \times \Omega$ as zero-set. Therefore, for every $C \in \mathcal{C}$ and $n \in \mathbb{N}$, we have

$$C \cap ([0, n] \times \Omega) \in \mathcal{H}.$$

The relation $\mathcal{C}_\sigma = \mathcal{H}_\sigma$ is then trivial.

To prove (*c*), we have only to note that, for every positive, adapted, continuous process X,

$$\{X > 0\} = \bigcup_{n \in \mathbb{N}} \left\{\frac{1}{n} - \left(X \wedge \frac{1}{n}\right) = 0\right\} \in \mathcal{C}_\sigma.$$

In the same way, we deduce $\mathcal{O}_\delta \supset \mathcal{C}$ from the relation

$$\{X = 0\} = \bigcap_{n \in \mathbb{N}} \left\{\frac{1}{n} - \left(X \wedge \frac{1}{n}\right) = 0\right\} \in \mathcal{O}_\delta. \quad \square$$

6.5 Proposition. *Let λ be a positive admissible measure on $\mathcal{P}$. Then, the following three properties hold for every $A \in \mathcal{P}$.*

(i) $\lambda(A) = \sup(\lambda(K) : K \in \mathcal{H}, \ K \subset A\}$;

(ii) $\lambda(A) = \sup\{\lambda(C) : C \in \mathcal{C}, \ C \subset A\}$;

(iii) $\lambda(A) = \inf\{\lambda(O) : O \in \mathcal{O}, \ O \supset A\}$.

Proof: In view of the relations between $\mathcal{C}$, $\mathcal{H}$ and $\mathcal{O}$, the three properties of the proposition are clearly equivalent. Let us then prove the third one.

Let us consider $]s, t] \subset \mathbb{R}^+$ and a sequence (φ_n) of continuous functions on $\mathbb{R}^+$, such that

$$\{u : \varphi_n(u) \neq 0\} =]s, t + \frac{1}{n}[.$$

Considering the processes X_n defined by

$$X_n(u, \omega) := 1_F(\omega)\, \varphi_n(u),$$

where $F \in \mathcal{F}_s$, we see that $]s, t] \times F \in \mathcal{O}_\delta$. We show in the same way that, for every $F \in \mathcal{F}_0$, $\{0\} \times F \in \mathcal{O}_\delta$. Recalling that the outer measure of A can be expressed as

$$\lambda^*(A) = \inf\{\sum_n \lambda(R_n) : R_n \in \mathcal{R}, \ \bigcup_n R_n \supset A\},$$

we can find a sequence (O_n) in $\mathcal{O}$ such that

$$\lambda(\bigcup_n O_n) \leqslant \sum_n \lambda(O_n) \leqslant \lambda^*(A) + \varepsilon.$$

This proves the proposition. $\square$

6.6 Theorem. (1) *Let T be a stopping time with the following property. There exists an increasing sequence (T_n) of stopping times with respect to $(\mathcal{F}_{t+})$ satisfying*

(a) $\lim_{n \uparrow \infty} T_n(\omega) = T(\omega)$ *for every* $\omega \in \Omega$,

(b) $T_n(\omega) < T(\omega)$ *for every* $\omega \in \{T > 0\}$.

Then, T is predictable and the family $\bigcup_{n \in \mathbb{N}} \mathcal{F}_{T_n}$ *generates the σ-algebra* $\mathcal{F}_{T^-}$.

(2) *Let T be a predictable stopping time. Then there exists an increasing sequence of stopping times for the filtration $(\mathcal{F}_{t+})$ with the properties*

(a′) $\lim_{n \uparrow \infty} T_n(\omega) = T(\omega)$ *a.s.,*

(b′) $T_n(\omega) < T(\omega)$ *for every* $\omega \in \{T > 0\}$.

(3) *If the "usual hypotheses" (see Definition 1.1) are verified by the stochastic basis $(\Omega, (\mathcal{F}_t), P)$, a stopping time T is predictable iff there exists a sequence (T_n) of stopping times with the properties (a) and (b) (called announcing sequence for T).*

Proof: Let us first remark that, if T is a stopping time and T' a real random variable with $T = T'$ a.s., then the "usual hypotheses" imply that T' is a stopping time. Since, in this case, the σ-algebras $\mathcal{F}_t$ and $\mathcal{F}_{t+}$ are equal for all t's, the conditions (a) and (b) together are clearly equivalent to the pair (a') and (b') of conditions. The third statement of the theorem is then a straightfordward consequence of the first two.

We now prove the first statement. Let (T_n) be a sequence of stopping times with the properties (a) and (b). Since

$$]0, T[\; = \bigcup_{n \in \mathbb{N}} \;]0, T_n],$$

we have then $]0, T[\in \mathcal{P}$, and T is predictable according to   4.10.

To prove the second statement, we first consider the special case $[T] \in \mathcal{C}$. Let X be a continuous adapted process, such that $[T] = \{X = 0\}$ and S_n be the beginning of the set $\left\{ X \leqslant \dfrac{1}{n} \right\}, n \in \mathbb{N}$. The sequence $(T_n) := (S_n \wedge n)$ is, according to Corollary 4.16, a sequence of predictable stopping times, which clearly satisfies (a') and (b').

Let us now consider any predictable stopping time T. According to Proposition 6.5, there exists an increasing sequence (K_n) in $\mathcal{H}$ with the two properties

(6.6.1) $K_n \subset [T]$

and

(6.6.2) $\mu_T([T]) = \sup_{n \in \mathbb{N}} \mu_T(K_n)$

where μ_T is the admissible measure defined in (6.2.2). Because of (6.6.2), we obtain

(6.6.3) $$P(\pi([T] - \bigcup_{n\in\mathbb{N}} K_n)) = 0\,,$$

where π denotes the canonical projection: $\mathbb{R}^+ \times \Omega \to \Omega$. Let us write S_n for the beginning of K_n. According to what we said above, S_n is predictable and there exists an increasing sequence $(S_n^P)_{p\in\mathbb{N}}$ of predictable stopping times such that

(6.6.4) $$S_n^P < S_n \text{ on } \{S_n > 0\}$$

and

(6.6.5) $$\lim_{p\uparrow\infty} S_n^P = S_n \text{ on } \Omega\,.$$

Let f be an increasing isomorphism from $\mathbb{R}^+$ onto $[0, 1[$. Extracting a subsequence, if necessary, we may assume the following

$$\forall n, \forall p, P\left\{f(S_n) - f(S_n^P) \geqslant \frac{1}{p}\right\} \leqslant \frac{1}{p}\cdot\frac{1}{2^n}\,.$$

Let us define

(6.6.6) $$T_n := (\sup_{k\leqslant n} S_k^k) \wedge (\inf_{k>n} S_k^n)\,.$$

The sequence (T_n) is clearly increasing, and T_n is a stopping time for the filtration $(\mathscr{F}_{t+})$ according to Corollary 4.4. We remark that, since (K_n) is increasing, the sequence (S_n) is decreasing and

(6.6.7) $$S := \inf_n S_n = T \quad \text{a.s.}$$

(with the extra property $S = S_n$ on $\{S_n < \infty\}$).

From the definition of S and T_n, we obtain

(6.6.8) $$\inf_{k\geqslant n} S_k^n \leqslant T_n \leqslant \inf_{k\geqslant n} S_k < S \leqslant S_m \quad \text{for all} \quad m$$

and

$$P\left\{f(S) - f(T_n) \geqslant \frac{1}{n}\right\} \leqslant \frac{1}{n}\sum_{k\geqslant n}\frac{1}{2^k} \leqslant \frac{1}{n}\cdot\frac{1}{2^{n-1}}\,.$$

As a consequence of the Borel-Cantelli lemma, the sequence (T_n) converges towards S, a.s. The second statement of the theorem then follows from the relations (6.6.7) and (6.6.8). □

6.7 Theorem *(Predictable section). We assume the "usual hypotheses" for the stochastic basis. Then, for every predictable set A and every $\varepsilon > 0$, there exists a predictable stopping time T with the properties*

$$[T] \subset A$$

and

$$P\{T < \infty\} \geqslant P(\pi(A)) - \varepsilon,$$

where π is the canonical projection $\mathbb{R}^+ \times \Omega \to \Omega$.

Proof: Let A be a predictable set. Considering A as an element of $\mathscr{B}_{\mathbb{R}^+} \otimes \mathscr{F}_\infty$ and applying a classical theorem (see [Del. 1] T. 37, chapter 1) there exists an $\mathscr{F}_\infty$-measurable mapping Z from $\pi(A)$ into $\mathbb{R}^+$, the graph of which lies in A. Let us then define the following measure on $\mathscr{B}_{\mathbb{R}^+} \otimes \mathscr{F}_\infty$.

(6.7.1) $$\mu(A) := \int 1_A(Z(\omega), \omega)\, P(d\omega).$$

This measure is clearly admissible and, according to Proposition 6.5, there exists $K \in \mathscr{H}$ such that

(6.7.2) $$K \subset A,\ \mu(A - K) \leqslant \varepsilon.$$

From the definition of μ, we obtain

$$\begin{aligned}\varepsilon \geqslant \mu(A - K) &= P(\{\omega : (Z(\omega), \omega) \in A \setminus K\}) \\ &= P(\{\omega : (Z(\omega), \omega) \in A\} - P(\{\omega : (Z(\omega), \omega) \in K\} \\ &= P(\pi(A)) - P(\pi(K)).\end{aligned}$$

The beginning of K is a predictable stopping time T included in K and clearly satisfies the conclusions of the theorem. □

6.8 Proposition. *We assume the "usual hypotheses" for the stochastic basis. Let T be a predictable stopping time and H an element of $\mathscr{F}_{T-}$. Then the stopping time T_H (see Proposition 5.4) is also predictable.*

For every $\mathscr{F}_{T-}$-measurable random variable Z, the process $Z \cdot 1_{[T,\infty[}$ is predictable.

Proof: Let (T^n) be an increasing sequence of stopping times which converge towards T a.s., with $T^n < T$ a.s. on $\{T > 0\}$. Using (5.3.2) (b), we may restrict ourselves to considering $H \in \bigcup_{n \in \mathbb{N}} \mathscr{F}_{T^n-}$. But, for $H \in \mathscr{F}_{T^n}$ and $k \geqslant n$, T_H^k is a stopping time (Proposition 5.4). We may then apply Theorem 6.6–(1) to the sequence $(T_H^k)_{k \in \mathbb{N}}$ to obtain the predictability of T_H.

To prove the second statement of the proposition, it is clearly sufficient to consider random variables Z of the form $Z = 1_H$ with $H \in \mathscr{F}_{T-}$. But, since $1_H \cdot 1_{[T,\infty[} = 1_{[T_H,\infty[}$, the predictability of $1_H \cdot 1_{[T,\infty[}$ follows from the predictability of T and Proposition 4.10. □

7. Decomposition theorems for stopping times

In this paragraph, the "usual hypotheses" are always assumed for the stochastic basis.

Decomposition theorem

7.1 Definition. A stopping time T is called *totally inaccesible* when, for every admissible measure on $\mathscr{G}$ and every predictable stopping time S, the following holds:

$$\mu([T] \cap [S]) = 0.$$

7.2 Proposition. *A stopping time T is totally inaccessible if, for every predictable stopping time S, we have*

$$P(\{S = T < \infty\}) = 0.$$

Proof: Let T be a totally inaccessible stopping time. Then, for every predictable stopping time S, we have

$$P(\{S = T < \infty\}) = \mu_T([T] \cap [S]) = 0.$$

Conversely, the equality $P(\pi([T] \cap [S])) = 0$ follows from $P(\{S = T < \infty\}) = 0$ and this equality implies $\mu([T] \cap [S]) = 0$ for every admissible measure μ. □

7.3 Theorem. *For every stopping time T, there exists one and (up to P-negligibility) only one pair (T_e, T_u) of stopping times with the properties*

(a) $[T] = [T_e] \cup [T_u]$, $[T_e] \cap [T_u] = \emptyset$.

(*b*) *T_u is totally inaccessible.*

(*c*) *There exists a sequence $(T_n)_{n \in \mathbb{N}}$ of predictable stopping times such that* $[T_e] \subset \bigcup_n [T_n]$

Proof: Let us consider the following family of subsets of

$$\mathbb{R}^+ \times \Omega: \mathscr{H} := \{[S] \cap [T]: S \text{ predictable stopping time}\}.$$

There exists a (up to P-negligibility) unique set $A \in \mathscr{H}^\sigma \subset \mathscr{G}$ such that

$$\mu_T(A) = \sup\{\mu_T(H): H \in \mathscr{H}\}.$$

Let T_e be the beginning of A and T_u the beginning of $[T] - A$. As A is contained in $[T]$, we have $[T_e] = A$ and $[T_u] = [T] - A$. According to the definition of A, the following property holds for T_u. For every predictable stopping time S, $\mu_T(T_u] \cap [S]) = 0$. As $[T_u] \subset [T]$, the last relation can be read $P\{T_u = S < \infty\} = 0$. This expresses the total inaccessibility of T_u.

The definition of A also implies the existence of denumerably many predictable stopping times $[T_n]$ such that $A = [T] - [T_u] \subset \bigcup [T_n]$.

It remains to prove the uniqueness. Suppose (T_e, T_u) and (T'_e, T'_u) are two decompositions of T with the properties (a), (b) and (c). The total inaccessibility of T_u and the existence of a denumerable family (T'_n) of predictable stopping times with $[T'_e] \subset \bigcup_n [T'_n]$ together imply

$$[T'_e] \setminus [T_e] = [T'_e] \cap [T_u] = \emptyset \ P.\ a.\ s.$$

In the same way, $[T_e] - [T'_e]$ is an evanescent set. This proves $P\{T_e = T'_e\} = 1$ and therefore $P\{T_u = T'_u\} = 1$. □

7.4 Definition. A stopping time T, the graph $[T]$ of which is included in the union of denumerably many graphs of predictable stopping times, is called *accessible.*

7.5 Examples. (1) In § 4.8, we gave an example of a non-predictable stopping time. Coming back to this example, we recall that, for every predictable A, $A \cap [0, T]$ is the trace on $[0, T]$ of a set $B \times \Omega$, where B is a Borel subset of $\mathbb{R}^+$. If A is a graph, then B reduces to a point or to $\emptyset$. If the probability P assigns measure zero to every $\omega \in \Omega$, then T is totally inaccessible.

(2) One naturally asks, therefore, about the existence of accessible stopping times which are not predictable. We give now a very simple example of a non-predictable accessible stopping time. We define $\Omega := \{1, 2\}$, $\mathcal{F}_t = \{\emptyset, \Omega\}$ if $t < 1$, $\mathcal{F}_t = \mathfrak{P}(\Omega)$ if $t \geqslant 1$, $T(\omega) = \omega$. As in the foregoing example, we remark that the trace of predictable sets on $[0, T]$ are of the form $[0, T] \cap (B \times \Omega)$, where B is a Borel subset of $\mathbb{R}^+$.

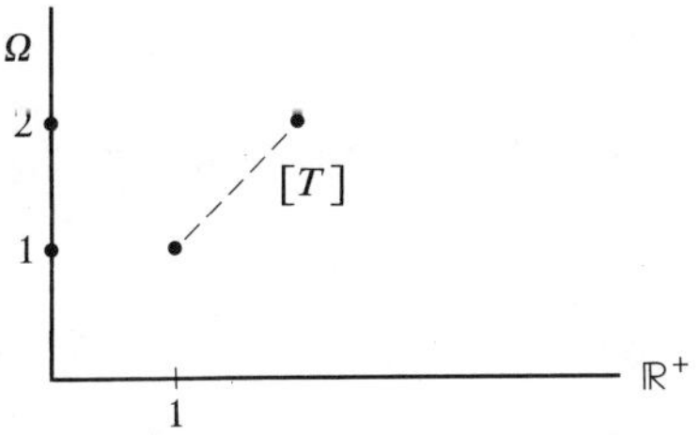

The graph of T is therefore not predictable but is included in the union of the graphs of the two constant, therefore predictable, stopping times $T_1 = 1$ and $T_1 = 2$.

Decomposition of R.R.C. processes

7.6 Theorem. *Every R.R.C. real process admits a representation of the type*

(7.6.1) $$X = Y + \sum_n \Delta X_{S_n} \cdot 1_{[S_n]} + \sum_n \Delta X_{T_n} \cdot 1_{[T_n]},$$

where Y is a left-continuous, adapted process, (T_n) (resp. (S_n)) is a sequence of totally inaccessible (resp. predictable) stopping times, and ΔX denotes the real optional process

$$(7.6.2) \qquad \Delta X(t, \omega) := X(t, \omega) - \lim_{s \uparrow t} X(s, \omega).$$

The sequences (S_n) and (T_n) can be chosen in such a way that X has no jump outside $\bigcup_n ([T_n] \cup [S_n])$ and all graphs $[S_n]$ and $[T_n]$ are mutually disjoint.

When X is predictable, the random variables ΔX_{T_n} in the above representation (7.6.1) *are negligible.*

Proof: For every $\varepsilon > 0$, the process X^{ε} defined by

$$(7.6.3) \qquad X^{\varepsilon}(t, \omega) := X((t - \varepsilon) \vee 0, \omega)$$

is clearly an R.R.C. process. From the relation

$$(7.6.4) \qquad \Delta X = \lim_{\varepsilon \downarrow 0} (X - X^{\varepsilon})$$

follows the optionality of ΔX.

Each path of X has only finitely many jumps greater than $\varepsilon > 0$ on every finite interval $[0, t]$. So, if we define

$$A_0^0 := \{|\Delta X| \geqslant 1\}$$
$$\vdots$$
$$A_n^0 := \left(\frac{1}{n+1} \leqslant |\Delta X| < \frac{1}{n}\right) \quad \text{for} \quad n \geqslant 1,$$
$$\vdots$$

the sets A_n^0 are optional with $[D_{A_n^0}] \subset A_n^0$. We can then define inductively

$$A_n^{k+1} := A_n^k - [D_{A_n^k}].$$

For every n, the set $\{t : (t, \omega) \in A_n^0\}$ is finite and therefore, $[D_{A_n^{k+1}}] \subset A_n^{k+1}$ and we have

$$\{|\Delta X| > 0\} = \bigcup_{n,k} [D_{A_n^k}].$$

Therefore, we have

$$\Delta X = \sum_n \Delta X_{\sigma_n} \cdot 1_{[\sigma_n]},$$

where (σ_n) is a denumerable family of stopping times with disjoint graphs. We set

$$Y := \lim_{\varepsilon \downarrow 0} X^{\varepsilon}.$$

Y is clearly an R.L.C. process.

Let us write T_n for the totally inaccessible part of σ_n and (S_k') for a denumerable family of predictable stopping times such that

$$\bigcup_n [\sigma_n] - [T_n] \subset \bigcup_k [S_k'].$$

If we define

$$H_n := \bigcap_{i \leqslant n-1} \{S_i' \neq S_n' < \infty\}$$

and

$$S_n := S'_{nH_n} = \begin{cases} S_n' & \text{on } H_n, \\ +\infty & \text{on } \complement H_n, \end{cases}$$

the S_n are stopping times with mutually disjoint graphs and, since

$$[S_n] = [S_n'] - \bigcup_{i \leqslant n-1} [S_i'],$$

they have predictable graphs and are, therefore, predictable.

The process Y and the thus defined stopping times (T_n) and (S_n) provide us with the desired representation.

Let us assume now that X is predictable. Then, the process $\Delta X = X - Y$ is itself predictable. The above sets A_n^k are predictable and the stopping times σ_n are predictable. □

7.7 Proposition. *A real R.R.C. process X is predictable iff the two following properties hold.*

(i) *For every predictable stopping time T, the random variable*

$$X_T \cdot 1_{\{T<\infty\}} \text{ is } \mathscr{F}_{T-}\text{-measurable}.$$

(ii) *For every totally inaccessible stopping time T, the set*

$$\{\Delta X \neq 0\} \cap [T] \text{ is evanescent.}$$

Proof: Let T be a predictable stopping time and (T_n) an increasing sequence of stopping times with

$$T_n < T \text{ a.s. on } \{T < \infty\}$$

and

$$\lim_n T_n = T \text{ a.s.}$$

The random variable

$$X_{T-} \cdot 1_{\{T<\infty\}} = \lim_{n\to\infty} X_{T_n} \cdot 1_{\{T<\infty\}}$$

is clearly $\mathscr{F}_{T-}$-measurable.

If we assume property (i), then the random variable

$$\Delta X_T = X_T \cdot 1_{\{T<\infty\}} - X_{T-} \cdot 1_{\{T<\infty\}}$$

is $\mathscr{F}_{T-}$-measurable and $\Delta X_T \cdot 1_{[T]}$ is a predictable process according to Proposition 6.8. Theorem 7.6 then shows clearly that (i) and (ii) together imply the predictability of X.

Conversely, let us assume that X is predictable. Theorem 5.5 expresses property (i). The last statement of Theorem 7.6 and the fact that, in the representation (7.6.1), X has no jump outside $\bigcup_n [S_n]$, together imply property (ii). □

7.8 Remark. We stated the two results 7.6 and 7.7 for real processes only. It should be noticed, however, that the statements and proofs of these propositions hold without any change for Banach-valued R.R.C. processes.

Exercises and supplements

Regularity of processes – Brownian and Poisson processes

E. 1. Let us consider $\Omega := x^{\mathbb{R}^+}$ with the canonical filtration $(\mathscr{F}_t)_{t\in\mathbb{R}^+}$ associated with the projections $(X_t)_{t\in\mathbb{R}^+}$. We write Ω_c for the subset of Ω consisting of continuous functions.

(1) Let $I \subset \mathbb{R}^+$. We write $\mathscr{G}_I$ for the σ-algebra of subsets of Ω generated by $\{X_s : s \in I\}$. Show that every $\mathscr{F}_\infty$-mesurable set belongs to some $\mathscr{G}_D$, where D is a denumerable subset of $\mathbb{R}^+$. Hence, deduce that an $\mathscr{F}_\infty$-measurable set F contains Ω_c iff there exists a denumerable dense D in $\mathbb{R}^+$ such that F contains all paths ω, the restriction of which to D is uniformly continuous on each bounded part of D.

(2) Let D be a denumerable dense subset of $\mathbb{R}^+$ and D_n an increasing family of subsets of D such that

(i) $$D = \bigcup_n D_n,$$

(ii) the intersection of D_n with every bounded interval $[a, b]$ is finite,

(iii) for every $D_n := \{t_0^n < t_1^n \ldots < t_j^n \ldots\}$, $\sup_j |t_{j+1}^n - t_j^n| \leqslant \frac{1}{2^n}$.

We set for every $\omega \in \Omega$ and $K > 0$,

$$\phi_D^K(\omega) := \sum_n \sup_{\substack{t_j \in D \\ t_j \leqslant K}} |\omega(t_{j+1}^n) - \omega(t_j^n)|.$$

Let P be a probability on $(\Omega, \mathscr{F}_\infty)$. Show that the property $P\{\omega : \phi_D^K(\omega) < \infty\} = 1$ for all K and D implies

$$P^*(\Omega_c) = 1.$$

E. 2. (1) Let $h \curvearrowright \varepsilon(h)$ and $h \curvearrowright \eta(h)$ be two real increasing functions defined on some $[0, \delta]$ and possessing the two properties

(i) $$\int_0^\delta \frac{\varepsilon(h)}{h}\, dh < \infty \quad \text{with} \quad \frac{\varepsilon(h)}{h} \quad \text{a monotone function of } h;$$

(ii) $$\int_0^\delta \frac{\eta(h)}{h^2}\, dh < \infty \quad \text{with} \quad \frac{\eta(h)}{h^2} \quad \text{a monotone function of } h.$$

Let P be a probability on $(\Omega := x^{\mathbb{R}^+}, \mathscr{F}_\infty)$ (same notations as in E. 1) and such that

$$P(\{|X_{t+h} - X_t| > \varepsilon(h)\}) \leqslant \eta(h) \quad \text{for all} \quad h \in [0, \delta]$$

Then

$$P^*(\Omega_c) = 1.$$

(2) Apply this to Brownian motion and prove, in particular, the existence of the canonical Brownian motion as defined in § 2.2.

E. 3. Let us consider $\Omega := x^{\mathbb{R}^+}$ and the canonical filtration $(\mathscr{F}_t)_{t \in \mathbb{R}^+}$ as in E. 1. For every denumerable subset I of $\mathbb{R}^+$, we call Ω_I the set of paths which, when restricted to D, have left and right limits in any $t \in \bar{I}$ and are right-continuous on I in every $t \in I$. We call Ω_r (resp. Ω_r') the set of regular paths, i.e. the set of ω which have a left limit and are right-continuous in every point $t \in \mathbb{R}^+$ (resp. which are increasing and right-continuous).

(1) Let P be a probability measure on $(x^{\mathbb{R}^+}, \mathscr{F}_\infty)$. Show that $P^*(\Omega_r) = 1$ if $P(\Omega_I) = 1$ for every denumerable $I \subset \mathbb{R}^+$.

(2) Let $x := \mathbb{N}$ and P be the law of the Poisson process. Show that $P^*(\Omega_r') = 1$.

(3) Letting $\Delta X_t(\omega) := X_t(\omega) - \lim_{s \uparrow t} X_s(\omega)$ for any $\omega \in \Omega_r$ and noting that

$$\sup_{0 \leqslant t < K} |\Delta X_t| = \lim_n \max_{0 \leqslant k \leqslant nK} \left| X\left(\frac{k}{n}\right) - X\left(\frac{k-1}{n}\right) \right|,$$

show that, for the law P of the Poisson process,

$$P^* \{\omega : \omega \in \Omega_r', \sup_t \Delta X(\omega) \leqslant 1\} = 1.$$

In other words, the paths of the canonical Poisson process have jumps of only magnitude one.

E. 4. Let $(\Omega_r, (\mathscr{F}_t)_{t \in \mathbb{R}^+}, P)$ be the canonical basis for the Poisson process starting at 0, where Ω_r is the set of regular functions (i.e., left limits and right continuous) from $\mathbb{R}^+$ into $\mathbb{N}$). Let T_i be the time of the ith jump of the process. More precisely,

$$T_1(\omega) := \inf\{t : X_t(\omega) \geqslant 1, t > 0\},$$
$$\vdots$$
$$T_n(\omega) := \inf\{t : t > T_{n-1}(\omega), X_t(\omega) > X_{T_{n-1}}(\omega)\}.$$

(1) Show that the T_i are stopping times.

(2) Using the independent increments property, show that, for every $t > 0$ and $s > 0$,

$$P\{T_1 > t + \tau\} = P\{T_1 > t\} \cdot P\{T_1 > s\},$$

and therefore

$$P\{T_1 > t\} = e^{-\lambda t} \quad \text{for some } \lambda .$$

(3) Show that the $T_1, T_2 - T_1, \ldots, T_n - T_{n-1}$. are independent random variables with the same distribution.

Hence, deduce that $P\{X_t = n\} = \left(\frac{\lambda t}{n!}\right)^n e^{-\lambda t}$ and, therefore, $\lambda = h$, where h is the parameter of the Poisson process.

Stopping times

E. 5. Let Ω be a set and $A \subset \Omega$ with $\emptyset \neq A \neq \Omega$. The σ-algebras $\mathscr{F}_0$ and $\mathscr{F}_1$ are as follows.

$$\mathscr{F}_0 := \{\emptyset, \Omega\},$$
$$\mathscr{F}_1 := \{A, \complement A, \emptyset, \Omega\}.$$

We define

$$\mathscr{F}_t := \begin{cases} \mathscr{F}_0 & \text{for} \quad t \in [0, 1[, \\ \mathscr{F}_1 & \text{for} \quad t \geqslant 1 . \end{cases}$$

Show that $1 + 1_A$ is a stopping time which is not predictable.

E. 6. We define

$$\Omega := \mathbb{R}^+, \ F_t := (\mathfrak{P}[0, t]) \cup \{]t, \infty[, \mathbb{R}^+\} \text{ for every } t \in \mathbb{R}^+, T(\omega) := \omega .$$

Show that $\mathscr{F}_{T-} \neq \mathscr{F}_T$, while $\mathscr{F}_t = \mathscr{F}_{t+} = \mathscr{F}_{t-}$ for every $t \in \mathbb{R}^+$.

E. 7. Let the canonical Poisson process be as in E. 4.

(1) Show that for every $t \in \mathbb{R}^+$ and every stopping time T with $t \geqslant T$, the following relation holds.

$$P(\{X_t \geqslant 1, X_T = 0\}) = P\{X_{t-T} \geqslant 1\} \cdot P\{X_T = 0\}$$

(Consider first a T with a denumerable number of values and then a decreasing sequence of stopping times converging to t).

(2) Let T_1 be the time of the first jump as defined in E. 4. Assuming that $T \leqslant T_1$, prove

$$P\{T < T_1\} = P\{X_{T_1 - T} \geqslant 1\} \cdot P\{X_T = 0\} .$$

Deduce from this inequality that $T \leqslant T_1$ implies $P\{0 < T < T_1\} = 0$.

(3) Prove that T_1 is totally inaccessible.

E. 8. Let T be a stopping time and Z a real random variable. Show that the conditions $Z \geqslant T$ and "Z is $\mathscr{F}_T$-measurable" together imply that Z is a stopping time.

E. 9. A decreasing sequence (T_n) of stopping times is called *stationary* when, for every ω, an integer $n(\omega)$ exists such that $T_{n(\omega)}(\omega) = \lim_n T_n(\omega)$. Show that the limit of a stationary decreasing sequence of predictable stopping times it itself a predictable stopping time.

E. 10. Let T be a stopping time and $A \in \mathscr{F}_\infty$. Prove that

$$A \cap [T = \infty] \in \mathscr{F}_{T-}.$$

E. 11. Let T_1 and T_2 be two stopping times. Prove the following equalities.

$$\{T_1 \leqslant T_2\} \cap \mathscr{F}_{T_1 \vee T_2} = \{T_1 \leqslant T_2\} \cap \mathscr{F}_{T_2}$$
$$\{T_1 < T_2\} \cap \mathscr{F}_{(T_1 \vee T_2)-} = \{T_1 < T_2\} \cap \mathscr{F}_{T_2-}$$

Optionality and predictability

E. 12. Show that $\mathscr{P}$ is generated by the family of processes

$$1_{[S,T]} \cdot 1_F,$$

where $S \leqslant T$ are stopping times, S being predictable and $F \in \mathscr{F}_{S-}$. Using example 4.8, show that, for non-predictable S and every $F \in \mathscr{F}_{S-}$, one may have

$$[S, T] \times F \notin \mathscr{P}.$$

E. 13. [Del. 1]. We assume the "usual hypotheses". Show that every real, adapted, right-continuous process X is optional. *Hint:* Every rightcontinuous evanescent process is optional. Call $\mathscr{A}$ the family of stopping times S for which there exists an optional process X^s such that the set

$$\{(t, \omega) : t \in [0, S(\omega)[, \, |X_t(\omega) - X_t^s(\omega)| \geqslant \varepsilon\}$$

is evanescent. Let $T :=$ ess sup $\mathscr{A}$ and $U :=$ beginning of

$$\{(t, \omega) : t > T(\omega), |X_T(\omega)| \geqslant \varepsilon\}.$$

Prove that $U \in \mathscr{A}$. Since, following the definition, we have $T < U$ on $\{U < \infty\}$, then $T = +\infty$. The possibility of approximating X through optional processes follows.

Markov processes

E. 14. Show that the Brownian motion, the Poisson-process are Markov processes. What are the semi-groups?

(*Answer:* Brownian motion $\Pi_t(x, dy) = \dfrac{1}{\sqrt{2\Pi t|C|}} \exp - \dfrac{(y-x) \cdot C^{-1}(y-x)}{2t} dy$

Poisson-process $\Pi_t(k, k+n) = e^{-\lambda t} \dfrac{(\lambda t)^n}{n!}$).

E. 15. Let X be the transport process of example 2.6 and (u_t) be the process of the speed. The laws of the τ_n are supposed exponential with parameters λ. Then the couple (X, u) is a Markov process.

Is X a Markov Process? (no!).

E. 16. Let X be a process on the stochastic basis $(\Omega, (\mathscr{F}_t), \mathfrak{A}, P)$ with increments independent of the past. It is called *stationary* if for every s and t the law of the variables $X_{s+t} - X_s$ is the same as the law of $X_t - X_0$. Let us denote this law by ν_t. Show that X is Markov process with semi-group $T_t f = \nu_t * f$. Conversely, if X is a Markov process with a "convolution semi-group" (i. e. such that $T_t f = \nu_t * f$, the family (ν_t) of probability-laws satisfying $\nu_{t+s} = \nu_t * \nu_s \, \forall s, t)$, X is a process with increments independent of the past.

E. 17. *Another form of the Markov property.* Let Ω be a subset of $\mathscr{X}^{\mathbb{R}^+}$ with the following property: $\forall s \in \mathbb{R}, \omega \in \Omega$ the function $t \curvearrowright \omega(t+s)$ is also an element of $\mathscr{X}^{\mathbb{R}^+}$. We call Θ_s the *translation operator*: $\omega \overset{\Theta_s}{\rightarrow} (\omega_{t+s})_{t \in \mathbb{R}}$. ξ being the canonical process on $\Omega\,(\xi_t(\omega) := \omega(t))$, we have by definition $X_t(\Theta_s(\omega)) = X_{t+s}(\omega)$.

Let $(\mathscr{F}_t)_{t \geqslant 0}$ be the filtration generated on Ω by ξ (i. e.: $\mathscr{F}_t :=$ σ-algebra generated by $\{\xi_s : s \leqslant t\}$). Assume we are given a semi-group (Π_t) of transitions as in 2.7 ($\forall f \in C_0(\mathscr{X})$ the function $x \curvearrowright \int f(y)\, \Pi_t(x, dy)$ belongs to $C_0(\mathscr{X})$) and that for every $x \in \mathscr{X}$ a probability P_x can be constructed on $(\Omega, \mathscr{F}_\infty)$, using the procedure described in 2.7, such that ξ is a Markov process associated with $(\Pi_t)_{t \geqslant 0}$ and starting from x at time zero.

(1°) Show the following: Writing $S(x, \Phi)$ for the expectation $E_x(\Phi)$ with respect to P_x of any bounded $\mathscr{F}_\infty$-measurable function $\mathscr{F}$ on Ω, for every and x

(1) $$E_x(\Phi \circ \Theta_t | \mathscr{F}_t) = S(\xi_t, \Phi)$$

where $\Phi \circ \Theta_t$ is the composition of mappings Θ_t and Φ.
This formula is usually written in the following slightly confusing form

(2) $$E_x(\Phi \circ \Theta | \mathscr{F}_t) = E_{\xi_t}(\Phi)$$

Hints: First prove this formula for functions Φ of the type $\Phi(\omega) = \varphi_0(\xi_0(\omega)), \ldots \varphi_n(\xi_{t_n}(\omega))$ $0 = t_0 < \ldots < t_n$ using iterated integrals as in (2.7.2) and use a classical extension procedure.

(2°) Show that formula (1) holds if $\mathscr{F}_t$ is replaced by $\mathscr{F}_{t+}$.

(3°) Prove the following Blumenthal zero-one law: $\forall F \in \mathscr{F}_0^+$ one has either $P_x(F) = 0$ or $P_x(F) = 1$.

(Note that $1_F \circ \Theta_0 = 1_F$.)

E. 18. *The strong Markov property.* We use the same notation and make the same assumptions as in E. 17. We assume, moreover, that the canonical process ξ on Ω is right-continuous (i. e. the probabilities P_x as defined in 2.7 are carried by the set of right-continuous paths in $\mathscr{X}^{\mathbb{R}^+}$). For every finite stopping time τ, Θ_τ denotes the transformation $\omega \curvearrowright \Theta_{\tau(\omega)}(\omega)$ in Ω.

(1° Prove that for every bounded $\mathscr{F}_\infty$-measurable function Φ on Ω every finite stopping time τ and $x \in \mathscr{X}$

$$(1) \qquad E_x(\Phi \circ \Theta_\tau | \mathscr{F}_\tau^+) = S(X_\tau, \Phi).$$

This formula, which expresses the so-called *strong Markov*-property, is usually written:

$$(2) \qquad E_x(\Phi \circ \Theta_\tau | \mathscr{F}_\tau^+) = E_{X_\tau}(\Phi)$$

(But notice here too that $E_{X_\tau}(\Phi)$ should be read as follows: take for every x the expectation $E_x(\Phi)$ and *then* substitute X_τ for x).

Hints: First prove the formula for stopping times τ, taking a denumerable set of values and writing $F = \bigcup_i F \cap \{\tau = t_i\}$ for every $F \in \mathscr{F}_\tau$. Use the properties of the semi-group $(\Pi_t)_{t \geqslant 0}$ to show that $x \curvearrowright S(x, \Phi)$ is continuous for every Φ depending on a finite number of coordinates of ω, approximate any τ by a decreasing sequence of stopping times with denumerably many values and use the right continuity of paths.

(2°) If τ can take the value $+\infty$, show that the following strong Markov-formula may be written.

$$(3) \qquad E(1_{\{\tau < \infty\}} \Phi \circ \Theta_\tau | \mathscr{F}_\tau^+) = 1_{\{\tau < \infty\}} E_{X_\tau}(\Phi)$$

E. 19. *The time needed for leaving a point.* Let ξ be a canonical right-continuous Markov process with a Feller semi-group of probality transitions as in E. 17. We call P_x the law of the process starting from x and set $\tau_1 := \inf\{t : t > 0\, \xi_t \neq x\}$.

Show that there exists $\lambda(x) \in \overline{\mathbb{R}}_+$ such that

$$P_x\{\tau_1 > t\} = e^{-\lambda(x)t}.$$

If $\lambda(x) = +\infty$, the state x is called *instantaneous.*
If $0 < \lambda(x) < \infty$, the state x is called *stable.*
If $\lambda(x) = 0$, the state x is called *absorbing.*

Hints: Use the strong Markov property to show that for every $t \geqslant 0$ and $h > 0$

$$P_x\{\tau_1 > t+h, \tau_1 > t\} = P_x\{\tau_1 > t\} P_x\{\tau_2 > h\}.$$

E. 20. *Moments of regular right continuous processes with stationary independent increments and bounded jumps.* Let $(X_t)_{t \geqslant 0}$ be a real, regular, right-continuous process. Define

$$\tau_N := \inf\{t : |X_t - X_0| > N\}.$$

Assume that the jumps of $t \to X_t(\omega)$ are, for all ω, bounded by C.

(1°) Prove that for every $\eta \geqslant 0$:

$$P\{\sup_{0\leqslant s\leqslant t}|X_s-X_0|>N+\eta+C\}\leqslant P\{\tau<t\sup_{\tau\leqslant s<t}|X_s-X_\tau|>\eta\}\leqslant$$
$$\leqslant P\{\sup_{0\leqslant s\leqslant t}|X_s-X_0|>N\}P\{\sup_{0\leqslant s\leqslant t}|X_s-X_0|>\eta\}$$

Hint: Use the strong Markov-property to write

$$P\{\tau<t\sup_{\tau\leqslant s\leqslant t}|X_s-X_\tau|>\eta\}\leqslant E\{1_{\{\tau<t\}}\cdot 1_{\{\sup_{\tau\leqslant s\leqslant t+\tau}/X_s-X_\tau/>\eta\}}\}\leqslant$$
$$\leqslant E\{1_{\{\tau<t\}}P\{\sup_{0\leqslant s\leqslant t}|X_s-X_0|>\eta\}\}$$

(2°) Choose η such that

$$P\{\sup_{0\leqslant s\leqslant t}|X_s-X_0|>\eta\}\leqslant\frac{1}{2}$$

to show

$$E\sup_{0\leqslant s\leqslant t}|X_s-X_0|^k\leqslant\sum_n n^k(\eta+C)^k\,2^{(1-n)}.$$

Conclusion: Under the hypothesis on X the random variables $\sup_{0\leqslant s\leqslant t}|X_t-X_0|$ have moments of all orders.

E. 21. *Regular, right-continuous processes with independent increments.* Let X be a regular, right-continuous process with increments independent of the past for a given filtration. Write $\nu_{t,t+u}(dx)$ for the law of $X_{t+u}-X_t$ and call $\tilde{X}^{x,u}$ the process with state $\mathbb{R}^d\times\mathbb{R}^+$ defined by

$$\tilde{X}_0^{x,u}=(x,u)\quad\tilde{X}_t^{x,u}=(X_t,u+t)$$

(1°) Show that the process X is a Markov process associated with the probability-transitions $(\tilde{\Pi}_t)_{t>0}$, i.e.,

$$\int_{\mathbb{R}^d\times\mathbb{R}^+}\varphi(y,v)\,\tilde{\Pi}_t(x,u;dy,dv)=\int\varphi(y,u+t)\,\nu_{t,t+u}(dy)$$

(2°) Apply the method of E. 20 to show that *every regular right-continuous process X with increments independent of the past and with bounded jumps is such that $\sup_{0\leqslant s\leqslant t}\|X_s-X_0\|$ has moments of all orders for every t.*

Historical and bibliographical notes

Stochastic processes were studied long before the theory presented here was established. Brownian motion is so-called on account of Brown who, in 1928, observed and described the motion of small particles suspended in water. The mathematical study was initiated by Bachelier (Bac] and carried on by many people. Wiener [Wie] proved, in particular, the path continuity. Many deep properties are due to P. Levy ([Lev. 1] and [Lev. 2]). An extensive exposition of the properties of Brownian motion may be found in the Ito-McKean book [ItM]. A short but substantial introduction to the main properties of Brownian motion can be found in chapter 12 of [Bre]. Much information is to be found in Feller's book [Fel].

The description of stochastic processes as families or random variables defined on a probability space starts with Kolmogorov's papers [Kol. 1], [Kol. 2] which are generally considered as founding modern probability theory. The fundamental "Kolmogorov's theorem", which shows how to define a probability law on a set of paths, has received many extensions under the name of projective-limit theorems for measures ([Boc], [Cho], [Met. 1], [Rau]).

The notion of a filtration of σ-algebras $(\mathscr{F}_t)$, representing the time before t, was considered systematically by Doob [Doo. 1], [Doo. 3]. In this and many other respects, which will be indicated in the following chapters, Doob's work is certainly one of the most significant pioneering works in what we call here the "dynamic theory of stochastic processes". His book [Doo. 1] contains most of the fundamental ideas leading to the notions and results presented here. The other main developments and fruitful new techniques since Doob's book was published are due to P.A. Meyer and the French school. Doob introduced various measurability conditions for processes, connected with the given filtration and made explicit the notion of a stopping time.[1]) It was P.A. Meyer, however, who introduced the notion of predictable sets, decomposition theorems for stopping times and many other important features of the present theory, which we shall meet all through this book. See [Mey. 1]. The present simplicity of the theory owes much to Dellacherie's book [Del. 1].

The present state of the theory contains many developments which are not included in the "kernel" which we present in this book. The two big volumes [DeM] give, on the contrary, a systematic exposition. For example, we have even omitted to mention the σ-algebra of accessible sets, which is between $\mathscr{P}$ and $\mathscr{G}$, and topics like "separability", which seem to us to have lost most of their interest.

Markov processes, introduced here to provide the reader with a field of applications of the theory, form in themselves an enormous subject. An efficient short introduction may be found in [DyY], [Bre], chap. 13 and [GiS] (from where we extracted the proof of E. 20 for processes with independent increments). Much more information on Markov processes may be found in (Chu. 3] (the classical theory of Markov chains), [Dyn], [BlG] (Markov processes in continuous time), [Ore. 2], [Rev] (Markov chains with general state-space).

[1]) See also Chung [ChD].

Chapter 2

Martingales and quasimartingales – Basic inequalities and convergence theorem – Application to stochastic algorithms

This chapter gives the basic definitions for Martingales and Quasimartingales and the elementary properties which follow from the basic fact, taken here as a definition, that quasimartingales (resp. martingales, resp. submartingales) have a "Doleans additive measure" with bounded variation, (resp. a positive "Doléans additive measure").

The convergence and regularity properties of paths are entirely and easily derived from this.

We assume once and for all that a stochastic basis $(\Omega, (\mathscr{F}_t)_{t \in \mathbb{R}^+}, \mathfrak{A}, P)$ is given and that $\mathscr{F}_\infty$ is the σ-algebra generated by $\bigcup_{t \in \mathbb{R}^+} \mathscr{F}_t$.

When we consider a process $(X_t)_{t \in I}$ indexed only by a subset of $\mathbb{R}^+$, such a process is said to be adapted (to the given filtration $\mathscr{F} := (\mathscr{F}_t)_{t \in \mathbb{R}^+}$) when X_t is $\mathscr{F}_t$-measurable for every $t \in I$.

8. Martingales, submartingales, supermartingales, quasimartingales: elementary properties

8.1 Definitions. Let $\mathbb{B}$ be a Banach space. A stochastic process X with values in $\mathbb{B}$ and index set $I \subset \bar{\mathbb{R}}^+$ is called a $(P, (\mathscr{F}_t)_{t \in I})$-*martingale* (or, more briefly, a martingale when P and $(\mathscr{F}_t)_{t \in I}$ are clear) when X is an adapted process with $E(\|X_t\|) < \infty$ for all $t \in I$ and when the following property holds

(M) For every $(s, t) \in I \times I$ with $s < t$, and every $F \in \mathscr{F}_s$,

$$E(1_F \cdot X_t) = E(1_F \cdot X_s).$$

Let us assume now that $\mathbb{B}$ has an order structure compatible with the vector-space structure and that this order is denoted by $\leqslant$ · A $\mathbb{B}$-valued process X with index set $I \subset \mathbb{R}^+$ will be called a *submartingale* (resp. a *supermartingale*) if for every $t \in I$, X_t is integrable and the following property M^+ (resp. M^-) holds.

(M^+) For every $(s, t) \in I \times I$ with $s < t$ and every $F \in \mathscr{F}_s$,

$$E(1_F \cdot X_t) \geqslant E(1_F \cdot X_s).$$

(resp. (M^-) $E(1_F \cdot X_t) \leqslant E(1_F \cdot X_s)$ for every $s < t$ and every $F \in \mathscr{F}_s$)

8.2 Examples. (1) The reader will easily prove that, if X is a process with independent increments (see §2) and if $(\mathscr{F}_t^X)$ is the filtration associated with X (i.e. $\mathscr{F}_t^X$ is, for every t, the σ-algebra generated by $\{X_s : s \leqslant t\}$), the process X is a $(\mathscr{F}_t^X)$-martingale when $E(X_t - X_s) = 0$ for every pair (s, t).

When $E(X_t - X_s) \geqslant 0$ for every couple (s, t), $s \leqslant t$, then X is a submartingale with respect to $(\mathscr{F}_t^X)$.

As a consequence, the canonical Brownian motion is a martingale with respect to its own filtration. The Poisson process N with parameter λ is a submartingale with respect to its own filtration $(\mathscr{F}_t^N)$, while $(N_t - \lambda t)_{t \in \mathbb{R}^+}$ is a martingale with respect to $\mathscr{F}_t^N$. When we consider other filtrations, this may no longer be the case. (see exercise E. 1. 1).

(2) Let P and P' be two probability measures on $(\Omega, (\mathscr{F}_t)_{t \in I})$. Let us assume that, for every $t \in I, P'$, restricted to $\mathscr{F}_t$, is absolutely continuous with respect to the restriction P_t of P to $\mathscr{F}_t$. If we denote by X_t the Radon-Nikodym density of P' with respect to P_t on the σ-algebra $\mathscr{F}_t$, it is readily seen that the family $(X_t)_{t \in I}$ of random variables is a $(P, (\mathscr{F}_t))$-martingale.

8.3 Remarks. An equivalent statement for the property (M) (resp. (M^+), resp. (M^-)) is the following, when X is real. For every $(s, t) \in I \times I$ with $s < t$,

(8.3.1) $$E(X_t | \mathscr{F}_s) = X_s \text{ (resp. } \geqslant X_s, \text{ resp. } \leqslant X_s) .$$

If X is Banach-valued, the property M can be expressed exactly in the same way by (8.3.1).

At this point, we remind the reader of the following argument to prove the existence of the conditional expectation for Banach-valued random variables. For every step function

$$f = \sum_{i=1}^{n} 1_{F_i} \cdot a_i, \; a_i \in \mathbb{B}, \; F_i \in \mathscr{F}_\infty \supset \mathscr{F}_s ,$$

we define

$$E(f | \mathscr{F}_s) = \sum_{i=1}^{n} E(1_{F_i} | \mathscr{F}_s) \cdot a_i .$$

This defines a continuous linear mapping from a dense linear subspace of $L^1_{\mathbb{B}}(\Omega, \mathscr{F}_\infty, P)$ into $L^1_{\mathbb{B}}(\Omega, \mathscr{F}_s, P)$, with norm $\leqslant 1$. This mapping can be uniquely extended into a linear operator from $L^1_{\mathbb{B}}(\Omega, \mathscr{F}_\infty, P)$ into $L^1_{\mathbb{B}}(\Omega, \mathscr{F}_s, P)$ with norm $= 1$. One thus defines the conditional expectation with respect to $\mathscr{F}_s$.

It should be emphasized that this conditional expectation is defined without using any theorem of the Radon-Nikodym type. For a general Banach space, such a theorem does not hold. More precisely, we shall say that $\mathbb{B}$ has the *Radon-Nikodym property* if, for every pair $\{\mu, \nu\}$ of bounded measures defined on a measurable space $(\Omega, \mathscr{F})$, with values in $\mathbb{B}$ and $\mathbb{R}^+$ respectively and such that $(F \in \mathscr{F}, \mu(F) = 0) \Rightarrow \nu(F) = 0$, then there exists a $\mathbb{B}$-valued, $\mathscr{F}$-measurable function f such that

$$\forall F \in \mathscr{F}, \; \nu(F) = \int_F f \, d\mu .$$

A large class of Banach spaces which has the Radon-Nikodym property (we write R. N. Property) is the class of separable reflexive spaces. An even larger class is the class of separable Banach spaces which are dual spaces of other Banach spaces.

8.4 The content associated with a process (see [Föll] or [MeP. 2]).

At the end of § 3 we introduced the ring of subsets of $\mathbb{R}^+ \times \Omega$, denoted by $\mathscr{A}$, generated by the family $\mathscr{R}$ of predictable rectangles.

For every index set $I \subset \mathbb{R}^+$, we now define the boolean ring $\mathscr{A}_I$ of subsets of $\mathbb{R}^+ \times \Omega$, generated by the following class $\mathscr{R}_I$.

$$\mathscr{R}_I := \{]s,t] \times F : (s,t) \in I \times I, s < t, F \in \mathscr{F}_s\} \cup \{[0, \inf I] \times F, F \in \mathscr{F}_0\}$$

If $I = \mathbb{R}^+$, then $\mathscr{A}_I$ and $\mathscr{A}$ are the same.

By a *content* on a boolean ring, we mean a real- or vector-valued additive function defined on this ring. Since $\mathscr{A}_I$ consists of the finite unions of sets of $\mathscr{R}_I$, and as $\mathscr{R}_I$ is a semi-ring[1]), it is easily seen that a content on $\mathscr{A}_I$ is uniquely determined by any additive function on $\mathscr{R}_I$.

In particular, with every process such that X_t is integrable for all $t \in I$, we associate the content λ_X on $\mathscr{A}_I$, uniquely determined by

$$\begin{cases} \lambda_X(]s,t] \times F) := E[1_F \cdot (X_t - X_s)], (s,t) \in I \times I, s \leqslant t, F \in \mathscr{F}_{s-}, \\ \lambda_X([0, \inf I] \times F) := 0, \ F \in \mathscr{F}_0 . \end{cases}$$

We make the following trivial remark. *The martingale property for an adapted process X is equivalent to the property* $\lambda_X = 0$, while $\lambda_X \geqslant 0$ (resp. $\leqslant 0$) is equivalent to the submartingale (resp. supermartingale) property.

Let us assume now that I possesses a larger element β and the Banach space $\mathbb{B}$ has the R. N. Property. Then, for every $t \in I$, the random variable X_t is entirely determined (up to P-equivalence) by λ_X and X_β by the relation

$$F \in \mathscr{F}_t, \int_F X_t \, dP = \int_F X_\beta \, dP - \lambda_X(]t, \beta] \times F) \tag{8.4.1}$$

and the condition

X_t is $\mathscr{F}_t$-measurable.

8.5 Proposition. *We assume* $\beta := \sup I \in I$. *Let* $(X_t)_{t \in I}$ *be a real adapted stochastic process. The content* λ_X *is of bounded variation* (*i. e.*, $\sup \{\sum_i |\lambda_X(A_i)| : (A_i)$ finite $\mathscr{A}_I$-partition of $[0,\beta] \times \Omega\} < \infty$) *iff X can be written*

$$X = M + X' - X'',$$

---

[1]) A semi-ring $\mathscr{J}$ is a class of subsets of a set closed under finite intersections and such that, for every $A, B \in \mathscr{J}$ with $A \subset B$, there exists a finite family of disjoint sets I_i in $\mathscr{J}$ such that $B \backslash A = \bigcup_i I_i$.

where M is a martingale and X' and X'' are two positive supermartingales with $X'_\beta = X''_\beta = 0$.

Proof: We recall the following known equivalence (see [Bou. 1], chap. 1 or [DuS]). "λ_X is of bounded variation iff λ_X is the difference of two positive contents $\lambda_X = \lambda_X^+ - \lambda_X^-$". We then define

$$M_t := E(X_\beta | \mathscr{F}_t) \text{ for } t \in I.$$

The processes X'' and X' are then defined by the formulas

$$\forall F \in \mathscr{F}_t, \int_F X''_t \, dP = \lambda_X^+ (]t, \beta] \times F),$$
$$\int_F X'_t \, dP = \lambda_X^- (]t, \beta] \times F),$$

X''_t and X'_t being $\mathscr{F}_t$-measurable for all $t \in I$. □

8.6 Definition. A Banach-valued process $(X_t)_{t \in I}$ is called a $(P, (\mathscr{F}_t)_{t \in I})$-*quasi-martingale* (or briefly, a quasi-martingale, if the stochastic basis is clear) when the associated content λ_X has a bounded variation on every set $]0, t] \times \Omega$, $t \in I$.

8.7 Remarks and example

Remark 1. The notion of quasi-martingale was first introduced by D. L. Fisk [Fis] and later studied by S. Orey [Ore. 1] who called quasi-martingales F-processes. The original definition was different from the one given here. For the equivalence of both definitions, the reader is referred to the exercise E. 1.2 at the end of this chapter.

Remerk 2. Let $(X_n)_{n \in \mathbb{N}}$ be a real process with index set $\mathbb{N}$. It is immediately seen that, in this case, for any $p \in \mathbb{N}$, the trace of $\mathscr{A}_\mathbb{N}$ on $]0, p] \times \Omega$ is a σ-ring, the elements of which are finite unions

$$\sum_{n \leq p-1}]n, n+1] \times F_n, \; F_n \in \mathscr{F}_n.$$

Since

$$\lambda_X(]n-1, n] \times F) = \int_F (X_n - X_{n-1}) \, dP,$$

the restriction of λ_X to $\mathscr{A} \; \cap]0, p] \times \Omega$ is σ-additive for all p. Since every finite σ-additive measure is of bounded variation, it follows that every real process $(X_n)_{n \in \mathbb{N}}$ such that $E(|X_n|) < \infty$, for every n, is a quasimartingale.

Remark 3. An important problem is to know whether or not the content λ_X, associated with a given process X, has a σ-additive extension to the σ-algebra $\mathscr{P}_I$ generated by $\mathscr{A}_I$. Since a real σ-additive measure is of finite variation on any set of finite measure, it is immediately seen that a real process X, for which λ_X has a σ-additive extension, is necessarily a quasimartingale. Sufficient conditions will be given in § 13. A trivial example is given by the process β^2, where β is a real Brownian motion: $\lambda_{\beta^2}(]s, t] \times F) = (t-s)\,P(F)$.

8.8 Proposition. (1) *Let $(X_t)_{t\in I}$ be a real submartingale. For every convex increasing function g, $(g(X_t))_{t\in I}$ is a submartingale.*

(2) *Let $(X_t)_{t\in I}$ be a $\mathbb{B}$-valued martingale and g a real function on $\mathbb{B}$ with the following property.*

At every $y \in \mathbb{B}$, there exists a linear form L_y such that

$$\forall y' \in \mathbb{B},\, g(y') - g(y) \geqslant L_y(y'-y), \text{ resp. } \leqslant).$$

Then the process $(g(X_t))_{t\in I}$ is a submartingale (resp. a supermartingale).

Proof: This proposition follows immediately from the Jensen inequality (cf [Bau]), namely,

$$E[g(X_t)|\mathcal{F}_s] \geqslant g(E(X_t|\mathcal{F}_s)) \text{ a.s. } \square$$

8.9 Proposition. *Let σ and τ be two stopping times taking their values in a finite subset of I, with $\sigma \leqslant \tau$. Then the set $]\sigma, \tau]$ belongs to $\mathcal{A}_I$ and*

$$\lambda_X(]\sigma, \tau]) = E(X_\tau - X_\sigma)$$

for every process $(X_t)_{t\in I}$ for which $E(X_t)$, $t \in I$, exists.

Proof: Let $\{t_0 < t_1 < \ldots < t_n\} \subset I$ be the set in which σ and τ take their values. It is enough to prove the following relations.

$$]t_0, \tau] \in \mathcal{A}_I$$
$$\lambda_X(]t_0, \tau]) = E(X_\tau - X_{t_0}).$$

We first notive that τ may be written as follows.

$$\tau = \sum_{i=1}^{n} (t_i - t_{i-1}) \,.\, 1_{\{\tau > t_{i-1}\}}.$$

Therefore, we obtain

$$\begin{aligned}\lambda_X(]t_0, \tau]) &= \lambda_X\Big(\bigcup_{i=1}^{n} (]t_{i-1}, t_i] \times \{\tau > t_{i-1}\}\Big)\\ &= \sum_{i=1}^{n} \int_{\{\tau > t_{i-1}\}} (X_{t_i} - X_{t_{i-1}})\, dP\\ &= \sum_{i=1}^{n} \int_{\{\tau = t_i\}} (X_{t_i} - X_{t_0})\, dP\\ &= E(X_\tau) - E(X_{t_0}). \quad \square\end{aligned}$$

For further references, we now prove two lemmas concerning processes indexed by $\mathbb{N}$. Analogous results hold for processes indexed by $\mathbb{R}^+$. For these extensions, the reader is referred to [MeP. 2] and exercise E. 1.5.

8.10 Lemma. *Let $\mathbb{B}$ be a Banach space and $(X_n)_{n\in\mathbb{N}}$ a $\mathbb{B}$-valued process with $E\|X_n\|<\infty$ for every $n\in\mathbb{N}$. Then $\mathscr{A}_{\mathbb{N}}$ is a boolean δ-ring (i.e. $\mathscr{A}_{\mathbb{N}}$ is a boolean ring, stable for denumerable intersections) and λ_X is a σ-additive content (i.e. a measure) on $\mathscr{A}_{\mathbb{N}}$ and, for every n, the following relation holds*

$$|\lambda_X|(]n,n+1]\times\Omega)=E\|E(X_{n+1}-X_n|\mathscr{F}_n)\|,$$

where $|\lambda_X|$ denotes the variation of λ_X.

Proof: For every n and $F\in\mathscr{F}_n$

$$\lambda(]n,n+1]\times F)=E\{1_F(X_{n+1}-X_n)\}=\int_F E(X_{n+1}-X_n|\mathscr{F}_n)\,dP.$$

The restriction of λ to $(]n,n+1]\times\Omega)\cap\mathscr{A}_{\mathbb{N}}=\{]n,n+1]\times F\colon F\in\mathscr{F}_n\}$ is therefore clearly a measure with density $E(X_{n+1}-X_n|\mathscr{F}_n)$ with respect to P. As is known for every measure with a (strongly measurable) density, the variation $|\lambda_X|$ of this measures admits the density $\|E(X_{n+1}-X_n|\mathscr{F}_n)\|$ The fact that $\mathscr{A}_{\mathbb{N}}$ is a δ-boolean ring, the elements of which are sets of the form

$$\sum_n\,]n,n+1]\times F_n,\ F_n\in\mathscr{F}_n$$

has already been noted in Remark 8.7–(2). □

8.11 Lemma. *We assume $(t_n)=I$ to be a monotone sequence in $\mathbb{R}_+$. Let $\mathbb{B}$ be a Banach space and $(X_{t_n})_{t_n\in I}$ a $\mathbb{B}$-valued quasi-martingale. Then, $(\|X_{t_n}\|)_{t_n\in I}$ is a real quasi-martingale and, moreover,*

$$|\lambda_{\|X\|}|(]0,t_n]\times\Omega)\leqslant 2|\lambda_X|(]0,t_n]\times\Omega)+\lambda_{\|X\|}(]0,t_n]\times\Omega)<\infty. \tag{8.11.1}$$

If $(X_{t_n})_{t_n\in I}$ is a martingale, then $(\|X_{t_n}\|)_{t_n\in I}$ is a submartingale.

Proof: We consider the case where (t_n) is increasing, the proof being the same when (t_n) is decreasing. Since

$$\begin{aligned}\|X_{t_n}\|&=\|E(X_{t_n}-X_{t_{n+1}}|\mathscr{F}_{t_n})+E(X_{t_{n+1}}|\mathscr{F}_{t_n})\|\\&\leqslant\|E(X_{t_n}-X_{t_{n+1}}|\mathscr{F}_{t_n})\|+E(\|X_{t_{n+1}}\||\mathscr{F}_{t_n}),\end{aligned}$$

we obtain, for every $F\in\mathscr{F}_{t_n}$,

$$\begin{aligned}-\lambda_{\|X\|}(]t_n,t_{n+1}]\times F)&=\int_F[\|X_{t_n}\|-E(\|X_{t_{n+1}}\||\mathscr{F}_{t_n})]\,dP\\&\leqslant\int_F\|E(X_{t_n}-X_{t_{n+1}}|\mathscr{F}_{t_n})\|\,dP.\end{aligned}$$

This inequality may be rewritten as follows.

(8.11.2) $$\lambda^-_{\|X\|} \leqslant |\lambda_X| .$$

From

$$\lambda^+_{\|X\|} - \lambda^-_{\|X\|} = \lambda_{\|X\|}$$

one obtains the inequality (8.11.1). If $(X_{t_n})_{t_n \in I}$ is a martingale, $\lambda_X = 0$ and (8.11.2) show that $\lambda^-_{\|X\|} = 0$. The real process $\|X\|$ is therefore a submartingale. □

For later use, we prove the following lemma.

8.12 Lemma. *Let $(X_t)_{t \in I}$ be a $\mathbb{B}$-valued quasimartingale on I; let J be a subset of $\mathbb{R}^+$ and $(\tau_j)_{j \in J}$ an increasing family of stopping times, each of which has finitely many values in I. Then the process $(Y_t)_{t \in J} := (X_{\tau_j})_{j \in J}$ is a quasimartingale whith respect to the filtration $(\mathscr{F}_{\tau_j})_{j \in J}$.*

Proof: Let us consider a finite disjoint family of rectangles $]s_k, t_k] \times F_k$ with $s_k \leqslant t_k \in J$, $F_k \in \mathscr{F}_{\tau_{s_k}}$, and $t_k \leqslant t \in J$. These rectangles are predictable with respect to the filtration $(\mathscr{F}_{\tau_j})_{j \in J}$ and included in $]0, t] \times \Omega$. The quasimartingale property for X, with respect to the filtration $(\mathscr{F}_t)_{t \in I}$ gives

$$\begin{aligned} - |\lambda_X|(]0, \tau_t]) &\leqslant \sum_k \|E\,1_{F_k}(X_{\tau_{t_k}} - X_{\tau_{s_k}})\| \\ &\leqslant |\lambda_X|(]0, \tau_t]) . \end{aligned}$$

The content associated with Y and $(\mathscr{F}_{\tau_j})_{j \in J}$ therefore has a variation on $]0, t] \times \Omega$ which is smaller than $|\lambda_X|(]0, \tau_t])$. □

9. Doob's inequalities for real quasimartingales and the almost sure convergence theorem

9.1 Theorem. *Let I be a denumerable index set with $a := \inf I \in I$ and $b := \sup I \in I$[1]. Let $(X_t)_{t \in I}$ be a real quasi-martingale. Then, for every $\alpha > 0$, the following inequalities hold.*

(9.1.1) $$\alpha \cdot P\,\{\sup_{t \in I} X_t > \alpha\} \leqslant \lambda^-_X\,(]a, b] \times \Omega) + \int_{\{\sup_{t \in I} X_t > \alpha\}} X_b\, dP ,$$

(9.1.2) $$\alpha \cdot P\,\{\inf_{t \in I} X_t < -\alpha\} \leqslant \lambda^+_X\,(]a, b] \times \Omega) - \int_{\{\inf_{t \in I} X_t < -\alpha\}} X_b\, dP .$$

In the particular case where $(X_t)_{t \in I}$ is a submartingale, these inequalities read

(9.1.3) $$\alpha \cdot P\,\{\sup_{t \in I} X_t > \alpha\} \leqslant \int_{\{\sup_{t \in I} X_t > \alpha\}} X_b\, dP \leqslant \int X_b^+\, dP ,$$

(9.1.4) $$\alpha \cdot P\,\{\inf_{t \in I} X_t < -\alpha\int \leqslant \int_{\{\inf_{t \in I} X_t \geqslant -\alpha\}} X_b\, dP - \int X_a\, dP .$$

[1]) It should be noticed that $I \subset \mathbb{R}^+$ and that b is allowed to be $+\infty$.

Proof: By (I_n) we denote an increasing sequence of finite subsets of I, such that $I = \bigcup_n I_n$. Since

$$\{\sup_{t \in I} X_t > \alpha\} = \bigcup_n \{\sup_{t \in I_n} X_t > \alpha\}$$

and

$$\{\inf_{t \in I} X_t < -\alpha\} = \bigcup_n \{\inf_{t \in I_n} X_t < -\alpha\}$$

it is clearly sufficient to prove the inequalities of the theorem for a finite I, with $b < \infty$. Let us denote by τ the beginning of the set

$$\{(t, \omega) : X_t(\omega) > \alpha,\ t \in I\}.$$

The random variable τ is a stopping time (according to Proposition 4.14) and we have

(9.1.5) $$\{\sup_{t \in I} X_t > \alpha\} = \{\tau \leqslant b\},$$

and since $X_\tau > \alpha$ on the set $\{\tau \leqslant b\}$,

$$\alpha \cdot P\{\sup_{t \in I} X_t > \alpha\} = \alpha \cdot P\{\tau \leqslant b\} \leqslant \int_{\{\tau \leqslant b\}} X_\tau \, dP.$$

From the relation

$$\int_{\{\tau \leqslant b\}} X_\tau \, dP = \int (X_{\tau \wedge b} - X_b) \, dP + \int_{\{\tau \leqslant b\}} X_b \, dP,$$

and from Proposition 8.9, we deduce

$$\alpha \cdot P\{\sup_{t \in I} X_t > \alpha\} \leqslant -\lambda_X(]\tau \wedge b, b]) + \int_{\{\tau \leqslant b\}} X_b \, dP.$$

This proves (9.1.1).

The inequality (9.1.2) follows immediately by applying (9.1.1) to the process $-X$. When X is a submartingale, λ_X is positive; therefore, λ_X^- is zero and (9.1.3) appears to be a trivial consequence of (9.1.1). Since in the case of a submartingale X,

$$\lambda_X^+(]a, b]) = E(X_b) - E(X_a) = \int (X_b - X_a) \, dP,$$

the inequality (9.1.4) follows readily from (9.1.2). □

9.2 Definition. Let $(\xi_t)_{t \in I}$ be a finite family of real numbers and a and b be two real numbers with $a < b$. The integer $N_{a,b}^I$, called the number of upcrossings of [a, b] by the sequence $(\xi_t)_{t \in I}$, is defined as the largest integer n with the following property.

There exists an increasing family $\{t_1 < t_2 \ldots < t_{2n}\}$ extracted from I, such that

$$\xi_{t_{2p-1}} \leqslant a, \text{ for } p = 1, \ldots, n;$$
$$\xi_{t_{2p}} \geqslant b, \text{ for } p = 1, \ldots, n.$$

9.3 Theorem. *Let I be a denumerable index set, with $I = \bigcup_n I_n$, where (I_n) is an* *increasing sequence of finite subsets of I. We define* $N^I_{a,b} = \sup_n N^{I_n}_{a,b}$. *We assume that* $(X_t)_{t\in I}$ *is a real quasi-martingale with*

(9.3.1) $$K := \sup_{t\in I} |\lambda_X|(]0, t]) < \infty$$

(1) *Then the following inequality holds.*

(9.3.2) $$E(N^I_{a,b}) = E(\sup_n N^{I_n}_{a,b}) \leqslant \frac{1}{b-a}(K + \sup_{t\in I} E(X_t - a)^-).$$

(2) *If we assume, moreover, that*

(9.3.3) $$\sup_{t\in I} E(X_t^-) < \infty,$$

there exists a P-negligible set $\Omega_0 \subset \Omega$ *with the following property.*

For every $\omega \in \Omega \setminus \Omega_0$ *and every monotone family* $(t_\alpha)_{\alpha\in J}$ *in I, with J totally ordered, the limit:* $\lim_\alpha X_{t_\alpha}(\omega)$ exists in $\bar{\mathbb{R}}$.

Proof: (1) Let I_n be an increasing sequence of finite subsets of I with

$$I = \bigcup_n I_n \text{ and } I_n = \{t_0 < \ldots < t_n\}.$$

We define inductively the pairs $(\tau_{2j+1}, \tau_{2j+2})$ as follows.

$$\begin{aligned}
\tau_1(\omega) &:= \inf\{t : t \in I_n, X_t(\omega) < a\} + t_n,\\
\tau_2(\omega) &:= \inf\{t : t \in I_n, t \geqslant \tau_1(\omega), X_t(\omega) > b\}\, t_n,\\
&\vdots\\
\tau_{2j+1}(\omega) &:= \inf\{t : t \in I_n, t > \tau_{2j}(\omega), X_t(\omega) < a\} \wedge t_n,\\
\tau_{2j+2}(\omega) &:= \inf\{: t \in I_n, t \geqslant \tau_{2j+1}(\omega), X_t(\omega) < b\} \wedge t_n.
\end{aligned}$$

The τ_k's form an increasing sequence of stopping times. Since

$$\sum_{j=1}^n \lambda_X(]\tau_{2j-1}, \tau_{2j}]) \leqslant |\lambda_X|(]0, t_n])$$

Proposition 8.9 implies

(9.3.4) $$\sum_{j=1}^n E(X_{\tau_{2j}} - X_{\tau_{2j-1}}) \leqslant K.$$

From the definition of the τ_k's it follows that

$$X_{\tau_{2j}} - X_{\tau_{2j-1}} > b - a \text{ for } 1 \leqslant j \leqslant N^{I_n}_{a,b},$$

and

$$X_{\tau_{2j}} - X_{\tau_{2j-1}} = 0 \text{ for } j > N^{I_n}_{a,b} + 1,$$

while

$$X_{\tau_{2j}} - X_{\tau_{2j-1}} = X_{t_n} - X_{\tau_{2j-1}} \geqslant (X_{t_n} - a) \wedge 0, \text{ for } j = N^{I_n}_{a,b} + 1.$$

Therefore,

(9.3.5) $$(b-a)\,E(N_{a,b}^{I_n}) - E(X_{t_n}-a)^- \leqslant K\,.$$

Since the sequence $(N_{a,b}^{I_n})_{n\in\mathbb{N}}$ of random variables is clearly increasing towards $N_{a,b}^{I}$, the inequality (9.3.2) follows from (9.3.5).

(2) Since $(X_t-a)^- \leqslant X_t^- + a^-$, the assumption (9.3.3) implies

$$E(N_{a,b}^{I}) < \infty$$

for every couple (a, b) with $a<b$. Therefore,

(9.3.6) $$P\Big(\bigcap_{\substack{a<b\\ a,b\in\mathbb{Q}}} \{N_{a,b}^{I}<\infty\}\Big) = 1\,.$$

We write

(9.3.7) $$\Omega_0 := \bigcup_{\substack{a<b\\ a,b\in\mathbb{Q}}} \{N_{a,b}^{I}=\infty\} \quad\text{and}\quad \alpha_\infty := \lim_{\alpha\in J}\alpha$$

If, for $\omega\in\Omega_0$, the family $(X_{t_\alpha}(\omega))_{\alpha\in J}$ does not converge when α goes to α_∞, there exist $a, b\in\mathbb{Q}$ with $a<b$ and

$$\liminf_{\alpha\uparrow\alpha_\infty} X_{t_\alpha}(\omega) < a < b < \limsup_{\alpha\uparrow\alpha_\infty} X_{t_\alpha}(\omega)\,.$$

This says that $\Omega-\Omega_0$ is in the convergence set of $(X_{t_\alpha})_{\alpha\in J}$, while (9.3.6) and (9.3.7) give $P(\Omega_0)=0$. □

9.4 Theorem *(Almost sure convergence of quasimartingales).* (1) *Let $(X_n)_{n\in\mathbb{N}}$ be a real adapted process which satisfies the following conditions.*

(i) $$\sum_n E(|E(X_{n+1}\,|\,\mathscr{F}_n) - X_n|) < \infty\,,$$

(ii) $$\sup_n E(X_n^-) < \infty\,.$$

Then the sequence $(X_n)_{n\in\mathbb{N}}$ converges a.s. towards an integrable random variable X_∞ and

$$E|X_\infty| \leqslant \liminf_n E|X_n| < \infty\,.$$

(2) *If $(X_n)_{n\in\mathbb{N}}$ is a real submartingale, conditions (i) and (ii) together are equivalent to the following condition.*

(j) $$\sup_n E|X_n| < \infty\,.$$

(3) *If $(X_n)_{n\in\mathbb{N}}$ is a real supermartingale, condition (i) is implied by (ii).*

Proof: (1) We first prove that (i) is equivalent to the condition

(9.4.1) $$\sup_n |\lambda_X|(]0,n]\times\Omega) < \infty\,.$$

For this purpose, let us write

$$F_n := \{E(X_{n+1}|\mathscr{F}_n) - X_n > 0\}$$

For every $F \in \mathscr{F}_n$ with $F \subset F_n$, we have

$$\lambda_X(]n, n+1] \times F) = E(1_F \cdot (X_{n+1} - X_n)) \geqslant 0\,,$$

while for every $F \in \mathscr{F}_n$ with $F \subset \complement F_n$,

$$\lambda_X(]n, n+1] \times F) = E(1_F \cdot (X_{n+1} - X_n)) \leqslant 0\,.$$

Therefore,

$$\begin{aligned} |\lambda_X|(]n, n+1] \times \Omega) &= \lambda_X^+(]n, n+1] \times \Omega) + \lambda_X^-(]n, n+1] \times \Omega) \\ &= E[1_{F_n} \cdot (X_{n+1} - X_n)] - E[1_{\complement F_n} \cdot (X_{n+1} - X_n)] \\ &= E(|E(X_{n+1}|\mathscr{F}_n) - X_n|)\,. \end{aligned}$$

This equality establishes the equivalence of (i) and (9.4.1). The convergence of (X_n) follows from Theorem 9.3.

Let us remark that condition (9.4.1) implies in particular that

$$\sup_n |\lambda_X(]0, \mathrm{n}] \times \Omega| = \sup_n |E(X_n)| < \infty.$$

Therefore, the conditions (i) and (ii) together imply

$$\sup_n E|X_n| < \infty$$

and the inequality

$$E|X_\infty| \leqslant \liminf_n E|X_n|$$

follows from the Fatou lemma.

(2) If $(X_n)_{n \in \mathbb{N}}$ is a submartingale, the measure λ_X is positive and (9.4.1) reduces to

$$\sup_n (E(X_n) - E(X_0)) < \infty\,.$$

The conditions (i) and (ii) are thus equivalent to

$$\sup_n E(X_n^-) < \infty \quad \text{and} \quad \sup_n E(X_n) < \infty\,,$$

which is clearly equivalent to the one condition

$$\sup_n E(|X_n|) < \infty\,.$$

(3) When $(X_n)_{n \in \mathbb{N}}$ is a supermartingale, the measure λ_X is negative and (i) reduces to

$$\sup_n (E(X_0) - E(X_n)) < \infty\,.$$

This inequality is clearly implied by (ii). □

The following result gives other helpful equivalent forms of the quasimartingale property.

9.5 Proposition. *Let $(X_n)_{n\in\mathbb{N}}$ be a real-valued process. For every n we set*

$$F_n := \{E(X_{n+1} - X_n | \mathcal{F}_n) > 0\}$$

(1) *The condition (i) of Theorem 9.4 is then equivalent to the following two*

(j) $$\sum_{n\geq 0} E(1_{F_n}(X_{n+1} - X_n)) < \infty$$

(jj) $$\sum_{n\geq 0} E(1_{F_n^c}(X_{n+1} - X_n)) > -\infty$$

(2) *If $\liminf_n E(X_n) > -\infty$, the condition (i) of Theorem 9.4 is implied by condition (j) above.*

Proof: (1) We have indeed

$$E(|E(X_{n+1} - X_n | \mathcal{F}_n)|) = E(1_{F_n}(X_{n+1} - X_n)) - E(1_{F_n^c}(X_{n+1} - X_n))$$

The equivalence of (i) and (j) + (jj) is therefore clear.

(2) From the equality

$$\sum_{n\leq r} E(1_{F_n}(X_{n+1} - X_n)) + E(1_{F_n^c}(X_{n+1} - X_n)) = \sum_{n\leq r} E(X_{n+1} - X_n)$$
$$= E(X_{r+1}) - E(X_0),$$

we immediately derive

$$\sum_{n\geq 0} E 1_{F_n^c}(X_{n+1} - X_n) \geq - \sum_{n\geq 0} E(1_{F_n}(X_{n+1} - X_n)) - E(X_0) + \liminf_n E(X_n)$$

The conclusion of (2) follows. □

The following result concerns vector-valued quasimartingales. For real valued quasimartingales it is an immediate consequence of inequalities (9.1.1) and (9.1.2). In the general Banach case it needs a special proof.

9.6 Theorem*. *Let I be a denumerable index set in $\mathbb{R}^+$ with $a := \inf I \in I$ and $b := \sup I \in I$. Let $\mathbb{B}$ be a Banach space and $(X_t)_{t\in I}$ a Banach valued quasi-martingale. Then, for every $\lambda > 0$, the following inequality holds.*

$$\lambda \cdot P\{\sup_{t\in I} \|X_t\| > \lambda\} \leq |\lambda_X|(]a, b] \times \Omega) + \int_{\{\sup_{t\in I} \|X_t\| > \lambda\}} \|X_b\| \, dP.$$

Proof: As in 9.1, it is sufficient to consider a finite set I. We define the stopping time with finitely many values,

$$\tau := \inf\{t : t \in I \, \|X_t\| > \lambda\} \wedge b$$

and notice that

(9.6.1) $$\{\sup_{t\in I} \|X_t\| > \lambda\} \subset \{\tau < b\} \cup \{\|X_b\| > \lambda\}.$$

We have

(9.6.2) $$|\lambda_X|(]\tau, b]) \leqslant |\lambda_X|(]a, b] \times \Omega).$$

If $F_1, .., F_k$ is any disjoint family in $\mathscr{F}$ and $x_1', \ldots, x_k'$ any family of continuous linear forms on $\mathbb{B}$ with $\|x_i'\| \leqslant 1$ for $i = 1, \ldots, k$, we obtain from the definition of $|\lambda_X|$

$$|\lambda_X|(]\tau, b]) \geqslant \sum_{i=1}^{k} < X_b - X_\tau, x_i' \cdot 1_{F_i} >$$

Therefore, it is possible to write, for any $\mathscr{F}_\tau$-measurable simple function φ, $\varphi := \sum_{i=1}^{k} x_i' \cdot 1_{F_i}$ with values in the dual space $\mathbb{B}'$ of $\mathbb{B}$ and $\|\varphi\| \leqslant 1$, that

$$|\lambda_X|(]\tau, b]) \geqslant E[< E(X_b - X_\tau | F_\tau), \varphi >]$$

and therefore

(9.6.3) $$|\lambda_X|(]\tau, b]) \geqslant E\|E(X_b - X_\tau|\mathscr{F}_\tau)\| \geqslant \int_{\{\tau<b\}} \|E(X_b|\mathscr{F}_\tau) - X_\tau\| \, dP.$$

From this inequality and (9.6.2), we derive

(9.6.4) $$\begin{aligned} |\lambda_X|(]a, b]) &\geqslant \int_{\{\tau<b\}} \|X_\tau\| \, dP - \int_{\{\tau<b\}} \|X_b\| \, dP \\ &\geqslant \lambda P\{\tau < b\} - \int_{\{\tau<b\}} \|X_b\| \, dP \end{aligned}$$

Using (9.6.1), we get

$$\begin{aligned} \lambda P\{\sup_{t\in I} \|X_t\| > \lambda\} &\leqslant |\lambda_X|(]a, b]) + \int_{\{\tau<b\}} \|X_b\| \, dP + \\ &\quad + \lambda P(\{\|X_b\| > \lambda\} - \{\tau < b\}) \\ &\leqslant |\lambda_X|(]a, b]) + \int_{\{\sup_{t\in I} \|X_t\| > \lambda\}} \|X_b\| \, dP, \end{aligned}$$

which is the equality of the theorem. □

9.7 Corollary. *For every right-continuous quasimartingale $(X_t)_{t\in\mathbb{R}^+}$ with values in a Banach space $\mathbb{B}$, the following inequality holds on any bounded interval $[0, b]$.*

$$\lambda P\{\sup_{t\in[0,b]} \|X_t\| > \lambda\} \leqslant |\lambda_X|(]0, b] \times \Omega) + \int_{\{\sup_{t\leqslant b} \|X_t\| > \lambda\}} \|X_b\| \, dP$$

Almost surely every path of X is bounded on any bounded interval $[0, b]$.

Proof: It is enough to consider a denumerable dense set in $\mathbb{R}^+$ and to apply (9.6).

10. Uniform integrability – Convergence in L^p – Regularity properties of trajectories

Uniform integrability

10.1 Theorem. *Let $(X_t)_{t\in I}$ be a positive submartingale with $I \subset \bar{\mathbb{R}}^+$. We assume that $b := \sup I \in I$ and denote by $\mathcal{T}_I{}^f$ the set of stopping times taking their values in a finite subset of I. Then, the random variables $\{X_\tau : \tau \in \mathcal{T}_I{}^f\}$ are uniformly integrable.*

Proof: Let τ be a stopping time with finitely many values $t_0 < t_1 \ldots < t_n$ in I. For every $F \in \mathcal{F}_\tau$ (in this particular case this amounts to saying $\{\tau = t_i\} \cap F \in \mathcal{F}_{t_i}$)

$$([0,b] \times F) \cap]\tau, b] = \sum_{i=0}^{n-1}]t_i, b] \times (F \cap \{\tau = t_i\}) \in \mathcal{A}_I,$$

and therefore,

$$\begin{aligned}\lambda_X(([0,b] \times F) \cap]\tau, b]) &= \sum_{i=0}^{n-1} E(1_{F \cap \{\tau = t_i\}}(X_b - X_{t_i})) \\ &= E(1_F \cdot (X_b - X_\tau)).\end{aligned}$$

If we consider in particular $F := \{X_\tau \geqslant \alpha\}$, the positivity of the content λ_X implies

$$0 \leqslant \lambda_X(([0,b] \times F) \cap]\tau, b]) = \int_{\{X_\tau \geqslant \alpha\}} (X_b - X_\tau)\, dP \tag{10.1.1}$$

Hence, we deduce

$$\int_{\{X_\tau \geqslant \alpha\}} X_b \, dP \geqslant \int_{\{X_\tau \geqslant \alpha\}} X_\tau \, dP, \tag{10.1.2}$$

and then

$$E(X_b) \geqslant \alpha P\{X_\tau \geqslant \alpha\}. \tag{10.1.3}$$

From the integrability of X_b follows, for every $\varepsilon > 0$, the existence of $\delta > 0$ such that

$$(F \in \mathcal{F}_\infty, P(F) \leqslant \delta) \Rightarrow \int_F X_b \, dP \leqslant \varepsilon. \tag{10.1.4}$$

By choosing $\alpha \geqslant \dfrac{E(X_b)}{\delta}$ we obtain from (10.1.3)

$$P\{X_\tau \geqslant \alpha\} \leqslant \delta \quad \text{for every} \quad \tau \in \mathcal{T}_I{}^f.$$

Therefore, from (10.1.2) and (10.1.4),

$$\int_{\{X_\tau \geqslant \alpha\}} X_b \, dP \leqslant \varepsilon \quad \text{for every} \quad \tau \in \mathcal{T}_I{}^f. \quad \square$$

10.2 Corollary. *Let $I \subset \bar{\mathbb{R}}^+$ with $b := \sup I \in I$, and let $(X_t)_{t\in I}$ be a Banach valued martingale. The random variables $\{X_\tau : \tau \in \mathcal{T}_I{}^f\}$ are uniformly integrable.*

Proof: We only have to consider the positive submartingale $(\|X_t\|)_{t\in I}$ and apply Theorem 10.1. □

10.3 Theorem. *Let $(t_n)_{n\in\mathbb{N}}$ be a decreasing family of real numbers with $a := \inf_n t_n$. We write $I := \{t_n\} \cup \{a\}$. Let $(X_t)_{t\in I}$ be a Banach-valued quasi-martingale. Then the random variables $\{X_t : t \in I\}$ are uniformly integrable.*

Proof: According to Lemma 8.11, we have only to prove the theorem for a positive quasi-martingale. Let us assume then that $(X_t)_{t\in I}$ is a positive quasi-martingale and write

$$K := \lambda_X^-(]a, t_0] \times \Omega + E(X_{t_0}).$$

From Doob's inequality, we may write for every $\alpha > 0$,

$$P\{\sup_{t\in I} X_t > \alpha\} \leqslant \frac{K}{\alpha}. \tag{10.3.1}$$

Given any $\varepsilon > 0$, any subsequence (t_n') in I and any increasing sequence (α_n) of positive numbers converging to $+\infty$, it is thus easy to recursively define a subsequence $(n_k)_{k\in\mathbb{N}}$ of $\mathbb{N}$ endowed with the following property.

$$\forall n \in \mathbb{N},\ \int_{\{X_{t_n'} > \alpha_{n_{k+1}}\}} X_{t_{n_k}'}\, dP \leqslant \frac{\varepsilon}{4}. \tag{10.3.2}$$

Let us now make the extra assumption that $(X_t)_{t\in I}$ is not uniformly integrable. In this case, there exists $\varepsilon > 0$, an increasing sequence (α_n) of positive numbers converging to $+\infty$ and a subsequence (t_n') of I, which can clearly be supposed decreasing, such that

$$\forall n \in \mathbb{N},\ \int_{\{X_{t_n'} > \alpha_n\}} X_{t_n'}\, dP \geqslant \varepsilon. \tag{10.3.3}$$

Then, using (10.3.2) and (10.3.3), we can construct a decreasing sequence (t_n'') in I having the following two properties.

$$\forall n \in \mathbb{R}, \forall k \in \mathbb{R},\ \int_{\{X_{t_n''} > \alpha_{k+1}\}} X_{t_k''}\, dP \leqslant \frac{\varepsilon}{4}, \tag{10.3.4}$$

$$\forall k \in \mathbb{R},\ \int_{\{X_{t_k''} > \alpha_k} X_{t_k''}\, dP \geqslant \varepsilon, \tag{10.3.5}$$

where $\varepsilon > 0$ and (α_k) is an increasing sequence of positive numbers, converging towards $+\infty$. Let us write

$$A_k :=]t_{k+1}'', t_k''] \times \{X_{t_{k+1}''} > \alpha_{k+1}''\} \in \mathcal{R}_I.$$

From (10.3.4) and (10.3.5) it follows that

$$\forall p,\ \sum_{k=1}^{p} \lambda_X(A_k) = \sum_{k=1}^{p} \int_{\{X_{t_{k+1}} > \alpha_{k+1}\}} (X_{t_k} - X_{t_{k+1}})\, dP \leqslant -3p\frac{\varepsilon}{4}$$

but this conradicts the property for λ_X to have a bounded variation on $]a, t_0] \times \Omega$, so the theorem is proved. □

L^p-convergence theorems ($p \geqslant 1$) for real martingales

10.4 Theorem. *Let $(X_n)_{n\in\mathbb{N}}$ be a real $(\mathscr{F}_n)$-martingale. Then, the following properties are equivalent.*

(a) *$(X_n)_{n\in\mathbb{N}}$ is uniformly integrable.*

(b) *There exists a random variable X such that*

$$X_n = E(X \mid \mathscr{F}_n) \text{ a.s. for all } n \in \mathbb{N}.$$

(c) *$(X_n)_{n\in\mathbb{N}}$ converges in $L^1_{\mathbb{R}}(\Omega, \mathfrak{A}, P)$ towards a random variable X_∞. If, as usual, $\mathscr{F}_\infty$ denotes the σ-algebra generated by $\bigcup_n \mathscr{F}_n$, the process $(X_n)_{n\in\bar{\mathbb{N}}}$ (where $\bar{\mathbb{N}} := \mathbb{N} \cup \{\infty\}$) is an $(\mathscr{F}_n)_{n\in\bar{\mathbb{N}}}$-martingale.*

Proof: We first remark that the assumption (a) implies $\sup_n E(|X_n|) < \infty$. The implication (a) ⇒ (c) then follows readily from Theorem 9.4 and the classical theorem which states that an a.s. convergent sequence of random variables converges in $L^1(\Omega, \mathfrak{A}, P)$ if it is uniformly integrable. The convergence in $L^1(\Omega, \mathfrak{A}, P)$ towards X_∞ implies

$$\forall F \in \mathscr{F}_n, \ \int_F X_n \, dP = \lim_{m\to\infty} \int_F X_m \, dP = \int_F X_\infty \, dP.$$

This equality says that

$$X_n = E(X_\infty \mid \mathscr{F}_n) \text{ a.s. for all } n \in \mathbb{N}.$$

Thus (c) ⇒ (b) and, moreover, $(X_n)_{n\in\bar{\mathbb{N}}}$ is an $(\mathscr{F}_n)_{n\in\bar{\mathbb{N}}}$-martingale.

Let us assume now that (b) is satisfied and let us define

$$X_\infty := E(X \mid \mathscr{F}_\infty).$$

The process $(X_n)_{n\in\bar{\mathbb{N}}}$ thus defined is clearly an $(\mathscr{F}_n)_{n\in\bar{\mathbb{N}}}$-martingale and, by applying Theorem 10.1 to $(|X_n|)_{n\in\bar{\mathbb{N}}}$, we immediately obtain the uniform integrability of the family $(X_n)_{n\in\bar{\mathbb{N}}}$.

This proves the theorem. □

To prove the L^P-convergence theorem for $p > 1$, we need an inequality which is again due to Doob.

10.5 Theorem. *Let $p \in \,]1, \infty[$ and $(X_n)_{n\in\mathbb{N}}$ be a positive submartingale with the property*

(10.5.1) $$\sup_n E(X_n^p) < +\infty.$$

Then $\sup_n X_n$ belongs to $L^P(\Omega, \mathscr{F}_\infty, P)$ and, moreover,

(10.5.2) $$\|\sup_n X_n\|_p \leqslant \frac{p}{p-1} \|X_\infty\|_P \leqslant \frac{p}{p-1} \cdot \sup_n \|X_n\|_P.$$

Proof: Since

$$\sup_n E(|X_n|) \leqslant \sup_n \{E(|X_n|^p)\}^{1/p} < \infty,$$

the submartingale (X_n) converges a. s. towards a random variable X_∞, as a consequence of Theorem 9.4–(2). The Fatou lemma implies

$$E(|X_\infty|^p) \leqslant \sup_n E\|X_n\|^p < \infty.$$

(10.5.1) also implies the uniform integrability of (X_n) and therefore, the convergence in L^1. Consequently $(X_n)_{n\in\overline{\mathbb{N}}}$ is a submartingale. Let us write

$$X^* := \sup_n X_n.$$

Doob's inequality (9.1.3) then gives

(10.5.3) $\alpha \cdot P(\{X^* > \alpha\}) \leqslant \int_{\{X^* > \alpha\}} X_\infty \, dP$ for all $\alpha > 0$.

Let μ be the probability law of the random variable X^*. We have

$$E[(X^*)^p] = \int_0^\infty y^p \mu(dy) = \int_0^\infty (\int_0^y p\, \xi^{p-1} d\xi)\mu(dy).$$

By applying twice the Fubinitheorem and (10.5.3), we obtain

$$E[(X^*)^p] = \int_0^\infty p\, \xi^{p-1} P\{X^* > \xi\}\, d\xi \leqslant \int_0^\infty p\, \xi^{p-2} (\int_{\{X^* > \xi\}} X_\infty \, dP)\, d\xi$$

$$\leqslant \int_0^\infty \frac{p}{p-1} y^{p-1}\, dF(y),$$

where F is the distribution function

$$F(y) := \int_{\{X^* \leqslant y\}} X_\infty \, dP.$$

The last inequality reads

$$E[(X^*)^p] \leqslant E\left[\frac{p}{p-1}(X^*)^{p-1} \cdot X_\infty\right].$$

From the Hölder inequality, we deduce that

$$E[(X^*)^p] \leqslant \frac{p}{p-1} [E[(X^*)^p]]^{\frac{p-1}{p}} [E(X_\infty^p)]^{1/p}$$

and therefore the inequality (10.5.2). □

10.6 Theorem. *Let* $(X_n)_{n\in\mathbb{N}}$ *be a real martingale for which there exists* $p > 1$ *with the property*

$$\sup_n E(|X_n|^p) < \infty.$$

Then the sequence (X_n) *converges in* $L^p(\Omega, \mathscr{F}_\infty, P)$ *towards a random variable* X_∞ *and for all* n,

$$X_n := E(X_\infty | \mathscr{F}_n) \text{ a.s.}$$

Prrof: Theorem 9.4 shows that the sequence (X_n) converges a.s. towards a random variable X_∞ but, from the domination property

$$\forall n, |X^n| \leqslant X^* := \sup_n |X_n|$$

and Theorem 10.5, the sequence (X_n) also converges towards X_∞ in $L^p(\Omega, \mathscr{F}_\infty, P)$. The last statement of the theorem follows from Theorem 10.4. □

Regularity properties of paths

We now come back to quasimartingales indexed by $\mathbb{R}^+$.

10.7 Theorem. (1) *Let* $(X_t)_{t \in \mathbb{R}^+}$ *be a real quasimartingale, which is regular right-continuous and such that*

(i) $$\sup_{t \in \mathbb{R}^+} |\lambda_X|(]0, t]) < \infty$$

(ii) $$\sup_t E(X_t^-) < \infty$$

Then $\lim_{t \to \infty} X_t(\omega)$ *exists a.s. and if we call* X_∞ *the limit-random variable*

$$E|X_\infty| \leqslant \liminf_t E|X_t| < \infty .$$

(2) *When* $(X_t)_{t \in \mathbb{R}^+}$ *is a real submartingale, conditions (i) and (ii) are together equivalent to the following condition.*

(j) $$\sup_t E|X_t| < \infty$$

(3) *If* $(X_t)_{t \in \mathbb{R}^+}$ *is a real supermartingale, condition (i) is implied by condition (ii).*

Proof: Part (1) of this theorem is a straightforward consequence of Theorem 9.3.2° and of the regularity property of X.

When X is a real submartingale, the hypothesis (i) reduces to $\sup_{t \in \mathbb{R}^+} E(X_t) < \infty$ and the equivalence is therefore clear.

When X is a supermartingale, the inequality (i) reduces to $\sup_{t \in \mathbb{R}^+} - E(X_t) < \infty$, which is clearly implied by (ii). □

10.8 Definition. Let $(X_t)_{t \in \mathbb{R}^+}$ be a process. We define

$$(10.8.1)\qquad \tilde{X}_{t^+}(\omega) := \begin{cases} \lim\limits_{\substack{s \downarrow t \\ s \in \mathbb{Q}}} X_s(\omega) & \text{when this right hand limit exists} \\ 0 & \text{otherwise,} \end{cases}$$

$$(10.8.2)\qquad \tilde{X}_{t^-}(\omega) := \begin{cases} \lim\limits_{\substack{s \uparrow t \\ s \in \mathbb{Q}}} X_s(\omega) & \text{when this left hand limit exists} \\ 0 & \text{otherwise.} \end{cases}$$

10.9 Theorem. *Let $(X_t)_{t \in \mathbb{R}^+}$ be a real quasimartingale (resp. martingale, resp. submartingale). Then*

(1) *The process $(\tilde{X}_{t^+})_{t \in \mathbb{R}^+}$ is a quasimartingale (resp. a martingale, resp. a submartingale) for the filtration $(\mathscr{F}_{t^+})_{t \in \mathbb{R}^+}$.*

(2) *If the function $t \rightsquigarrow |\lambda_X|(]0, t])$ is right-continuous and if $\mathscr{F}_{t^+} = \mathscr{F}_t$ for every t, one has*

$$\tilde{X}_{t^+} = X_t \text{ a.s. for every } t.$$

Proof: We remark first that, on every bounded interval $[0, K]$,

$$\sup_{t \in [0, K]} E|X_t| \leqslant \lambda_{|X|}(]0, K] \times \Omega) + E|X_0| < \infty$$

As a consequence of Theorem 9.4, there exists a set $\Omega_K \subset \Omega$ with $P(\Omega_K) = 0$ and with the property that for every $\omega \notin \Omega_K$ and $t \in [0, K[$, the limits in (10.8.1) and (10.8.2) exist. Accordingly, we have

$$(10.9.1)\qquad \tilde{X}_{t^+} = \lim_{\substack{s \downarrow t \\ s \in \mathbb{Q}}} X_s \quad \text{a.s.}$$

and

$$(10.9.2)\qquad \tilde{X}_{t^-} = \lim_{\substack{s \uparrow t \\ s \in \mathbb{Q}}} X_s \quad \text{a.s.}$$

For the martingale X on $[0, K]$, let us now consider the decomposition

$$X_t = X'_t - X''_t + M_t$$

as in Proposition 8.5, where X' and X'' are positive supermartingales and M is a martingale with $X'_k = X''_k = 0$, $M_k = X_k$.

We recall that for every $t \in [0, K]$,

$$|\lambda_X|(]0, t]) = \lambda_X^+(]0, t]) + \lambda_X^-(]0, t]),$$

where λ_X^+ and λ_X^- are positive contents with

$$\lambda_X^+(]0, t]) = E(X''_0 - X''_t),$$
$$\lambda_X^-(]0, t]) = E(X'_0 - X'_t).$$

Then since

$$\lambda_X^+(]s,t]) \leqslant |\lambda_X|(]s,t])$$

and

$$\lambda_X^-(]s,t]) \leqslant |\lambda_X|(]s,t]),$$

the right continuity of $t \rightsquigarrow |\lambda_X|(]0,t])$ implies the right-continuity of both functions $t \rightsquigarrow E(X_t')$ and $t \rightsquigarrow E(X_t'')$, while $t \rightsquigarrow E(M_t)$ is a constant function. This shows that, to prove the theorem, we can restrict ourselves to considering only supermartingales.

Proof of statement (1): Let X be a supermartingale, (s_n) a decreasing sequence of rational numbers which converges to s, and (t_n) a decreasing sequence of rational numbers which decreases to t, with $s<t$ and $s_n<t_n$ for all n. Since $(X_{s_n})_{n\in\mathbb{N}}$ (resp. $(X_{t_n})_{n\in\mathbb{N}}$) converges towards $\tilde{X}_{s^+}$ (resp. $\tilde{X}_{t^+}$) a.s., according to the above remark, we may write, using Theorem 10.3,

(10.9.3) $$\lim_{n\to\infty} E|X_{s_n} - \tilde{X}_{s^+}| = 0$$

and

(10.9.4) $$\lim_{n\to\infty} E|X_{t_n} - \tilde{X}_{t^+}| = 0.$$

Since X is a supermartingale, the sequences $(\int_F X_{s_n}\,dP)_{n\in\mathbb{N}}$ and $(\int_F X_{t_n}\,dP)_{n\in\mathbb{N}}$ are increasing sequences for every $F\in\mathscr{F}_{s^+}$ with $\int_F X_{t_n}\,dP \geqslant \int_F X_{t^+}\,dP$.

Because of (10.9.3) and (10.9.4), we may therefore write

$$\int_F \tilde{X}_{s^+}\,dP = \lim_n \int_F X_{s_n}\,dP \geqslant \lim \int_F X_{t_n}\,dP = \int_F \tilde{X}_{t^+}\,dP$$

This proves that $(\tilde{X}_{t^+})_{t\in\mathbb{R}^+}$ is a supermartingale. If X is a submartingale, then $-X$ is a supermartingale and the above result shows that $(-\tilde{X}_{t^+})_{t\in\mathbb{R}^+}$ is a supermartingale, and therefore that $\tilde{X}_{t^+}$ is a submartingale.

Proof of statement (2): We consider a supermartingale for which $t \rightsquigarrow E(X_t)$ is a right continuous function.

Let (t_n) be a decreasing sequence with $\lim_n t_n = t$.

Because of (10.8.4) and the supermartingale property of X, we have

$$\int_F \tilde{X}_{t^+}\,dP \leqslant \lim_n \int_F X_{t_n}\,dP \leqslant \int_F X_t\,dP.$$

Assuming $\mathscr{F}_{t^+} = \mathscr{F}_t$ we therefore find that

$$\tilde{X}_{t^+} \leqslant X_t \quad \text{a.s.}$$

and the inequality $\tilde{X}_{t^+} = X_t$ a.s. holds if and only if $E(X_t - \tilde{X}_{t^+}) = 0$. However, this is true with the hypotheses of statement (2) since $E(X_t - \tilde{X}_{t^+}) = \lim_n E(X_{t_n} - \tilde{X}_{t^+}) = 0$. □

10.10 Corollary. *Let us assume that the filtration $(\mathcal{F}_t)_{t\in\mathbb{R}^+}$ is right-continuous, and consider a real quasimartingale X with the property that the function $t \rightsquigarrow |\lambda_X|(]0,t])$ on $\mathbb{R}^+$ is also right continuous. (In particular this is true for every martingale X). Then there exists a regular quasimartingale $\tilde{X}$ with the property $\tilde{X}_t = X_t$* a.s. for all t.

Proof: We consider $\tilde{X}$ as defined in Theorem 10.9. It follows immediately from the definition of $\tilde{X}$ that $\tilde{X}_t$ is F_{t^+}-measurable and therefore $\mathcal{F}_t$-measurable. Theorem 10.9 says that $\tilde{X}_t = X_t$ a.s.

10.11 Remarks

Remark 1. Statements for left limits are given as exercises at the end of the chapter (exercises E. 2.2 and E. 2.3).

Remark 2. When the filtration $(\mathcal{F}_t)_{t\in\mathbb{R}^+}$ is right-continuous every quasimartingale X with respect to this filtration, such that $t \rightsquigarrow E(X_t)$ is right continuous, admits a modification $\tilde{X}$ (i.e.: $\tilde{X}$ is a process defined on the same stochastic basis with $X_t = \tilde{X}_t$ a.s. for all $t \in \mathbb{R}^+$) which is regular right-continuous. This leads to the following immediate extension of Theorem 10.5.

10.12 Theorem. *Let $p \in]1, \infty[$ and $(X_t)_{t\in\mathbb{R}^+}$ be a positive submartingale with the property.*

(10.12.1) $$\sup_{t\in\mathbb{R}^+} E|X_t|^p < \infty$$

Then

$$\|\sup_t X_t\|_p \leqslant \frac{p}{p-1}\|X_\infty\|_p \leqslant \frac{p}{p-1}\sup_{t\in\mathbb{R}^+}\|X_t\|_p$$

11. Convergence theorems for vector-valued quasimartingales*

11.1 Theorem.* *Let $\mathbb{B}$ be a Banach space with the R. N. property (See. 8.3) and $(X_n)_{n\in\mathbb{N}}$ a $\mathbb{B}$-valued process, for which the following inequality holds*

(i) $$\sum_n E(\|E(X_{n+1}|\mathcal{F}_n) - X_n\|) < \infty$$

Then there exists a martingale M and a quasi-martingale V with the following properties.

(11.1.1) $$X = M + V.$$

(11.1.2) $$\sum_n E(\|E(V_{n+1}|\mathcal{F}_n) - V_n\|) = \sum_n E(\|E(X_{n+1}|\mathcal{F}_n) - X_n\|) < \infty.$$

(11.1.3) $$\lim_{n\to\infty} E(\|V_n\|) = 0.$$

(11.1.4) $\lim_{n\to\infty} V_n = 0$ a.s.

Proof: Following Lemma 8.10, λ_X is a measure on $\mathscr{A}_{\mathbb{N}}$ and condition (i) says that the total variation $|\lambda_X|(\mathbb{R}^+ \times \Omega)$ of this measure is finite. For every n, let us define the measure α_n on $\mathscr{F}_n$

(11.1.5) $\alpha_n(F) = -\lambda_X(]n, \infty[\times F)$.

We write V_n for the $\mathscr{F}_n$-measurable Radon-Nikodym density of α_n with respect to P. The process V is clearly an adapted process with $\lambda_V = \lambda_X$. We set $M := X - V$. As $\lambda_M = \lambda_{X-V} = \lambda_X - \lambda_V = 0$, the process M is a martingale and (11.12) is immediate.

From (11.1.5), we deduce

(11.1.6) $\forall F \in \bigcup_n \mathscr{F}_n, \ \lim_n E(1_F \cdot V_n) = \lim_n - \lambda_X(]n, \infty[\times F) = 0$.

Since

$$\|\lambda_X(]n, \infty[\times F)\| \leqslant |\lambda_X|(]n, \infty[\times \Omega),$$

the convergence in (11.1.6) is uniform in F, which means

$$\lim_{n\to\infty} E(\|V_n\|) = 0.$$

It now follows from Lemma 8.11 that the quasi-martingale $\|V\|$ satisfies the assumptions of Theorem 9.4. Therefore, $\lim_n \|V_n\|$ exists almost surely with

$$E(\lim_n \|V_n\|) \leqslant \liminf_n E(\|V_n\|) = 0.$$

The property (11.1.4) follows. □

11.2 Theorem*(*J. Neveu – A. and C. Ionescu-Tulcea*). *Let $\mathbb{B}$ be any Banach space, X_∞ an $\mathscr{F}_\infty$-measurable $\mathbb{B}$-valued random variable and $X_n := E(X_\infty | \mathscr{F}_n)$.*

Then the sequence (X_n) converges a.s. towards X_∞ for the norm of $\mathbb{B}$. The convergence also holds in $L^1_{\mathbb{B}}(\Omega, \mathscr{F}_\infty, P)$.

Proof: Since $(X_n)_{n\in\bar{\mathbb{N}}}$ is a martingale, $(\|X_n\|)_{n\in\bar{\mathbb{N}}}$ is a positive submartingale (Lemma 8.11). Therefore, we may write

(11.2.1) $0 \leqslant \int \|X_\infty\| \, dP - \int \|X_n\| \, dP \leqslant \int \|X_\infty\| \, dP - \lim_n \int \|X_n\| \, dP$,

and, according to Theorem 10.1, the family $\{\|X_n\|\}_{n\in\mathbb{N}}$ is uniformly integrable. Since, by Theorem 9.4, the positive submartingale $(\|X_n\|)_{n\in\mathbb{N}}$ converges a.s. to a positive random variable Y, we also have

(11.2.2) $\lim_n E(\|X_n\|) = E(Y)$.

Let $\mathbb{B}'$ denote the Banach dual space of $\mathbb{B}$. The convergence theorem 10.4 implies the convergence of the real martingale $<X_n, x'>$ towards $<X_\infty, x'>$ in $L^1_{\mathbb{R}}(\Omega, \mathscr{F}_\infty, P)$ for every $x' \in \mathbb{B}'$. Let us consider any $\mathbb{B}'$-valued random variable of the following form

(11.2.3) $$\phi(\omega) := \sum_{k=1}^{n} 1_{F_i} \cdot x_i', \quad x_i' \in \mathbb{B}', \quad F_i \in \mathscr{F}_\infty .$$

Then the sequence $<X_n, \phi>$ clearly converges to $<X_\infty, \phi>$ in $L^1_{\mathbb{R}}(\Omega, \mathscr{F}_\infty, P)$ and, since

$$E(\|X_\infty\|) = \sup_{\|\phi\|\infty \leqslant 1} |E < X_\infty, \phi > | ,$$

where ϕ ranges over all $\mathbb{B}'$-valued random variables of the form (11.2.3) with uniform essential norm less than 1, one may write

(11.2.4) $$E\|X_\infty\| \leqslant \lim_n E\|X_n\| = E(Y) .$$

The formulas (11.2.1) and (11.2.4) and the Fatou lemma together imply

$$0 \leqslant \int \|X_\infty\| \, dP - \lim_n \int \|X_n\| \, dP \leqslant \int \|X_\infty\| \, dP - \int Y \, dP \leqslant 0 .$$

Therefore,

$$Y = \|X_\infty\| \quad \text{a.s.}$$

The same argument holds for the martingale $(X_n - y)_{n \in \bar{\mathbb{N}}}$, whenever $y \in \mathbb{B}$, so we have

$$\lim_{n \to \infty} \|X_n - y\| = \|X_\infty - y\| \quad \text{a.s.}$$

The random variables $(X_n)_{n \in \mathbb{N}}$, taking a.s. their values in a separable subspace of $\mathbb{B}$ (this follows from the assumption of strong measurability of the X_n's), we may, without loss of generality, assume that $\mathbb{B}$ is separable. Let $\mathbb{B}_0$ be a denumerable dense set in $\mathbb{B}$, and let Ω_0 be a P-negligible set such that

$$\forall \omega \notin \Omega_0, \; \forall y \in \mathbb{B}_0, \; \lim_n \|X_n(\omega) - y\| = \|X_\infty(\omega) - y\| .$$

Hence, it is straightforward to show that

$$\forall \omega \notin \Omega_0, \; \lim_n \|X_n(\omega) - X_\infty(\omega)\| = 0 .$$

The sequence (X_n) thus converges a.s. to X_∞ and the uniform integrability of the family $(\|X_n\|)_{n \in \mathbb{N}}$ implies that the convergence also occurs in $L^1_{\mathbb{B}}(\Omega, \mathscr{F}_\infty, P)$. □

11.3 Theorem. *Let $\mathbb{B}$ be a Banach space and $(X_n)_{n \in \mathbb{N}}$ an $(\mathscr{F}_n)$-martingale with state-space $\mathbb{B}$. We consider the three following properties for (X_n).*

(a) $(X_n)_{n \in \mathbb{N}}$ is uniformly integrable.

(b) *There exists a random variable X such that*

$$X_n := E(X \mid \mathscr{F}_n) \quad \textit{a.s. for all} \quad n \in \mathbb{N} .$$

(c) $(X_n)_{n\in\mathbb{N}}$ *converges in* $L^1(\Omega, \mathscr{F}_\infty, P)$ *towards a random variable* X_∞. *Then the following implications hold.*

$$\text{(b)} \Leftrightarrow \text{(c)} \Rightarrow \text{(a)}.$$

When (b) *or* (c) *is satisfied, the sequence* (X_n) *also converges almost surely.*

Proof: The implications (c) $\Rightarrow$ (b) $\Rightarrow$ (a) are derived exactly as in Theorem 10.4. The implication (b) $\Rightarrow$ (c) and the last statement of the theorem are already part of Theorem 11.2. □

11.4 Remark. The difference with the real case is clearly that the equivalence (a) $\Leftrightarrow$ (c) does not hold for a general Banach space $\mathbb{B}$.

The class of Banach spaces for which this equivalence is true is precisely the class of Banach spaces having the Radon-Nikodym property. This will follow from the results in § 11.7 and 11.8.

To begin with, we state and prove a lemma in measure theory. Since this lemma is not absolutely classical, we prove it for the sake of clarity.

11.5 Lemma. *Let* $\mathbb{B}$ *be a Banach space,* $\mathscr{A}$ *an algebra of subsets of a set* Ω *and* m *a* $\mathbb{B}$*-valued content, defined on* $\mathscr{A}$*, with bounded variation. We denote the variation on a set* A *by* $|m|(A)$. *Then there exist two contents* m_σ *and* m_f *with the following properties.*

(1) m_σ *is* σ*-additive with* σ*-additive finite variation* $|m_\sigma|$.

(2) m_f *has a purely finitely additive variation* $|m_f|$ *(i.e. for every* σ*-additive positive content* μ *on* $\mathscr{A}$*, the inequality*

$$\mu \leqslant |m_f| \quad \textit{implies} \quad \mu = 0).$$

(3) $|m| = |m_\sigma| + |m_f|$.

Proof: We use the classical decomposition of a real content in the complete Rieß-space of real contents with finite variation (see [Bou. 1]). This means we can write $|m| = \mu + \nu$, where μ is σ-additive, ν is purely finitely additive and this decomposition is unique. As $|m|$ is positive, so are μ and ν. From the decomposition property follows the existence of a sequence $(A_n)_{n\in\mathbb{N}}$ in $\mathscr{A}$, such that

$$\forall n \in \mathbb{N},\ \mu(A_n) + \nu(\Omega - A_n) \leqslant \frac{1}{n}.$$

Since

$$\|m(B \cap (A_n \Delta A_{n+p})\| \leqslant |m|(A_n \Delta A_{n+p}) \leqslant \frac{2}{n} \text{ for all } p, n \text{ and } B \in \mathscr{A},$$

we can define

$$m_f(B) := \lim_{n\to\infty} m(B \cap A_n) \tag{11.5.1}$$

and

(11.5.2) $$m_\sigma(B) := m(B) - m_f(B) = \lim_{n\to\infty} m(B \cap (A - A_n)).$$

Expressing $|m_\sigma|(B)$ with the help of $\mathscr{A}$-partitions of B and using the definition (11.5.2), we can immediately see that $|m_\sigma| = \mu$. Similarly, we get $|m_f| = \nu$. The σ-additivity of $|m_\sigma|$ implies the σ-additivity of m_σ.

Let us remark now that, for every $x' \in \mathbb{B}'$ (the dual space of $\mathbb{B}$) $<m, x'> = <m_\sigma, x'> + <m_f, x'>$, the real content $<m_\sigma, x'>$ being σ-additive while $<m_f, x'>$ is purely finitely additive. The uniqueness of the decomposition for m follows immediately from the uniqueness of the decomposition für real contents with bounded variation. □

11.6 Proposition. *Let $\mathbb{B}$ be a Banach space with the R.N. property and $(M_n)_{n\in\mathbb{N}}$ a $\mathbb{B}$-valued martingale, such that* $\sup_n E(\|M_n\|) < \infty$. *Then there exists a decomposition*

$$M_n = M_n^\sigma + M_n^f$$

with the following properties.

(1) $(M_n^\sigma)_{n\in\mathbb{N}}$ *is a $\mathbb{B}$-valued martingale and there exists a $\mathbb{B}$-valued random variable M_∞^σ such that*

$$M_n^\sigma = E(M_\infty^\sigma \mid \mathscr{F}_n) \quad \text{a.s.}$$

(2) $(M_n^f)_{n\in\mathbb{N}}$ *is a martingale and*

$$\lim_{n\to\infty} M_n^f = 0 \quad a.s.$$

(3) *Let $M_n'^\sigma + M_n'^f$ be another decomposition of M_n with the properties* (1) *and* (2). *Then*

$$M_n^\sigma = M_n'^\sigma \quad and \quad M_n^f = M_n'^f \quad a.s. for all \quad n.$$

Proof: We define a content α on $\bigcup_n \mathscr{F}_n$ by setting

$$\forall F \in \mathscr{F}_n,\ \alpha(F) = \int_F M_n\, dP.$$

We know from Theorem 9.4 that the real submartingale $(\|M_n\|)_{n\in\mathbb{N}}$ converges a.s.

We note that the variation of the content α_n, α_n being the restriction of α to $\mathscr{F}_n$, is given by

$$\forall F \in \mathscr{F}_n, |\alpha_n|(F) = \int_F \|M_n\|\, dP$$

and the sequence $(|\alpha_n|(F))_{m\leqslant n}$, $F \in \mathscr{F}_m$ is increasing with $|\alpha|(F) = \lim_{n\to\infty} |\alpha_n|(F)$. We may therefore write, for every $F \in \bigcup_n \mathscr{F}_n$,

(11.6.1) $$\int_F \lim_n \|M_n\|\, dP \leqslant \lim_n \int_F \|M_n\|\, dP = |\alpha|(F) = \sup_n \int_F \|M_n\|\, dP.$$

Let us denote by $\alpha = \alpha_\sigma + \alpha_f$ the decomposition of α given by Lemma 11.5. For every $F \in \bigcup_n \mathscr{F}_n$, with $P(F) = 0$, we may write $|\alpha|(F) = 0$ and therefore

$$|\alpha_\sigma|(F) = 0\,.$$

From the R.N. property of $\mathbb{B}$, we derive the existence of a Radon-Nikodym density M_∞^σ for α_σ with respect to P. We define

(11.6.2) $$M_n^\sigma = E(M_\infty^\sigma \,|\, \mathscr{F}_n)$$

and

(11.6.3) $$M_n^f = M_n - M_n^\sigma$$

From (11.6.2) it follows that $\alpha_\sigma(F) = \int_F M_n^\sigma \, dP$ for every $F \in \mathscr{F}_n$ and therefore

$$\alpha_f(F) = \int_F M_n^f \, dP\,.$$

This says that α_f is the content on $\bigcup_n \mathscr{F}_n$ associated with the martingale M^f. Applying the inequalities (11.6.1) to this martingale, we get

(11.6.4) $$\int_F \lim_n \|M_n^f\| \, dP \leqslant |\alpha_f|(F)\,.$$

But, α_f being purely finitely additive, the σ-additive content $F \rightsquigarrow \int_F \lim_n \|M_n^f\| \, dP$ has to be zero. In other words,

(11.6.5) $$\lim_n \|M_n^f\| = 0\,.$$

To prove the uniqueness of the decomposition, let us now consider any decomposition $M_n = M_n'^\sigma + M_n'^f$ with the properties (1) and (2). We have

$$M_n'^f - M_n^f = M_n^\sigma - M_n'^\sigma$$

and the martingale $M'^f - M^f$ has therefore an associated content on $\bigcup_n \mathscr{F}_n$ which, at the same time, is σ-additive and purely finitely additive. Since this content is given by $F \rightsquigarrow \int_F (M_\infty^\sigma - M_\infty'^\sigma)\, dP$, we see that $M_\infty^\sigma - M_\infty'^\sigma = 0$ a.s. Statement (3) of the theorem follows immediately. □

If the Banach space $\mathbb{B}$ has the R.N. property, the analogue of the Theorem 9.4 can be stated for $\mathbb{B}$-valued quasi-martingale.

11.7 Theorem.* *Let $\mathbb{B}$ be a Banach space with the R.N. property, and $(X_n)_{n\in\mathbb{N}}$ a $\mathbb{B}$-valued process with the following properties* (i) *and* (ii).

(i) $$\sum_n E\,\|E(X_{n+1}\,|\,\mathscr{F}_n) - X_n\| < \infty\,.$$

(ii) $$\sup_n E\,\|X_n\| < \infty\,.$$

Then the sequence (X_n) *converges a.s. to an integrable random variable* X_∞ *and, moreover,*

$$E\,\|X_\infty\| \leqslant \liminf_n E\,\|X_n\|.$$

Proof: As a consequence of theorems 11.1 and 11.6, we may write

$$X_n = X_n^\sigma + X_n^f + V_n,$$

where

$$X_n^\sigma = E(X_\infty^\sigma \mid \mathscr{F}_n)$$

and

$$\lim_{n\to\infty} \|X_n^f\| = \lim_{n\to\infty} \|V_n\| = 0 \quad \text{a.s.}$$

Since (X_n^σ) converges to X_∞^σ, according to Theorem 11.2, we immediately have the first statement of the theorem. The last inequality follows from the Fatou lemma. □

11.8 Corollary. *Let* $\mathbb{B}$ *be a Banach space with the R.N. property, and* $(X_n)_{n\in\mathbb{N}}$ *a* $\mathbb{B}$*-valued martingale. Then the following statements are equivalent.*

(a) $(X_n)_{n\in\mathbb{N}}$ is *uniformly integrable.*

(b) *There exists a Random variable* X *such that*

$$X_n = E(X \mid \mathscr{F}_n) \quad \textit{a.s. for every} \quad n \in \mathbb{R}.$$

(c) *The sequence* $(X_n)_{n\in\mathbb{N}}$ *converges in* $L^1_{\mathbb{B}}(\Omega, \mathscr{F}_\infty, P)$ *to a random variable* X_∞.
When these statements are true, the sequence $(X_n)_{n\in\mathbb{N}}$ *also converges to* X_∞ *a.s.*

Proof: This corollary follows immediately from 11.3 and 11.7. □

12. A typical application of quasimartingale convergence theorems: convergence of stochastic algorithms

12.1 Description of the algorithms

Robbins and Monro ([RoM]) and later Kiefer and Wolfowitz [KiW], [Wol] have introduced a class of stochastic algorithms of the type

$$(12.1.1) \qquad \Theta_{n+1} = \Theta_n - \gamma_n V_n(\Theta_n, Y_{n+1}),$$

where the Y_n are observed $\mathbb{R}^d$-valued random variables, V_n are given functions on $\mathbb{R}^k \times \mathbb{R}^d$ with values in $\mathbb{R}^k$, and (γ_n) is a decreasing family of positive numbers. The goal of the algorithms is to provide estimates Θ_n of a quantity tied with the random variables Y_n. This generalizes the situation where, (Y_n) being a sequence of independent identically distributed random variables Y_n with unknown mean α,

$$\Theta_n := \frac{1}{n} \sum_{k \leqslant n} Y_k ,$$

is an estimate of α which converges towards α when $n \to \infty$. According to the law of large numbers, this estimate satisfies the recurrence equation

$$\Theta_{n+1} = \Theta_n - \frac{1}{n+1} (\Theta_n - Y_{n+1}) ,$$

which is an equation of type (12.1.1).

We consider a filtration $(\mathscr{F}_n)$ for which the processes $(\Theta_n)_{n \in \mathbb{N}}$ and $(Y_n)_{n \in \mathbb{N}}$ are adapted.

The class of algorithms considered is the one for which the following hypotheses (A_1) and (A_2) are satisfied.

(A_1) There exists a function $\bar{V}_n$ on $\mathbb{R}^k$ such that for every n,

$$E(V_n(\Theta_n, Y_{n+1}) | \mathscr{F}_n) = \bar{V}_n(\Theta_n) ,$$

(A_2) There exists a positive function S_n^2 on $\mathbb{R}^k$ such that for every n,

$$E(V_n^2(\Theta, Y_{n+1}) | \mathscr{F}_n) = S_n^2(\Theta_n) .$$

Robbins-Monro case

In the case first treated by R. Robbins and H. Monro, for every value x of a real "control parameter", $Y(x)$ is a real random variable with unknown expectation $M(x)$. One starts with an initial value Θ_0 and at each step n, one has a trial which leads to the observation of a random variable $V_n(\Theta_n, Y_{n+1})$, whose conditional law with respect to $\mathscr{F}_n$ is the law of the random variable $Y(\Theta_n) - \alpha$, where α is a given vector. values of Θ_n in the successive trials is defined by

$$\Theta_{n+1} = \Theta_n - \gamma_n (Y(\Theta_n) - \alpha) .$$

It will be proven that, under suitable assumptions about the γ_n's and M, the sequence (Θ_n) converges a.s. to the solution of the (unknown!) equation

$$M(\Theta) = \alpha .$$

One of the first examples to be considered was the following: x is the quantity of powder used to make tubes of a certain type explode. The observed event is either explosion or no explosion, described by a random variable $Y(x)$ taking the value 1 or 0. The problem is to determine Θ^* such that $E(Y(\Theta^*)) = P\{Y(\Theta^*) = 1\} = \alpha$, a given probability of explosion, without knowing the function $M: M(x) := P\{Y(x) = 1\}$. The above algorithm describes an empirical way of successively approximating Θ^*. To summarize: the Robbins-Monro case is the case where a family $\mu(x, dy)$ of probability laws on $\mathbb{R}^k$ is given (depending on $x \in \mathbb{R}^k$) and where, in the hypothesis (A_1) and (A_2) above, we set

(12.1.2) $$\bar{V}(x) := M(x) - \alpha := \int y\,\mu(x, dy) - \alpha, \quad \bar{V}_n(x) := \bar{V}(x) \quad \text{for all} \quad n.$$

(12.1.3) $$S_n^2(x) := \int (y-\alpha)^2\,\mu(x, dy) \quad \text{and} \quad S_n^2(x) := S^2(x) \quad \text{for all} \quad n.$$

Kiefer-Wolfowitz case

In this case, we have a family of real random variables $Y(x)$ as in the previous case, x being real, and, in addition to the sequence (γ_n), another decreasing sequence (c_n) of positive numbers. The algorithm is roughly described by

(12.1.4) $$\Theta_{n+1} = \Theta_n - \gamma_n \frac{1}{c_n} (Y(\Theta_n - c_n) - Y(\Theta_n + c_n)).$$

This means precisely that if $\mu(x, dy)$ denotes the probability law of the real random variable $Y(x)$, the functions $\bar{V}_n$ in (A_1) and (A_2) are given by

(12.1.5) $$\bar{V}_n(x) := \frac{1}{c_n} [M(x - c_n) - M(x + c_n)],$$

where M is, as above, defined by

(12.1.6) $$M(x) := \int y\,\mu(x, dy).$$

If $x \in \mathbb{R}^k$ and $\mu(x, dy)$ denotes a probability law on $\mathbb{R}$, the algorithm (12.1.4) is replaced by

(12.1.7) $$\Theta_{n+1} = \Theta_n - \gamma_n \frac{1}{c_n} \sum_{i=1}^{k} (Y_n^{i,1} - Y_n^{i,2})\,\vec{e}_i,$$

where $Y_n^{i,1}$ (resp. $Y_n^{i,2}$) is a real random variable which, depending on the previous experiments $(Y^{i,j}_\ell, \ell < n)$, has the law $\mu(\Theta_n - c_n\vec{e}_i, dy)$ (resp $\mu(\Theta_n + c_n\vec{e}_i, dy)$) and where $(\vec{e}_i)_{i=1\ldots k}$ is a basis of $\mathbb{R}^k$. In this case,

(12.1.8) $$\bar{V}_n(x) := \frac{1}{c_n} \sum_{i=1}^{k} [M(x - c_n\vec{e}_i) - M(x + c_n\vec{e}_i)]\,\vec{e}_i.$$

The Kiefer-Wolfowitz algorithm is intended to give a convergent estimate of the solution Θ of the problem

$$M(\Theta) = \max_x M(x).$$

If the function M indeed is differentiable and strictly concave we have $\bar{V}(x) := \lim_n \bar{V}_n(x) = -\operatorname{Grad} M(x)$, and the point Θ solution of this maximum problem can be characterized by the property $(x - \Theta) \cdot \bar{V}(x) > 0$ for every $x \neq \Theta$ and every n. It will be proved that, under suitable assumptions about (γ_n), (c_n), (S_n^2) and $\bar{V}_n(x)$, the sequence (Θ_n) of solutions of the recurrence equation (12.1.1) converge a.s. towards a Θ which satisfies the inequality $(x - \Theta) \cdot \bar{V}(x) > 0$ for all $x \neq \Theta$ and n, and therefore towards the solution of the maximum problem.

12.2 Theorem. *We assume the hypotheses* (A_1), (A_2) *as well as the following.*

(H_1) *There exists* $\Theta^* \in \mathbb{R}^k$ *such that* $\forall \varepsilon > 0$

$$\liminf_n \inf_{\varepsilon < \|x-\Theta^*\| < \frac{1}{\varepsilon}} (x - \Theta^*) \cdot \bar{V}^n(x) > 0$$

(denoting by $x \cdot y$ *the scalar product in* $\mathbb{R}^k$*).*

(H_2) *There exists a sequence* (δ_n) *of positive numbers bounded by a number* K *such that*

(i) $$\sum_{n \geq 0} \gamma_n = +\infty$$

(ii) $$\sum_{n \geq 0} \frac{\gamma_n^2}{\delta_n^2} < \infty$$

(iii) $$S_n^2(x) \leq \frac{1}{\delta_n^2}\left[1 + \|x - \Theta^*\|^2\right]$$

Then, for every initial condition Θ_0*, the sequence* (Θ_n) *solution of* (12.1.1) *converges a. s. and in* L^1 *towards* Θ^*.

Proof: We set

(12.2.1) $$T_n := \Theta_n - \Theta^* .$$

The equation (12.1.1) can then be written

(12.2.2) $$T_{n+1} = T_n - \gamma_n V_n(\Theta_n, Y_{n+1}) .$$

Using (A_1) and (A_2), we get

(12.2.3) $$E(\|T_{n+1}\|^2 \mid \mathscr{F}_n) = \|T_n\|^2 - 2\gamma_n T_n \cdot \bar{V}_n(\Theta_n) + \gamma_n^2 S_n^2(\Theta_n) .$$

and following (H_2) (iii), we get

$$E(\|T_{n+1}\|^2 \mid \mathscr{F}_n) - \|T^n\|^2 \leq \frac{\gamma_n^2}{\delta_n^2}(1 + \|T_n\|^2) - 2\gamma_n T_n \cdot \bar{\mathrm{V}}_n(\Theta_n)$$

or

(12.2.4) $$E(\|T_{n+1}\|^2 \mid \mathscr{F}_n) - \left(1 + \frac{\gamma_n^2}{\delta_n^2}\right)\|T_n\|^2 \leq \frac{\gamma_n^2}{\delta_n^2} - 2\gamma_n T_n \cdot \bar{V}_n(\Theta_n)$$

We define

$$\Pi_n := \prod_{k=1}^{n-1}\left(1 + \frac{\gamma_k^2}{\delta_k^2}\right)$$

and

$$T_n' := \left(\frac{1}{\Pi_n}\right)^{\frac{1}{2}} T_n$$

With this change of random variables, inequality (12.2.4) can be written

(12.2.5) $$E(\|T'_{n+1}\|^2 \mid \mathscr{F}_n) - \|T'_n\|^2 \leqslant \Pi_{n+1}^{-1} \frac{\gamma_n^2}{\delta_n^2} - 2\gamma_n T_n \cdot \bar{V}_n(\Theta_n)$$

From the hypothesis (H_2) (ii) follows the convergence of the sequence (Π_n) (take the logarithm!) as well as the property $\sum_{n \geqslant 0} \Pi_{n+1}^{-1} \frac{\gamma_n^2}{\delta_n^2} < \infty$.

(H_1) implies $\liminf E(T_n \cdot \bar{V}_n(\Theta_n)) \geqslant 0$

Therefore, if $F_n := \{E(\|T'_{n+1}\|^2 - \|T'_n\|^2 \mid \mathscr{F}_n > 0\}$, we find from (12.2.5) that

(12.2.6) $$\sum_{n \geqslant 0} E(1_{F_n}(\|T'_{n+1}\|^2 - \|T'_n\|^2)) < \infty .$$

Since the $\|T'_n\|^2$ are positive random variables, Proposition 9.5 and Theorem 9.4 together imply the a.s. convergence of the sequence $\|T'_n\|^2$ towards some limit Z. We are left to prove that this limit is zero. The same property will hold for $\|T_n\|^2$.

From inequality (12.2.5) and $\sum_n \Pi_{n+1} \frac{\gamma_n^2}{\delta_n^2} < \infty$ we derive

$$E(\|T'_{n+1}\|^2) \leqslant \sum_{k \geqslant 0} \Pi_{k+1} \frac{\gamma_k^2}{\delta_k^2} < \infty$$

and

(12.2.7) $$\sup_n E\|T_n\|^2 < \infty .$$

Inequality (12.2.3) and property (H_1) imply

$$0 \leqslant \sum_{n=1}^{\infty} 2\gamma_n \cdot E(T_n \cdot \bar{V}_n(\Theta_n))$$

$$\leqslant \sum_{n \geqslant 0} E(\|T_n\|^2) - E(\|T_{n+1}\|^2) + \sum_{n \geqslant 0} \frac{\gamma_n^2}{\delta_n^2}[1 + E(\|T_n\|^2)]$$

$$\leqslant E(\|T_0\|^2) + \sum_{n \geqslant 0} \frac{\gamma_n^2}{\delta_n^2}(1 + \sup_k E(\|T_k\|^2)) < \infty$$

From $\sum_{n>0} \gamma_n = +\infty$ (hypothesis (H_2) (i)) there follows the existence of an increasing sequence $(n_k)_{k>0}$ of integers such that $\lim_k E(T_{n_k} \cdot \bar{V}_{n_k}(\Theta_{n_k})) = 0$.

On account of (H_1), this implies

$$\liminf_k \|T_{n_k}\| \leqslant \varepsilon \quad \text{a.s. for every} \quad \varepsilon > 0,$$

that is,

(12.2.8) $$\liminf_k \|T_{n_k}\| = 0 \quad \text{a.s.}$$

Since the sequence $(\|T_n\|)_{n \in \mathbb{N}}$ converges almost surely, this relation implies

(12.2.9) $\lim_n \|T_n\| = 0$ a.s.

The relations (12.2.9) and (12.2.7) together imply

$$\lim_n E\|T_n\| = 0.$$

The proof of the theorem is therefore complete. □

12.3 Corollary 1. *We consider the Robbins-Monro algorithm associated with the family $(\mu(x,.) : x \in \mathbb{R}^k)$ of probability laws on $\mathbb{R}^k$. We assume*

(a) *The equation $M(\Theta^*) - \alpha = 0$ has a unique solution Θ^* and for every*

$$\varepsilon > 0 \qquad \inf_{\varepsilon \leqslant \|x-\Theta^*\| \leqslant \frac{1}{\varepsilon}} (x - \Theta^*)(M(x) - \alpha) > 0$$

(b) *There exists $K > 0$ such that*

$$S^2(x) \leqslant K(1 + \|x - \Theta^*\|^2) \quad \textit{for all} \quad x \in \mathbb{R}^k$$

(c) *The sequence (γ_n) of positive numbers satisfies*

$$\sum_n \gamma_n = +\infty, \quad \sum_n \gamma_n^2 < \infty.$$

Then the sequence (Θ_n) converges a.s. and in L^1 towards Θ^*.

Proof. This corollary is a clear and immediate consequence of the above theorem. □

12.4 Corollary 2. *We consider the Kiefer-Wolfowitz algorithm associated with the family $\{\mu(x), dy) : x \in \mathbb{R}^k\}$ of probability-laws on $\mathbb{R}$. With the definitions* (12.1.4), (12.1.5), (12.1.8) *we assume the following.*

(a) *The function M is twice differentiable with bounded second derivative such that for some point Θ^* : $(x - \Theta^*)$.* Grad $M(x) < 0$ *for all $x \neq \Theta^*$*

(b) *There exists $K > 0$ such that $S^2(x) := \int y^2 \mu(x, dy) \leqslant K(1 + \|x\|^2)$ for all $x \in \mathbb{R}^k$*

(c) *The decreasing sequences (γ_n) and (c_n) of positive numbers satisfy*

$$\sum_n \gamma_n = +\infty, \quad \sum_n \frac{\gamma_n^2}{c_n^2} < \infty, \quad \sum_n \gamma_n c_n < \infty$$

Then the sequence (Θ_n) converges a.s. and in L^1 towards Θ^*, which is the unique solution of the maximum problem $M(\Theta^*) = \max_{x \in \mathbb{R}^k} M(x)$.

Proof. We set $\tilde{V}_n := V_n + \bar{V}(\Theta_n) - \bar{V}_n(\Theta_n)$, where V_n, $\bar{V}_n$ and $\bar{V}$ are as in § 12.1. Then $E(\tilde{V}_n | \mathscr{F}_n) = \bar{V}(\Theta_n) = -2\,\text{Grad}\, M(\Theta_n)$ and, using the boundedness of second order derivatives of M, $E(\|\tilde{V}_n\|^2 | \mathscr{F}_n) \leqslant 2E(\|V_n\|^2 | \mathscr{F}_n) + 2K_1^2 c_n^2$ for some constant K_1. Therefore, sor some constant K_2,

(12.4.1) $$E(\|\tilde{V}_n\|^2 \mid \mathscr{F}_n) \leqslant \frac{K_2}{c_n^2}(1 + \|\Theta_n\|^2)$$

Let us now consider the following algorithms.

(12.4.2) $$\tilde{\Theta}_{n+1}^k = \tilde{\Theta}_n^k - \gamma_n \tilde{V}_n, n \geqslant k, \tilde{\Theta}_k^k = \Theta_k.$$

For every $n \geqslant k$,

(12.4.3) $$\Theta_{n+1} = \tilde{\Theta}_{n+1}^k + \sum_{k<p\leqslant n} \gamma_p(\bar{V}(\Theta_p) - \bar{V}_p(\Theta_p))$$

with $\|\bar{V}(\Theta_n) - \bar{V}_n(\Theta_n)\| \leqslant K_1 c_n$

Equality (12.4.3) therefore implies

(12.4.4) $$\sup_{n\geq k} \|\Theta_n - \tilde{\Theta}_n^k\| \leqslant K_1 \sum_{k\leqslant n} \gamma_n c_n < \infty$$

The inequality (12.4.1) and the uniform boundedness of $\|\Theta_n - \tilde{\Theta}_n^k\|$ together show that

(12.4.5) $$E(\|\tilde{V}_n^2\| \mid \mathscr{F}_n) \leqslant \frac{K}{c_n^2}(1 + \|\tilde{\Theta}_n^k\|^2) \quad \text{for some constant} \quad K.$$

The algorithm (12.4.2) satisfies the hypothesis of Theorem 12.2 and therefore converges a.s. and in L^1 towards Θ^*. From (12.4.4) and $\lim_k \sum_{n\geqslant k} \gamma_n c_n = 0$ one easily finds the a.s. convergence of Θ_n to Θ^k, and the convergence in L^1 as well. □

Exercises and supplements

1. On the concept of quasimartingale

E. 1.1. *The martingale property under a change of filtration.* Let β be the canonical Brownian motion on its canonical basis $(\Omega_c, (\mathscr{C}_t)_{t\in\mathbb{R}^+}, P)$ (see 2.2). For every t, we denote by $\mathscr{F}_t$ the σ-algebra which is generated by $\mathscr{C}_t \cup \mathscr{F}_1$.

(a) For $0 \leqslant s \leqslant t \leqslant 1$ express the conditional law of β_t with respect to $\mathscr{F}_s$.

(b) One shows that β_t is not a martingale with respect to the filtration $(\mathscr{F}_s)_{s\in\mathbb{R}}$ but that the process M defined by

$$M_t := \beta_t - \int_0^{t\wedge 1} \frac{\beta_1 - \beta_u}{1-u} du \quad \text{is a Martingale.}$$

E. 1.2. *Quasimartingales as defined by D. L. Fisk.* Let $(X_t)_{t\in I}$ be a real – or Banach – valued process. Prove the equivalence of the following statements (i) and (ii).

(i) $$\forall t, \sum_{i\in\mathbb{N}} E(|E(X_{t_{i+1}} \mid \mathscr{F}_{t_i}) - X_{t_i}|) \leqslant K(t) < \infty$$

$$\forall t_0 < t_1 \ldots < t_n < \ldots < t,$$

and

(ii) $$|\lambda_X|([0,t]\times\Omega)<\infty,\ K(t),\ \forall t).$$

The original definition of a quasimartingale (see D. L. Fisk [Fis], S. Orey [Ore. 1]) was the following: $(X_t)_{t\in I}$ is a quasimartingale if (i) is satisfied. Compare this definition with the one taken here (see [MeP. 2]).

E. 1.3. *A Riesz decomposition of quasimartingales.* Let $I\in\bar{\mathbb{R}}^+$ with $\beta:=\sup I\in I$. Let $(X_t)_{t\in I}$ be a quasimartingale on the stochastic basis $(\Omega,(\mathscr{F}_t)_{t\in I},P)$.

(1) Prove that X can be written as the difference $X'-X''$ of two submartingales and that there exists only one such decomposition with the following two properties.

(i) $$X''_\beta=0,$$

(ii) For every representation $X=Y'-Y''$ of X as a difference of two submartingales with $Y''_\beta=0$, one has $Y'\geqslant X'$ and $Y''\geqslant X''$ (*Hint:* X' can be defined as having the property

$$\forall F\in\mathscr{F}_t,\ \forall t\in I,\ \int_F X'_t\,dP=\int_F X_\beta\,dP-\lambda_X^+(]t,\beta]\times F)$$

(2) Prove too that X can be written as the difference $X=X'-X''$ of two positive supermartingales and that there exists only one such decomposition with the following two properties.

(j) $$X'_\beta=X^+_\beta,\ X''_\beta=X^-_\beta,$$

(jj) For every representation $X=Y'-Y''$ of X as a difference of two positive supermartingales with $Y'_\beta=Y^+_\beta$, $Y''_\beta=Y^-_\beta$, one has $Y'\geqslant X'$, $Y''\geqslant X''$. (*Hint:* X' can be defined as having the property

$$\forall t\in I,\ \forall F\in\mathscr{F}_t,\ \int_F X'_t\,dP=\int_F X^+_\beta\,dP+\lambda_X^-(]t,\beta]\times F)$$

E. 1. 4. Prove that neither of the two properties (9.3.1) and (9.3.3) implies the other.

E. 1.5. Let $(X_t)_{t\in I}$ be a Banach-valued quasimartingale. Prove that $(\|X_t\|)_{t\in I}$ is a positive quasimartingale (when $I=\mathbb{N}$, this is Lemma 8.10).

2. Inequalities, convergence and regularity properties

E. 2.1. *Semimartingales with Index set* $\mathbb{N}$ *and their convergence property.* We will say that a real process $(X_n)_{n\in\mathbb{N}}$ is a semimartingale with respect to the filtration $(\mathscr{F}_n)_{n\in\mathbb{N}}$ if it is adapted to this filtration and if the following property holds.

[S.M.] $$\sum_{n\in\mathbb{N}}|E(X_{n+1}-X_n|\mathscr{F}_n)|<\infty$$

Prove the following. If $(X_n)_{n\in\mathbb{N}}$ is a semimartingale such that $\sup_n E(|X_{n\wedge\tau}|)<\infty$

for every stopping time τ, there exists a random variable X_∞ such that

$$X_\infty = \lim_{n\to\infty} X_n .$$

(*Hint:* For every $k \in \mathbb{N}$ consider the stopping time

$$\tau_k := \inf\{n: \sum_{0\leqslant \ell\leqslant n} |E(X_{\ell+1} - X_\ell \mid \mathscr{F}_\ell)| > k\}$$

and the process

$$X_n^k(\omega) := \begin{cases} X_n(\omega) & \text{if } n \leqslant \tau_k(\omega) \\ X_{\tau_k(\omega)}(\omega) & \text{if } n > \tau_k(\omega). \end{cases}$$

E. 2.2. *Left-hand modification of a positive supermartingale.* Let $(X_t)_{t\in\mathbb{R}^+}$. be a positive supermartingale with respect to the filtration $(\mathscr{F}_t)_{t\in\mathbb{R}^+}$. We define $(\tilde{X}_{t^-})_{t\in\mathbb{R}^+}$ as in 10.8.

(1) Prove that $(\tilde{X}_{t^-})_{t\in\mathbb{R}^+}$ is a supermartingale with respect to $(\mathscr{F}_{t^-})_{t\in\mathbb{R}^+}$. (*Hint:* For every increasing sequence (t_n) converging to t one can write

$$X_{t_n} = E(X_t|\mathscr{F}_{t_n}) + X_{t_n} - E(X_t|\mathscr{F}_{t_n})$$

and derive

$$E(X_t|\mathscr{F}_{t^-}) \leqslant \tilde{X}_{t^-} .$$

For $s < t_n \uparrow t$ and $F \in \mathscr{F}_{s^-}$ one then shows that

$$\int_F \tilde{X}_{s^-}\, dP \geqslant \int_F X_s\, dP \geqslant \lim_n \int_F X_{t_n}\, dP \geqslant \int_F \tilde{X}_{t^-}\, dP)$$

(2) Prove that the relation $\tilde{X}_{t^-} = E(X_t|\mathscr{F}_{t^-})$ a.s. holds for every t if and only if $t \frown E(X_t)$ is left-continuous. This property is also equivalent to the uniform integrability of the family $\{X_s : s < t\}$.

(*Hint:* One uses the fact that, for the positive random variables X_{t_n}, the convergence to $\tilde{X}_{t^-}$ in L^1 of the sequence $\{X_{t_n} : t_n \uparrow t\}$ is equivalent with the relation $\lim_n E(X_{t_n}) \leqslant E(\tilde{X}_{t^-})$.

(3) Give an example of a positive supermartingale such that $t \frown E(X_t)$ is continuous and $(\tilde{X}_{t^-})_{t\in\mathbb{R}^+}$ is not a modification of $(X_t)_{t\in\mathbb{R}^+}$.

E. 2.3. *Lefthand modification of a quasimartingale.* Let I be a closed subinterval of $\bar{\mathbb{R}}^+$ with $b := \sup I$ and $(X_t)_{t\in\mathbb{R}^+\cap I}$ a quasimartingale with respect to the filtration $(\mathscr{F}_t)_{t\in I}$. One defines $(\tilde{X}_{t^-})_{t\in\mathbb{R}^+\cap I}$ as in 10.8.

(1) Prove that $(X_{t^-})_{t\in I\cap\mathbb{R}^+}$ is a quasimartingale with respect to $(\mathscr{F}_{t^-})_{t\in I}$.

(2) Prove that $\tilde{X}_{t^-} = E(X_t|\mathscr{F}_{t^-})$ a.s. for every $t \in I$ if and only if $t \frown |\lambda_X|(]t, b])$ is left-continuous.

(*Hint:* Use exercises E. 1.3.2° and E. 2.2).

E. 2.4. *Some convergence condition for stochastic algorithms*. One considers the stochastic algorithm defined by the equation (12.1.1) with the hypotheses (A_1) and (A_2) of §12.1 and the following.

(a) There exists $\Theta \in \mathbb{R}^k$ such that

$$(x-\Theta)\cdot \bar{V}_n(x) > 0 \quad \text{for every} \quad n \in \mathbb{N} \quad \text{and} \quad x \neq \Theta .$$

(b) There exists a bounded sequence (δ_n) with the properties

$$\text{(b. 1)} \qquad \sum_n \lambda_n \delta_n = +\infty\,, \ \sum_n \frac{\gamma_n^2}{\delta_n^2} < \infty \quad \text{and} \quad \lim_{n\to\infty} \frac{\gamma_n}{\delta_n^3} = 0\,,$$

$$\text{(b. 2)} \qquad \delta_n \|x-\Theta\|^2 \leqslant (x-\Theta)\cdot \bar{V}_n(x) \quad \text{for every} \quad x \neq \Theta\,,$$

$$\text{(b. 3)} \qquad S_n^2(x) \leqslant \frac{1}{\delta_n^2}[1+\|x-\Theta\|^2] \quad \text{for every} \quad x \in \mathbb{R}^k .$$

Prove that Θ_n converges a.s. in L^2 towards Θ. (*Hint:* Start with equation 12.2.3 to derive

$$E(\|T_{n+1}\|^2 | \mathscr{F}_n) \leqslant \|T_n\|^2 \left(1 - 2\gamma_n \delta_n + \frac{\gamma_n^2}{\delta_n^2}\right) + \frac{\gamma_n^2}{\delta_n^2},$$

and therefore

$$E(\|T_{n+1}\|^2 | \mathscr{F}_n) - \|T_n\|^2 \leqslant \frac{\gamma_n^2}{\delta_n^2}.$$

Deduce from the quasimartingale convergence theorem the a.s. convergence of $\|T_n\|^2$ and use the relation

$$E(\|T_{n+1}\|^2) \leqslant E\|T_n\|^2 \left(1 - 2\gamma_n \delta_n + \frac{\gamma_n^2}{\delta_n^2}\right) + \frac{\gamma_n^2}{\delta_n^2},$$

to prove that $\lim_n E\|T_n\|^2 = 0$).

E. 2.5. *Lyapunov functions for a recurrence equation* (see [Buc], [BuJ]). We consider the recurrence equation

$$\text{(a)} \qquad \Theta_{n+1} = f_n(\Theta_n, Y_n)\,,$$

where the Θ_n's (resp. the Y_n's) are $\mathbb{R}^d$-valued (resp. $\mathbb{R}^k$-valued) random variables and f_n is, for every n, a mapping from $\mathbb{R}^d \times \mathbb{R}^k$ into $\mathbb{R}^d$.

We call a function L on $\mathbb{R}^d \times \mathbb{N}$ a Lyapunov function for the equation (a) if the following conditions are satisfied.

(L_1) There exists two increasing positive functions α and β on $\mathbb{R}^+$ with the properties

$$\text{(i)} \qquad L(0,n) = \beta(0) = 0 \quad \text{for every} \quad n \in \mathbb{N}\,,$$

$$\text{(ii)} \qquad \alpha(\|x\|) \leqslant L(x,n) \leqslant \beta(\|x\|) \quad \text{for every} \quad n \in \mathbb{N},\ x \in \mathbb{R}^d .$$

(L_2) For every sequence (Θ_n) of random variables satisfying (a) the process $(L(\Theta_n, n))_{n \geqslant 0}$ is a supermartingale.

Show that, if a Lyapunov function L exists, then for every $\varrho > 0$, $\varepsilon > 0$ there exists $\eta > 0$ and $\delta > 0$ such that

$$P\{\|\Theta_0\| \geqslant \delta\} \leqslant \eta \Rightarrow P\{\sup_n \|\Theta_n\| \geqslant \varepsilon\} \leqslant \varrho$$

(This is called stochastic stability around 0. *Hint:* One shows that, for every $\varrho > 0$, $\varepsilon > 0$ and $\lambda > \alpha(\varepsilon)$ (setting $X_n := L(\Theta_n, n)$),

$$P\{\sup_n \|\Theta_n\| \geqslant \varepsilon\} \leqslant \frac{1}{\alpha(\varepsilon)} E(X_0 \wedge \lambda) \leqslant \frac{1}{\alpha(\varepsilon)} E(\beta(\|\Theta_0\|) \wedge \lambda)).$$

E. 2.6. *Lyapunov functions, continued.* We consider equation (a) of E. 2.5 and assume the existence of a function L on $\mathbb{R}^d \times \mathbb{N}$ with the property (L_1) of E. 2.5 and

(L_3) for some $r > 0$ and every sequence $(\Theta_n)_{n \in \mathbb{N}}$ of random variables satisfying (a),

$$E(L(\Theta_n, n) | \mathscr{F}_{n+1}) - L(\Theta_{n+1}, n-1) \leqslant -\gamma_n \|\Theta_{n-1}\|^r + U_n,$$

where $\gamma_n > 0$, $\sum_n \gamma_n = +\infty$, and the positive random variables U_n are such that $\sum_n E(U_n) < \infty$.

Prove that $(L_n(\Theta))_{n \in \mathbb{N}}$ is a quasimartingale and, if α is strictly increasing in a neighborhood of 0, for every choice of the initial condition Θ_0 one has

$$\lim_{n \to \infty} \|\Theta_n\| = 0,$$

(almost sure asymptotic stability around 0).

(*Hint:* Use the quasimartingale convergence theorem).

E. 2.7. *Lyapunov functions: examples.* (1) Take for the recurrence equation

$$\Theta_{n+1} = \Theta_n - \gamma_n V_n(\Theta_n, Y_n),$$

as in (12.1.1) with functions $\bar{V}_n$ and S_n^2 defined as in § 12.1.

We make the following assumptions.

(j) $\exists \Theta \in \mathbb{R}^d, \forall x \in \mathbb{R}^d, (x - \Theta) \cdot \bar{V}_n(\Theta) \geqslant \delta_n \|x - \Theta\|^r$ for some $r > 0$

and a bounded strictly positive sequence (δ_n).

(jj) $S_n^2 \leqslant d_n^2$ for some sequence (d_n),

(jjj) $\sum_n \gamma_n \delta_n = +\infty, \sum_n \gamma_n^2 d_n^2 < \infty.$

One shows that the result of E. 2.6 can be applied with the function $L(x, n) = \|x - \Theta\|^2$ for all n.

(2) (cf. [BuJ]). Consider the recurrence equation

(e) $$\Theta_{n+1} = A_n \Theta_n,$$

where the A_n are random independant matrices such that there exist matrices $B > 0$ and $C < 0$ with the property: $E(A_n' B A_n) - B < C$ for all n.

Show that $L(x, n) = x \cdot Bx$ defines a Lyapunov function and that equation (e) is stochastically stable around 0.

Prove that if, moreover, $x \cdot Cx \leqslant -\gamma \|x\|^{2'}$ the equation is also a. s. asymptotically stable around 0.

3. Martingales with filtering index set and Radon-Nikodym derivates

E. 3.1. Let I be a set endowed with an order structure $<$ such that $\forall s, t \in I$ there exists $u \in I$ such that $s < u$ and $t < u$. Let $(\mathscr{F}_s)_{s \in I}$ be a family of σ-algebras of subsets of Ω with the property: $s < t \Rightarrow \mathscr{F}_s \subset \mathscr{F}_t$ Further, let $\mathscr{F}$ be a σ-algebra containing all the σ-algebra's $\mathscr{F}_t$, $t \in I$ and P a probability law on $(\Omega, \mathscr{F})$. We call a $\mathbb{B}$-valued process X a generalized martingale when, for every pair (s, t) with $s < t$, the equality $E(X_t | \mathscr{F}_s) = X_s$ holds.

(a) Let us assume, that there exists a $\mathbb{B}$-valued random variable X such that $X_t = E(X | \mathscr{F}_t)$ for every $t \in I$. Prove that the generalized sequence $(X_t)_{t \in I}$ converges in $L^1_{\mathbb{B}}(\Omega, \mathscr{F}, P)$. (The convergence is trivial, when X is $\mathscr{F}_t$-measurable for some $t \in I$. Observe then, that a dense subset of $L^1_{\mathbb{B}}(\Omega, \mathscr{F}_\infty, P)$ is provided by the set of random variables with the last measurability property).

(b) Let $(X, \mathscr{B}, P)$ be a probability space and I the family of finite σ-sub-algebras of $\mathscr{B}$ with the inclusion order. Let μ be a $\mathbb{B}$-valued measure on $\mathscr{B}$. For every $\Pi \in I$, we define $\mathscr{F}_\pi = \Pi$ and

$$X_\pi := \sum_{A \in \mathring{\Pi}} \frac{\mu(A)}{P(A)} \quad \left(\text{with} \frac{\mu(A)}{P(A)} := 0 \quad \text{when} \quad P(A) = 0\right),$$

where $\mathring{\Pi}$ is the set of atoms of Π.

Prove that if the implication "$P(A) = 0 \Rightarrow \mu(A) = 0$" holds, then $(X_\pi)_{\pi \in I}$ is a martingale.

(c) With the help of (a) and (b), show that the following two statements are equivalent.

(i) Every uniformly integrable generalized $\mathbb{B}$-valued martingale converges in $L^1_{\mathbb{B}}(\Omega, \mathscr{F}, P)$ for every $(\Omega, \mathscr{F}, \mathscr{F}_t)_{t \in I}, P)$ and every index set I.

(ii) $\mathbb{B}$ has the R. N. property.

4. Applications to Markov processes

E. 4.1. *The infinitesimal generator and the Dynkin formula.* Let X be a $\mathbb{R}^d$-valued Markov process associated with the semi-group $(T_t)_{t \geqslant 0}$ (see section 2.7) operating on $C_0(\mathbb{R}^d)$. We call $\mathscr{D}_{\mathscr{G}}$ the subset of $C_0(\mathbb{R}^d)$ constituted by the functions f for which

$\lim_{t\downarrow 0} \dfrac{T_t f - f}{t}$ exists in the Banach space $C_0(\mathbb{R}^d)$ (with the norm of uniform convergence) and we set, for $f \in \mathscr{D}_{\mathscr{G}}$,

$$\mathscr{G}f = \lim_{t\downarrow 0} \frac{T_t f - f}{t}$$

$\mathscr{G}$ is called infinitesimal generator.
Assume X is right-continuous and $P[X_0 = x] = 1$.

Show that, for every $\varphi \in \mathscr{D}_{\mathscr{G}}$, the process M^φ defined by

$$M_t^\varphi := \varphi(X_t) - \varphi(x) - \int_0^t \mathscr{G}f(X_s)\, ds$$

is a martingale (the "*Dynkin Formula*").

(Set $\Phi(F, t) = E(1_F \varphi(X_t))$ for every $t \geqslant t'$ and $F \in \mathscr{F}_{t'}$ and show that

$\lim \frac{1}{h}[\Phi(F, s+h) - \Phi(F, s)] = E(1_F \mathscr{G}f(X_s)) \forall s \geqslant 0$ and therefore that

$$\Phi(F, t) - \Phi(F, t') - \int_{t'}^t E(1_F \mathscr{G}f(X_s))\, ds = 0).$$

E. 4.2. *Regularity of processes with independent increments.* Let X be a process with increments "independent of the past" (see section 2.5).

(1°) Assume that for every $s < t$, $E|X_t - X_s| < \infty$, and set $a(t) = E(X_t)$. Show that the process $(X_t - a(t))_{t \geqslant 0}$ is a martingale and conclude that, under the usual assumptions about the filtration $(\mathscr{F})$, there is an R.R.C. version of X whenever a(.) itself is right-continuous with left limits at every t.

(2°) Make no assumptions about moments of X but assume X to be stochastically continuous (i.e. $\forall t \lim_{h\downarrow 0} \text{Prob}\,|X_{t+h} - X_t| = 0$) and to have "stationary increments" (i.e. the law of $X_t - X_0$ is the same as the law of $X_{t+r} - X_r$ for every r: see chapter I, exercise E. 16). Then one shows that X has an R.R.C. version. To do this one may proceed as follows (see books by Breiman [Bre], chap. 14 and Doob [Doo. 1]).

(a) The problem can be reduced to the case when X is real valued and $X_0 = 0$.

(b) For every φ bounded and continuous on $\mathbb{R}$, denote by Y^φ the R.R.C. version of the martingale $E(\varphi(X_T)|\mathscr{F}_t)_{t<T}$ for T fixed and use the Markov property of X to show that $Y_t^\varphi = \Theta\ (\tau - t, X_t)$, where the function Θ is defined by $\Theta(t, x) = E(\varphi(X_t) + x)$.

(c) Take φ continuous and strictly increasing such that $\lim_{x\to+\infty} \varphi(x) = -\lim_{x\to-\infty} \varphi(x) = \alpha$. Note that the mapping $(t, x) \curvearrowright \Theta(t, x)$ is continuous and that the equation

$$y = \Theta(\tau - t, \Psi(y, t))$$

defines Ψ uniquely as a continuous mapping from $]-\alpha, +\alpha[\times [0, T]$ into $\mathbb{R}$. Apply this to Y_t.

Historical and bibliographical notes

The mathematical notion of martingale is mentioned in P. Levy's book [Lev. 1] and in a work by Ville [Vil], where the concept of stopping time is implicit. The modern impuls given to the theory is due to Doob [Doo. 1,3]. An early systematic exposition is in [Mey. 1], which contains Doob's inequalities, convergence and regularity properties.

The content λ_X of a process X was introduced by C. Doleans [Dol. 1] in connection with Meyer's decomposition theorem, studied in the next chapter. The characterization of quasimartingales, introduced by Fisk [Fis] (see also the F-process of Orey [Ore. 1]) as the processes for which λ_X has finite variation, was made by Föllmer [Föll] and Metivier-Pellaumail [MeP. 2] (see also Stricker [Str. 1 and 2] and Pop-Stojanovic [Pop] for the Banach-valued processes). The formulation of Doob's inequalities with the help of the variation of λ_X was given in [Met. 4].

The method for proving regularity and convergence properties for quasimartingales, based on Doob's inequality and upcrossings of intervals, is used in Doob's book. This method does not apply to martingales indexed by a non-totally ordered set, which have been studied by K. Krickeberg [Kri. 1], [Kri. 2] and K. Krickeberg and Pauc [KrP]. In this case the convergence in probability only holds, in general, assuming the boundedness of first order moments. Extra conditions (of "Vitali type" [KrP]) are to be added to obtain the a. s. convergence.

Convergence theorems for Banach-valued martingales abound in papers written in the 60's [Cht. 1, 2, 3, 4] [Met. 1, 2, 3], [IoT. 1], [Sca]. The convergence theorem for Banach-valued quasimartingales is stated and proved here as in [Met. 4].

Relations between martingale theory and Radon Nikodym derivatives was already noted by J. L. Doob. It is systematically studied in [KrP]. In the case of densities of vector-valued measures, see [Cht. 4] and [Met. 3].

The theory of stochastic algorithms is an enormous subject. The interested reader may consult the expositions in the books by Kushner and Clark [KuC], Has'Minskii and Nevelson [HaN] or Schmetterer [Schm]. The proof of convergence given here, using the quasimartingale convergence theorem is very close to one due to Gladysev [Gla]. The reader interested in a survey of the theory with many very practical applications is referred to part II of [Lan].

The theory of martingales indexed by $\mathbb{N}$ gave birth to a number of papers providing deep and useful inequalities ([Bur. 3], [Dar], [Dav. 2]...). The reader will find much information in Garsia's lecture notes [Gar] and also in [Nev. 6], where applications to the optimal stopping-time problem are also considered. This latter book also contains an extensive bibliography on the subject. Proofs of Burkholder, Davis, Fefferman inequalities are also contained in [MeP. 1], chap. 5.

Further results on the convergence of martingales may be found in [Chw. 1, 2, 3].

More information on the immediate relation between martingales and Markov processes (Dynkin formula – E. 4.1), which is presented in a rather elementary way, may be found in [DyY]. See also [Dyn], chap. 5.

Chapter 3

Quasimartingales from class [L. D] – Predictable and dual predictable projection of processes

This chapter is devoted to some of the deepest theorems of the general theory of stochastic processes, which we interpret systematically as expressing relations between some type of processes and admissible measures which they generate either on $\mathscr{P}$ or on the σ-algebra $\mathscr{B}_{\mathbb{R}^+} \otimes \mathfrak{A}$ of measurable sets.

The first section gives a necessary and sufficient condition for a quasimartingale to admit a σ-additive Doleans measure and derives, as an immediate consequence, the stopping theorem for quasimartingales.

The following section (§14), introduces the notion of predictable projection of processes (due to Dellacherie) and the dual notion of dual predictable projection of admissible measures which plays an ever increasing role in the study of processes. A predictable measure is a measure which coincides with its predictable projection.

In §15, the predictable measures are characterized as admissible measures generated by predictable processes with finite variation. The fundamental Doob-Meyer decomposition theorem appears to be an immediate translation of this characterization.

We conclude the chapter with examples. Some of them, such as the example of point processes, illustrate the point of considering vector-valued processes. However, as in the preceeding chapters, we have organized things in such a way that the reader can disregard anything concerning vector-valued processes if he is not interested in processes or measures which are not real-valued.

13. Doleans measure of an [L. D] – Quasimartingale

This paragraph is devoted to the study of the σ-additivity of the content λ_X associated to a quasimartingale X.

We shall always assume throughout this paragraph that the processes considered are all defined on a stochastic basis $(\Omega, (\mathscr{F}_t)_{t\in\mathbb{R}^+}, \mathfrak{A}, P)$ *which satisfies the "usual hypotheses"*.

13.1 Definition. For every subset I of $\mathbb{R}^+$ we denote by $\mathscr{T}_I$ (resp. $\mathscr{T}_I^f$) the set of stopping times which take their values in I (resp. in a finite subset of I).

Let X be a quasimartingale with index set $\mathbb{R}^+$. We say that X belongs to the class [D] on I when the family $\{X_\tau : \tau \in \mathcal{T}_I^f\}$ of random variables is uniformly integrable. We call briefly class [D] the class [D] on $\mathbb{R}^+$.

We say that X belongs locally to the class [D] or, briefly, belongs to the class [L. D], when X belongs to the class [D] on every bounded interval $[0, \alpha] \subset \mathbb{R}$.

We recall (see §6.3) that a subset A of $\mathbb{R}^+ \times \Omega$ is said to be *bounded*, when there is a $t \in \mathbb{R}^+$ such that $A \subset [0, t] \times \Omega$.

13.2 Remark. (1) The notion of classes [D] and [L. D] was introduced by P.A. Meyer for supermartingales [Mey. 1]. He proved that supermartingales of class [L. D] are precisely those which admit a Doob-Meyer decomposition. This will be proved in §15, where it will turn out as tightly related to the σ-additivity property of λ_X.

In Meyer's original definition, a process X was said to be of class [D] on I when the family $\{X_\tau : \tau \in \mathcal{T}_I\}$ is uniformly integrable. Theorem 12.3, will show that this property of X is equivalent to the one used here to define the class [D].

(2) It follows immediately from Theorem 10.1 that every positive submartingale belongs to the class [L. D].

(3) Remark on vector-valued measures: As is well known (see, for example, [DuS]) every vector-valued measure which is defined on a σ-algebra is bounded. If a positive content λ_X on $\mathcal{A}$ admits a σ-additive extension to every σ-algebra $\mathcal{P} \cap [0,t] \times \Omega$, $t \in \mathbb{R}^+$, there is an $\overline{\mathbb{R}}^+$-valued extension to $\mathcal{P}$. When λ_X takes its values in a vector space $\mathbb{B}$, the σ-additive extension, when it exists, is in general defined only on the bounded elements of $\mathcal{P}$.

13.3 Theorem. *Let $(X_t)_{t \in \mathbb{R}^+}$ be a right-continuous quasimartingale with values in a Banach space $\mathbb{B}$.*

(1) *Let $\alpha \in \mathbb{R}^+$. The content λ_X can be extended into a $\mathbb{B}$-valued, σ-additive measure on $\mathcal{P} \cap ([0, \alpha] \times \Omega)$ if and only if X belongs to class [D] on $[0, \alpha]$. This extension has bounded variations. (We also denote the extension of λ_X by λ_X).*

(2) *The content λ_X can be extended into a $\mathbb{B}$-valued, σ-additive measure on $\mathcal{P}$ with bounded variation if and only if X belongs to the class [L. D] and $|\lambda_X|(\mathbb{R}^+ \times \Omega) < \infty$.*

(3) *A quasimartingale belongs to the class [D] on the interval I, if and only if $\{X_\tau : \tau \in \mathcal{T}_I\}$ is uniformly integrable.*

(4) *If X belongs to the class [D] on the interval I, one can write, for every two I-valued stopping times σ and τ with $\sigma \leqslant \tau$,*

$$\lambda_X(]\sigma, \tau]) = E(X_\tau - X_\sigma).$$

Proof: The proof is divided into the following steps.

Step 1: If X belongs to the class [D] on $[0, \alpha]$, then λ_X and $|\lambda_X|$ have a σ-additive extension to $\mathcal{P}_{[0,\alpha]} := \mathcal{P} \bigcap ([0, \alpha] \times \Omega)$.

Step 2: If $|\lambda_X|$ has a σ-additive extension to $\mathcal{P}_{[0,\alpha]}$, then $\lambda_X(]\sigma, \tau]) = E(X_\tau - X_\sigma)$ for every $\sigma, \tau \in \mathcal{T}_{[0,\alpha]}$ with $\sigma \leqslant \tau$.

Step 3: If $|\lambda_X|$ has a σ-additive extension to $\mathscr{P}_{[0,\alpha]}$, the family $\{X_\tau : \tau \in \mathscr{T}_{[0,\alpha]}\}$ is uniformly integrable.

These three statements clearly contain statements (1), (3) and (4) of the theorem.

Statement (2) is easily derived as follows. If X belongs to the class [L. D], then $|\lambda_X|$ has a σ-additive, $\bar{\mathbb{R}}^+$-valued extension to $\mathscr{P}$, which is bounded iff $|\lambda_X|(\mathbb{R}^+ \times \Omega) = \infty$. The existence of a bounded σ-additive extension for $|\lambda_X|$ implies the existence of a σ-additive extension for λ_X. This follows indeed from the continuity of the mapping $A \curvearrowright \lambda_X(A) \in \mathbb{B}$ when we consider on $\mathscr{P}$ the distance $\delta(A, A') := |\lambda_X|(A \triangle A')$, and from the density of $\mathscr{A}$ in $\mathscr{P}$ for this distance.

Proof of step 1: According to the above reasoning, we have only to prove that the positive content $|\lambda_X|$ has a σ-additive, bounded extension to $\mathscr{P}_{[0,\alpha]}$. The boundedness will follow immediately from the quasimartingale property of X. To prove the σ-additivity, we need the following lemma.

Lemma 1: *Let μ be a positive content on $\mathscr{A} \cap ([0,\alpha] \times \Omega)$ with the following two properties.*

(i) $$\forall s < \alpha, \quad \lim_{s' \downarrow s} \mu(]s, s'] \times \Omega) = 0 .$$

(ii) *For every decreasing sequence (H_n) of elements of $\mathscr{F}_\alpha$ with empty intersection, the following relation holds.*

$$\lim_{n\to\infty} \Big(\sup_{\substack{A \in \mathscr{A} \\ A \subset [0,\alpha] \times H_n}} \mu(A) \Big) = 0$$

Then μ has a unique σ-additive extension to $\mathscr{P}_{[0,\alpha]}$

Proof of Lemma 1: We know from measure theory we have only to prove the σ-additivity of the content μ on $\mathscr{A}_{[0,\alpha]} := \mathscr{A} \cap ([0,\alpha] \times \Omega)$.

Let us consider the class $\mathscr{C}$ of subsets $[s,t] \times F$ of $[0,\alpha] \times \Omega$ with $F \in \mathscr{F}_s$ and of finite unions of such subsets.

As a consequence of (i) and of the additivity of μ, we may write for every $]s,t] \times F \in \mathscr{R}$, $t \leqslant \alpha$,

$$\mu(]s,t] \times F) = \lim_n \mu(]s + \frac{1}{n}, t] \times F) .$$

The class $\mathscr{C}$ therefore has the following property.

(*) $$\forall A \in \mathscr{A}_{[0,\alpha]}, \ \forall \varepsilon > 0, \ \exists\ C \in \mathscr{C} \text{ and } \ \exists A' \in \mathscr{A}_{[0,\alpha]}$$

such that

$$A \supset C \supset A' \text{ and } \quad \mu(A - A') \leqslant \varepsilon .$$

From an elementary and classical argument in measure theory, it is easily seen that property (*) and the following (**) imply the σ-additivity of the content μ on $\mathscr{A}_{[0,\alpha]}$.

(**) For every decreasing sequence (C_n) in $\mathscr{C}$ with empty intersection and every decreasing sequence (A_n) in $\mathscr{A}_{[0,\alpha]}$ with $A_n \subset C_n$ for every n, the following holds.

$$\lim_n \mu(A_n) = 0.$$

We are thus actually left to prove that property (ii) implies (**) for μ.

Let us then consider sequences (C_n) and (A_n) as in (**) and set

$$H_k := \{\omega : C_k \cap (\mathbb{R}^+ \times \{\omega\}) \neq \emptyset\}.$$

Since the ω-section $:= \{t : (t, \omega) \in C_k\}$ of the set C_k is, for each ω, a finite union of compact intervals, and as these ω-sections, $k \in \mathbb{N}$, have an empty intersection, there exsists, for every ω, *a* k such that $\omega \notin H_k$. In other words, the intersection of the decreasing sequence (H_k) is empty. From property (ii) and the inclusions $A_k \subset C_k \subset [0, \alpha] \times H_k$ it follows that

$$\lim_n \mu(A_n) = 0,$$

so we have the property (**) for μ.

The proof of step 1 will now consist in showing that $|\lambda_X|$ fulfills conditions (i) and (ii).

Lemma 2: *The following properties* (j) *and* (jj) *are sufficient for* $|\lambda_X|$ *to fulfill conditions* (i) *and* (ii) *of Lemma 1.*

(j) $$\forall s < \alpha,\ F \in \mathscr{F}_s.$$

$$\lim_{s' \downarrow s} \|\lambda_X(]s, s'] \times F)\| = 0.$$

(jj) *For every decreasing sequence* (H_n) *in* $\mathscr{F}_\alpha$ *with an empty intersection the following holds.*

$$\lim_{n \to \infty} \sup \{\|\lambda_X(]\sigma, \tau])\| :]\sigma, \tau] \subset [0, \alpha] \times H_n,\ \sigma, \tau \in \mathscr{T}^f_{[0,\alpha]}\} = 0.$$

Proof of Lemma 2: For every $\varepsilon > 0$, it is possible to find a partition $\{R_k : k = 1 \dots r, R_k \in \mathscr{R}\}$ of $[0, \alpha] \times \Omega$ with the property

$$|\lambda_X|(]0, \alpha] \times \Omega) \geqslant \sum_{k=1}^{r} \|\lambda_X(R_k)\| \geqslant |\lambda_X|(]0, \alpha] \times \Omega) - \varepsilon.$$

From the definition of the variation $|\lambda_X|$ it follows that for every

$$A \subset [0, \alpha] \times \Omega,\ A \in \mathscr{A},$$

(13.3.1) $$|\lambda_X|(A) \geqslant \sum_{k=1}^{r} \|\lambda_X(A \cap R_k)\| \geqslant |\lambda_X|(A) - \varepsilon.$$

Let us apply this to $A =]s, s'] \times \Omega$. From (j) and (13.3.1) it clearly follows that

$$\limsup_{s' \downarrow s} |\lambda_X|(]s, s'] \times \Omega) \leqslant \varepsilon ,$$

which is property (i) of Lemma 1.

Let us now consider $A \in \mathscr{A}_{[0,\alpha]}$. Since A is a finite union of predictable rectangles with $A \subset [0, \alpha] \times H$, there exist two stopping times $\sigma, \tau \in \mathscr{T}^{f}_{[0,\alpha]}$ with $\sigma \leqslant \tau$ such that $A \subset]\sigma, \tau] \subset [0, \alpha] \times H$. Therefore, $|\lambda_X|(A) \leqslant |\lambda_X|(]\sigma, \tau])$.

Let (H_n) be a sequence as in (ii). The latter inequality and property (jj) imply, using (13.3.1), that

$$\lim_n \Big(\sup_{\substack{A \in \mathscr{A} \\ A \subset [0,\alpha) \times H}} |\lambda_X|(A)\Big)$$

$$\leqslant \varepsilon + \limsup_n \left\{ \sum_{k=1}^{r} \|\lambda_X(]\sigma, \tau] \cap R_k)\| : \sigma, \tau \in \mathscr{T}^{f}_{[0,\alpha]} \sigma \leqslant \tau ,\ ]\sigma, \tau] \subset H_n \right\}$$

$$\leqslant \varepsilon .$$

This proves property (ii) and therefore Lemma 2.

End of Step 1: We now prove that the condition for X to belong to the class [L. D] and the right continuity of X together imply properties (j) and (jj) for $|\lambda_X|$.

Property (j) follows immediately from

$$\lim_{s' \downarrow s} \|\lambda_X(]s, s'] \times F)\| = \lim_{s' \downarrow s} \|\int_F (X_{s'} - X_s)\, dP\| = 0 .$$

We come to the proof of (jj). Let σ, τ and (H_n) be as in the statement of this condition. As in Proposition 8.9, we may write

$$\|\lambda_X(]\sigma, \tau])\| = \|E(X_\tau - X_\sigma)\| = \|\int_{H_n} (X_\tau - X_\sigma)\, dP\|$$

$$\leqslant \int_{H_n} \|X_\tau\|\, dP + \int_{H_n} \|X_\sigma\|\, dP .$$

Since the family $\{X_\tau : \tau \in \mathscr{T}^{f}_{[0,\alpha]}\}$ is uniformly integrable, we immediately get

$$\lim_n \int_{H_n} \|X_\tau\|\, dP = \lim_n \int_{H_n} \|X_\sigma\|\, dP = 0 ,$$

and therefore property (jj).

Proof of step 2: If $|\lambda_X|$ has a σ-additive extension to $\mathscr{P}_{[0,\alpha]}$ then $\lambda_X(]\sigma, \tau]) = E(X_\tau - X_\sigma)$ for every $\sigma, \tau \in \mathscr{T}^{f}_{[0,\alpha]}$ with $\sigma \leqslant \tau$.

For finite valued stopping times σ and τ, this result has already been proved in 8.9. We clearly have to prove it only for $\sigma = 0$ because $\lambda_X(]\sigma, \tau]) = \lambda_X]0, \tau] - \lambda_X]0, \sigma]$.

We consider the decreasing sequence (τ_n) of stopping times in $\mathscr{T}^{f}_{[0,\alpha]}$

$$\tau_n := \sum_{k=1}^{2^n-1} \frac{k+1}{2^n}\alpha \cdot 1_{]\frac{k\alpha}{2^n} < \tau \leqslant \frac{k+1}{2^n}\alpha]},$$

which converges towards τ.

We know from Lemma 8.12 that if, for a decreasing sequence of positive numbers (t_n) converging to a, we set $y_{t_n} := X_{\tau_n}$, $Y_a := X_\tau$, $\mathscr{G}_{t_n} := \mathscr{F}_{\tau_n}$ and $\mathscr{G}_a = F_\tau$, then the process $(Y_u)_{u\in I}$, where $I := \{t_n : n \geqslant 0\} \cup \{a\}$, is a quasimartingale with respect to the filtration $(\mathscr{G}_u)_{u\in I}$. Theorem 10.3, tells us that the random variables form a uniformly integrable family. This, with the right continuity of X, implies

$$\lim_n E(X_{\tau_n}) = E(X_\tau)$$

and therefore, on account of the σ-additivity of λ_X,

$$\lambda_X(]0,\tau]) = \lim_n \lambda_X(]0,\tau_n]) = \lim_n (E(X_{\tau_n}) - E(X_0)) = E(X_\tau) - E(X_0).$$

This proves the claim of Step 2.

Proof of Step 3: If $|\lambda_X|$ has a σ-additive extension to $\mathscr{P}_{[0,\alpha]}$ the family $\{X_\tau : \tau \in \mathscr{T}_{[0,\alpha]}$ is uniformly integrable$\}$.

We define

$$\tau_n := \inf\{t : \|X_t\| \geqslant n\} \wedge \alpha$$

As a consequence of Corollary 9.7, we have

$$\lim_{n\to\infty} P(\{\tau_n < \alpha\}) = 0. \tag{13.3.3}$$

We turn now to the proof of the uniform integrability property, that is to the proof of

$$\lim_n \sup_{\sigma\in\mathscr{T}_{[0,\alpha]}} \int_{\{\|X_\sigma\| \geqslant n\}} \|X_\sigma\|\, dP = 0. \tag{13.3.4}$$

We set

$$H_\sigma^n := \{\|X_\sigma\| \geqslant n\}.$$

Since H_σ^n belongs to $\mathscr{F}_\sigma$, the variable $\sigma_n := \sigma \cdot 1_{H_\sigma^n} + (\alpha - \sigma) 1_{(H_\sigma^n)^c}$ is a stopping time and

$$\lambda_X(]\sigma_n, \alpha]) = \int_{\{\sigma_n < \alpha\}} (X_\alpha - X_\sigma)\, dP$$
$$= \int_{\{\|X_\sigma\| \geqslant n\}} (X_\alpha - X_\sigma)\, dP.$$

Reasoning as above (for example in the proof of 9.6.3), we get

$$|\lambda_X|(]\sigma_n, \alpha]) \geqslant \int_{\{\|X_\sigma\| \geqslant n\}} \|X_\alpha - X_\sigma\|\, dP,$$

and therefore

$$\int_{\{\|X_\sigma\| \geqslant n\}} \|X_\sigma\|\, dP \leqslant \int_{\{\|X_\sigma\| \geqslant n\}} \|X_\alpha\|\, dP + |\lambda_X|(]\sigma_n, \alpha]). \tag{13.3.5}$$

We remark then that for every stopping time $\sigma \leqslant \alpha$ and every n,

$$\{\|X_\sigma\| \geqslant n\} \subset \{\tau_n < \alpha\} \cup \{\|X_\alpha\| \geqslant n\}$$

and

$$]\sigma_n, \alpha] \subset]\tau_n, \alpha].$$

As a consequence of (13.3.5), we get

$$\int_{\{\|X_\sigma\| \geqslant n\}} \|X_\sigma\| \, dP \leqslant \int_{\{\tau_n < \alpha\}} \|X_\alpha\| \, dP + \int_{\{\|X_\alpha\| \geqslant n\}} \|X_\alpha\| \, dP + |\lambda_X|(]\tau_n, \alpha]).$$

Equation (13.3.4) now follows immediately from (13.3.3). This completes the proof of Theorem 13.3.

13.4 Definition. The σ-additive measure λ_X, the existence of which follows from Theorem 13.3 for every quasimartingale of class [L. D], is called the *Doléans measure* of the quasimartingale X.

According to Remark 13.2.3, this measure is defined on $\mathscr{P}$ when it is positive (it takes its values in $\mathbb{R}^+$) and in general on the δ-ring of bounded predictable sets when it is vector-valued.

13.5 Theorem *(Stopping Theorem). For every right-continuous martingale (resp. real submartingale) $(X_t)_{t \in \mathbb{R}^+}$ of class [D] on the interval $I \subset \mathbb{R}^+$ and for any two I-valued stopping times σ and τ with $\sigma \leqslant \tau$, we have*

$$E(X_\tau | \mathscr{F}_\sigma) = X_\sigma$$
$$(\text{resp.} \quad E(X_\tau | \mathscr{F}_\sigma) \geqslant X_\sigma).$$

(We recall that on account of Theorem 10.1 every martingale and every positive submartingale on a closed interval I is of class [D] on I).[1])

Proof of Theorem 13.5: For every $H \in \mathscr{F}_\sigma$, $\sigma' := \sigma(\tau - \sigma) 1_{H^c}$ is a stopping time and one may write

$$\lambda_X(]\sigma', \tau]) = \int_H (X_\tau - X_\sigma) \, dP.$$

When X is a martingale (resp. a submartingale), we have

$$\lambda_X(]\sigma', \tau]) = 0 \quad (\text{resp.} \quad \lambda_X(]\sigma', \tau]) \geqslant 0).$$

This yields the theorem. □

---

[1]) The importance of the assumption "X of class [D] on I is underlined by the following counter-example. Take for X the Brownian motion starting at 0, set $\sigma := 0$ and $\tau := \inf\{t : |X_t| > 1\}$.

13.6 Theorem. *For every right-continuous martingale* $(M_t)_{t \in \mathbb{R}^+}$ *and every two bounded stopping times* σ *and* τ *with* $\sigma \leqslant \tau$ *and* σ *predictable, the following relation holds.*

(13.6.1) $$E(X_\tau | \mathscr{F}_{\sigma^-}) = X_{\sigma^-}.$$

If M *is a uniformly integrable martingale on* $\mathbb{R}^+$, *the formula* 13.6.1 *holds for every pair* (σ, τ) of stopping times, where σ is predictable.

Proof: We assume M uniformly integrable on the interval I (which is always the case if I is bounded). In this case, M can be extended into a martingale on the closed interval $\bar{I}$ according to the martingale convergence theorem and it is of class [D] on $\bar{I}$.

If σ is a predictable I-valued stopping time, it is possible to find an increasing sequence (σ_n) of stopping times such that $\sigma_n < \sigma$ on $\{\sigma > 0\}$ and

$$\sigma = \lim_n \sigma_n$$

According to Theorem 13.5,

$$X_{\sigma_n} = E(X_\sigma | \mathscr{F}_{\sigma_n}) = E(X_\tau | \mathscr{F}_{\sigma_n})$$

and the convergence theorem (Theorem 9.4 in the real case and 11.2 in the vector case) shows that the sequence $(X_{\sigma_n})_{n \in \mathbb{N}}$ converges a.s. and in $L^1(\Omega, \mathscr{F}_{\sigma^-}, P)$ to X_{σ^-} and that $E(X_\tau | \mathscr{F}_{\sigma^-}) = X_{\sigma^-}$.

This proves the theorem.

13.7 Corollary 1. *Every predictable right-continuous martingale* M *is continuous.*

Proof: It is enough to prove that, for any stopping time τ and any $\alpha > 0$, $E(\|\Delta M_{\tau \wedge \alpha}\|) - 0$. Since M is assumed to be predictable, the random variable $\Delta M_{\tau \wedge \alpha}$ is $\mathscr{F}_{(\tau \wedge \alpha)^-}$-measurable (Theorem 5.5).

We have thus only to prove that, for every $H \in \mathscr{F}_{(\tau \wedge \alpha)^-}$, $E(1_H \cdot \Delta M_{\tau \wedge \alpha}) = 0$ but this follows immediately from (13.6.1), which gives $E(M_{\tau \wedge \alpha} | \mathscr{F}_{(\tau \wedge \alpha)^-}) = M_{(\tau \wedge \alpha)^-}$.

13.8 Corollary 2. If T is a stopping time and h an integrable $\mathscr{F}_T$-measurable random variable with $E(h | \mathscr{F}_{T^-}) = 0$ the process $h 1_{[T, \infty[}$ is a right continuous martingale. In particular, if M is a martingale, which is uniformly integrable on $[0, \alpha]$ and T is a predictable stopping time with values in $[0, \alpha]$, $\Delta M_T 1_{[T, \infty[}$ is a martingale.

Proof: Let us set $X := h 1_{[T, \infty[}$ and $X_\infty := \lim_{t \to \infty} h 1_{[T, \infty[} = h 1_{\{T < \infty\}}$ we have to prove that for every t and $F \in \mathscr{F}_t$

(13.8.1) $$E(1_F X_t) = E(1_F h 1_{\{T < \infty\}})$$

But

(13.8.2) $$E(1_F X_t) = E(1_{F \cap \{T \leqslant t\}} h)$$

Since the condition $E(h|\mathscr{F}_{T-}) = 0$ and $F \cap \{t < T < \infty\} \in \mathscr{F}_{T-}$ imply $E(1_{F \cap \{t<T<\infty\}} h) = 0$, the relation (13.8.2) gives immediately (13.8.1).

If M is a martingale, which is uniformly integrable on $[0, \alpha]$ and T is a predictable stopping time with values in $[0, \alpha]$, Theorem 13.6 shows that $E(\Delta M_T|\mathscr{F}_{T-}) = 0$ and the corollary follows. □

13.9 Remark on weak $\mathbb{B}'$-valued quasimartingales. Let $\mathbb{B}'$ be the dual Banach space of a separable Banach space $\mathbb{B}$. We may consider families $(X_t)_{t\in\mathbb{R}^+}$ of mappings from Ω in $\mathbb{B}'$ which are only weakly measurable and adapted, that is, such that the real function $\omega \frown\!\!\!\rightarrow \langle x, X_t(\omega)\rangle$ is $\mathscr{F}_t$-measurable for every $t \in \mathbb{R}^+$ and $x \in \mathbb{B}$. Because of the separability of $\mathbb{B}$, $\|X_t\|$ is a real random variable for each t and, if $E\|X_t\| < \infty$, for each $F \in \mathfrak{A}$, one may define the weak integral $\int_F X_t\, dP$ as being the unique element of $\mathbb{B}'$ such that $\langle x, \int_F X_t\, dP\rangle = \int_F \langle x, X_t\rangle\, dP$ for every $x \in \mathbb{B}$ (the existence of such an element in $\mathbb{B}'$ is trivial because of the continuity of $x \frown\!\!\!\rightarrow \int_F \langle x, X_t\rangle\, dP$). We can define a *weak $\mathbb{B}'$-valued quasimartingale* as being a process X_t with the property that the $\mathbb{B}'$-valued associated Doléans measure has bounded variation (for the norm of $\mathbb{B}'$). It is a *weak martingale* if $\lambda_X = 0$.

A weak $\mathbb{B}'$-valued quasimartingale is said of class [L. D] if the family $\{\|X_\tau\|;\ T \in \mathscr{T}_I^f\}$ of random variables is uniformly integrable for any interval I.

Inspection of the proof of Theorem 13.3 shows that the statement of the theorem holds also for a right-continuous (for the norm of $\mathbb{B}'$), weak $\mathbb{B}'$-valued quasimartingale of class [L. D].

14. Predictable projection of a process and dual predictable projection of an admissible measure

In this paragraph as in the preceeding ones, the "usual hypotheses" are always assumed for the stochastic basis $(\Omega, (\mathscr{F}_t)_{t\in\mathbb{R}^+}, \mathfrak{A}, P)$.

14.1 Introduction. In this paragraph, we define a projection $X \frown\!\!\!\rightarrow X^{\mathscr{P}}$ from the space of real measurable processes onto the space of real predictable processes, which possesses the following property. For every predictable stopping time T,

(14.1.1) $$\int X\, d\mu_T = \int X^{\mathscr{P}}\, d\mu_T,$$

where μ_T is the admissible measure defined in §6.2. That is, for every $A \in \mathscr{B}_{\mathbb{R}^+} \otimes \mathscr{A}$,

$$\mu(A) := \int 1_A(T(\omega), \omega)\, P(d\omega).$$

The projection $X \frown\!\!\!\rightarrow X^{\mathscr{P}}$ will be called a *predictable projection*.

The relation (14.1.1) can therefore be written as

(14.1.2) $E(X_T \cdot 1_{\{T<\infty\}}) = E(X_T^{\mathscr{P}} \cdot 1_{\{T<\infty\}})$.

If $\tilde{\mathscr{A}}$ is a σ-algebra of measurable sets in $\mathbb{R}^+ \times \Omega$, containing $\mathscr{P}$, the projection $X \curvearrowright X^{\mathscr{P}}$ admits a dual mapping $\mu \curvearrowright \mu^{\mathscr{P}}$, which associates to every admissible measure μ on $\tilde{\mathscr{A}}$ a measure $\mu^{\mathscr{P}}$ on $\mathscr{B}_{\mathbb{R}^+} \otimes \mathfrak{A}$. That is,

(14.1.3) $\int X\, d\mu^{\mathscr{P}} := \int X^{\mathscr{P}}\, d\mu$.

We will call $\mu^{\mathscr{P}}$ the *dual predictable projection* of μ.

The relation (14.1.1) shows immediately that, when T is predictable, $\mu_T = \mu_T^{\mathscr{P}}$. The admissible measures on $\mathscr{B}_{\mathbb{R}^+} \otimes \mathfrak{A}$ with this property will be precisely called *predictable* and characterized in §15.

14.2 Theorem. *For every bounded, real measurable process X, there exists (up to indistinguishability) a unique predictable process $X^{\mathscr{P}}$ such that, for every predictable stopping time T,*

(14.2.1) $E(X_T 1_{\{T<\infty\}}) = E(X_T 1_{\{T<\infty\}}^{\mathscr{P}})$.

For every bounded real predictable process Z,

(14.2.2) $(ZX)^{\mathscr{P}} = ZX^{\mathscr{P}}$.

Proof: We first prove the uniqueness.

Let X^1 and X^2 be two bounded measurable processes and Y^1 and Y^2 two bounded predictable processes such that, for $i = 1, 2$, and every predictable stopping time T, one has

(14.2.3) $E(X_T^i 1_{\{T<\infty\}}) = E(Y_T^i 1_{\{T<\infty\}})$.

We prove that if $X^1 \geqslant X^2$, the set $\{Y^1 < Y^2\}$ is an evanescent set. Otherwise, we could apply the section theorem, Theorem 6.7, and there would exist a predictable stopping time T with $[T] \subset \{Y^1 < Y^2\}$ and $P\{T<\infty\} > 0$. This would imply

(14.2.4) $E(Y_T^1 1_{\{T<\infty\}}) < E(Y_T^2 1_{\{T<\infty\}})$

but since the inequality $X^1 \geqslant X^2$ implies

$$E(X_T^1 1_{\{T<\infty\}} \geqslant E(X_T^2 1_{\{T<\infty\}}),$$

the inequality (14.2.4) would contradict (14.2.3) and $X^1 \geqslant X^2$. Therefore, one has $Y^1 \geqslant Y^2$ outside an evanescent set as soon as $X^1 \geqslant X^2$ holds. The uniqueness of $X^{\mathscr{P}}$, if it exists, is thus proved up to indistinguishability.

We prove now the existence of $X^{\mathscr{P}}$ for every bounded positive measurable X. Let $\mathscr{H}$ be the class of all positive bounded measurable processes X for which a predictable process $X^{\mathscr{P}}$ exists with the properties (14.2.1) and (14.2.2). It follows immediately from the definition that $\mathscr{H}$ is stable for all the linear combinations with positive coefficients and, if (X_n) is an increasing bounded sequence in $\mathscr{H}$, the process $\sup_n X_n$ also belongs to $\mathscr{H}$ with $(\sup_n X_n)^{\mathscr{P}} = \sup_n X_n^{\mathscr{P}}$.

If $X > Y$ and X and Y belong to $\mathscr{H}$, the same is true for $X - Y$. To prove that $\mathscr{H}$ contains all the positive bounded measurable processes, it is therefore enough to prove that every process of the form $X = 1_{]s,t] \times F}$ with $s < t$ and $F \in \mathfrak{A}$ is in $\mathscr{H}$.

For such an F, we consider the right-continuous martingale M such that

(14.2.5) $$M_t = E(1_F | \mathscr{F}_t) \quad \text{a.s.}$$

and we let M^- denote the process

$$M_t^- := \lim_{\substack{s \uparrow t \\ s < t}} M_s$$

and

(14.2.6) $$Y := 1_{]s,t]} M^- .$$

We show now that $Y = X^{\mathscr{P}}$. We will actually prove that for every bounded predictable process Z, we have

(14.2.7) $$ZY = (ZX)^{\mathscr{P}}.$$

This will prove the theorem for positive bounded measurable processes. The result for real bounded measurable processes can then be easily obtained by representing any bounded measurable process as the difference of two positive ones.

The adapted left-continuous process Y is predictable and according to Theorem 13.6, we have, for every predictable stopping time T,

$$M_T^- = E(1_F | \mathscr{F}_{T^-}) .$$

The $\mathscr{F}_{T^-}$-measurability of T and Z_T (Theorem 7.6) implies that

$$\begin{aligned} E(Z_T Y_T 1_{\{T < \infty\}}) &= E(M_T 1_{\{s < T \leqslant t\}} Z_T) \\ &= E(1_{\{s < T \leqslant t\}} Z_T E(1_F | \mathscr{F}_{T^-})) \\ &= E(1_F 1_{\{s < T \leqslant t\}} Z_T) \\ &= E(X_T Z_T 1_{\{T < \infty\}}) . \end{aligned}$$

This exactly proves (14.2.7)and therefore the theorem. □

14.3 Definition. Let X be a bounded measurable real process. The process $X^{\mathscr{P}}$ defined in Theorem 14.2 is called the *predictable projection* of X.

14.4 Remark. (1) The measures μ_T in (14.1.1) are probability measures as soon as $P\{T < \infty\} = 1$ and the defining relation of $X^{\mathscr{P}}$ says that $X^{\mathscr{P}}$ is a common conditional expectation of X with respect to the σ-algebra $\mathscr{P}$ for all the probability laws μ_T, where T is predictable.

In the language of statistics, one would say that $\mathscr{P}$ is an exhaustive sub σ-algebra of $\mathscr{B}_{\mathbb{R}^+} \otimes \mathfrak{A}$ for the family $\{\mu_T : T \text{ predictable}, P\{T < \infty\} = 1\}$ of probability laws.

(2) Analogously one could define the *optional projection* of a measurable process by considering the optional process $X^{\mathscr{G}}$ with the property

$$E(X_T \cdot 1_{\{T<\infty\}}) = E(X_T^{\mathscr{G}} \cdot 1_{\{T<\infty\}})$$

for every stopping time.

The method of proving the uniqueness and existence of $X^{\mathscr{G}}$ is entirely analogous to the one used here for $X^{\mathscr{P}}$. The difference is that, for processes X of the form $X = 1_{]s,t] \times F}$, we have $X^{\mathscr{G}} = 1_{]s,t]} M$ where M is the right-continuous martingale (14.2.5).

14.5 Examples. (1) From the definition it follows immediately that for every totally inaccessible stopping time S, $(1_{[S]})^{\mathscr{P}} = 0$.

(2) Let T be a predictable stopping time, Y an $\mathfrak{A}$-measurable random variable and M^- the left-continuous process associated with the martingale $(E(Y \mid \mathscr{F}_t))_{t \in \mathbb{R}^+}$. We leave it as an exercise to prove that if $X := 1_{[0,T]} Y$, we have

(14.5.1) $$X^{\mathscr{P}} = 1_{[0,T]} M^- .$$

14.6 Proposition. (1) *The mapping $X \curvearrowright X^{\mathscr{P}}$ is linear and for every increasing sequence $(X_n)_{n \in \mathbb{N}}$ of positive bounded measurable processes, we have*

$$(\sup_n X_n)^{\mathscr{P}} = \sup_n (X_n)^{\mathscr{P}} .$$

(2) *Let $\mathscr{A}$ be a sub-σ-algebra of $\mathscr{B}_{\mathbb{R}^+} \otimes \mathfrak{A}$ with $\mathscr{A} \supset \mathscr{P}$ and E a topological vector space. For every E-valued admissible measure μ on $\mathscr{A}$*), there exists a unique E-valued (admissible) measure μ on $\mathscr{B}_{\mathbb{R}^+} \otimes \mathfrak{A}$ such that, for every bounded real measurable process X, the following relation holds.*

(14.6.1) $$\int X^{\mathscr{P}} d\mu = \int X \, d\mu^{\mathscr{P}} .$$

Proof: (1) The first statement of the proposition follows immediately from the uniqueness of $X^{\mathscr{P}}$ and the continuity property of the mathematical expectation.

(2) The first statement of the proposition shows that the mapping $X \curvearrowright \int X^{\mathscr{P}} d\mu$ possesses the properties of an integral, namely the integral with respect to the measure $A \curvearrowright \int 1_A^{\mathscr{P}} d\mu$ on $\mathscr{B}_{\mathbb{R}^+} \otimes \mathfrak{A}$. □

14.7 Definition. Let μ be an admissible E-valued measure on $\mathscr{A} \supset \mathscr{P}$. The measure $\mu^{\mathscr{P}}$ defined by (14.6.1) on $\mathscr{B}_{\mathbb{R}^+} \otimes \mathfrak{A}$ is called the *dual predictable projection* of μ.

When μ coincides with the restriction of $\mu^{\mathscr{P}}$ to $\mathscr{A}$, μ is said to be *predictable*.

Theorem 14.9 provides us with an important example of a predictable measure. It will turn out that all positive predictable measures are of this type. To prove this theorem, we need the following auxiliary lemma.

---

*) One can also consider measures, which are defined on the δ-ring $\mathscr{A} \cap \bigcup_{\alpha \in +} [0, \alpha] \times \Omega$ of bounded predictable sets, to include the case of unbounded vector measures (see Remark 13.2.3). In this case, μ is defined on $\bigcup_{\alpha} \mathscr{B}_{[0,\alpha]} \otimes \mathfrak{A}$.

14.8 Lemma. *Let A be an adapted, increasing, right-continuous process with $A_0 = 0$ and M be a bounded right-continuous martingale. Then, for every finite stopping time T such that $E(A_T) < \infty$,*

$$E\Big(\int_0^T M_t \, dA_t\Big) = E(M_T A_T).$$

When M is positive, then whether it is bounded or not, this relation holds without assuming $E(A_T) < \infty$.

Proof: Let (I_n) be an increasing sequence of finite subsets of $\mathbb{R}^+$, the union of which is dense in $\mathbb{R}^+$. More precisely, let $I_n = \{0 < t_n^1 < \ldots < t_n^{k_n}\}$.

The martingale property for M implies that

$$\begin{aligned} &E\left[\sum_{j=1}^{k_n-1} M_{t_n^{j+1} \wedge T} (A_{t_n^{j+1} \wedge T} - A_{t_n^j \wedge T})\right] \\ &= E(M_{t_n^{k_n} \wedge T} A_{t_n^{k_n} \wedge T}) - \sum_{j=1}^{k_n-1} E[A_{t_n^j \wedge T} (M_{t_n^{j+1} \wedge T} - M_{t_n^j \wedge T})] \\ &= E M_{t_n^{k_n} \wedge T} A_{t_n^{k_n} \wedge T} \\ &= E(M_T A_{t_n^{k_n} \wedge T}). \end{aligned}$$

When n goes to infinity, the left side of the formula converges to $E\Big(\int_0^T M \, dA_t\Big)$, while the right side tends to $E(M_T A_T)$. □

14.9 Theorem. *Let A be a right-continuous, measurable, increasing process with $A_0 = 0$ and $E(A_t) < \infty$ for every $t \in \mathbb{R}^+$. Then the measure μ_A on $\mathscr{B}_{\mathbb{R}^+} \otimes \mathfrak{A}$ is predictable if and only if A is predictable.*

(Note that we do not assume that A is adapted).

Proof: (1) We first prove the implication$A$ predictable $\Rightarrow \mu_A$ predictable. We have only to prove that $\int X \, d\mu_A = \int X^{\mathscr{P}} d\mu_A$ for every X of the form $X = 1_{]0, t] \times F}$, when $F \in \mathfrak{A}$. In this case, $X^{\mathscr{P}} = M^- 1_{]0, t]}$, where M^- is the left-continuous process of the martingale $(E(F \mid \mathscr{F}_t))_{t \in \mathbb{R}^+}$. We have thus to prove that for every $t \in \mathbb{R}^+$ and $F \in \mathfrak{A}$,

$$E(1_F A_t) = E\Big(\int_0^t M_s^- \, dA_s\Big), \tag{14.9.1}$$

which can be written

$$E(M_t A_t) = E \int_0^t M_s^- \, dA_s. \tag{14.9.2}$$

According to Lemma 14.9 this is equivalent to

(14.9.3) $$E\left(\int_0^t (M_s - M_s^- \, dA_s\right) = 0 .$$

We have thus reduced the proof to showing that (14.9.3) holds. However $\int_0^t (M_s - M_s^-)dA_s = \sum_{s<t} \Delta M_s \Delta A_s$, and according to Theorem 7.6, there exists a sequence (S_n) of predictable stopping times such that

(14.9.4) $$\int_0^t (M_s - M_s^-)dA_s = \sum_n \Delta M_{S_n} \Delta A_{S_n} 1_{\{S_n \leqslant t\}} \quad \text{a.s.}$$

Theorem 5.5 says that $\Delta A_{S_n^-}$ is $\mathscr{F}_{S_n^-}$-measurable which, in view of Theorem 13.6, implies

(14.9.5) $$E(\Delta A_{S_n} \Delta M_{S_n}) = 0 .$$

Since the martingale M is bounded by one, we may interchange expectation and summation on the right hand side of (14.9.4), so equality (14.9.3) follows from (14.9.4) and (14.9.5).

(2) We now prove that μ_A predictable $\Rightarrow$ A predictable.

We use the characterization of predictable processes given gy Theorem 7.7.

Let T be a predictable stopping time and Y a bounded, real random variable with $E(Y \mid \mathscr{F}_{T^-}) = 0$. As seen in (14.5.2), the predictable projection of $Y 1_{[0,T]}$ is the process $1_{[0,T]} M^-$, where M^- is the left-continuous process of the martingale $(E(Y \mid \mathscr{F}_t))_{t \in \mathbb{R}^+}$. As

$$M^-_{t \wedge T} = E(Y \mid \mathscr{F}_{(t \wedge T)^-}) \quad \text{(Theorem 13.6)},$$

the process $1_{[0,T]}\; M^-$ is evanescent since $E(Y \mid \mathscr{F}_{T^-}) = 0$.

The definition of μ_A and the predictability of this measure show that

$$E(A_T \; Y) = \int 1_{[0,T]} Y \, d\mu_A = \int 1_{[0,T]} M^- \, d\mu_A = 0 .$$

The equation $E(A_T \; Y) = 0$, which holds for every bounded real random variable possessing the property $E(Y \mid \mathscr{F}_{T^-}) = 0$, and the completness of the σ-algebra $\mathscr{F}_{T^-}$ together imply the $\mathscr{F}_{T^-}$-measurability of A_T. This satisfies condition (i) of Proposition 7.7.

Now let T be a totally inaccessible stopping time. Then

$$E(A_T - A_{T^-}) = \mu_A([T]) = E\left(\int 1^{\mathscr{P}}_{[T]} \, d\mu_A\right).$$

As we noted in (14.5.1), we have $1^{\mathscr{P}}_{[T]} = 0$ and therefore $E(A_T - A_{T^-}) = 0$ for every totally inaccessible stopping time. The process A thus has the property (ii) of Proposition 7.7, so it is predictable. □

15. The predictable F. V. process of an admissible measure on $\mathscr{P}$ and the Doob-Meyer decomposition of a quasimartingale

As in the previous sections of this chapter, we assume the "usual hypotheses" for the stochastic basis.

15.1 Random measures – Primitive processes – F. V. processes

Let $\mathbb{B}$ be a Banach-space with norm $\|.\|$, and I an interval in $\mathbb{R}^+$. A family $\{v(\omega.,): \omega \in \Omega\}$ of $\mathbb{B}$-valued Borel measures on $\mathscr{B}_I$ is called a *$\mathbb{B}$-valued random measure*, when, for every $B \in \mathscr{B}_I$, the mapping $\omega \curvearrowright v(\omega, B)$ is a random variable. The process F_v: $F_v(t, \omega) := v(\omega, [0, t])$ is called the *primitive process of the random measure* v.

If $\mathbb{B}'$ denotes the Banach dual space of $\mathbb{B}$, a family $\{v(\omega,.): \omega \in \Omega\}$ of $\mathbb{B}'$-valued measures on $\mathscr{B}_I$ will be called a *weak $\mathbb{B}'$-valued random measure* if for every $x \in \mathbb{B}$, $\langle x, v(\omega,.)\rangle$ is a real random measure. The process F_v is then only a *weak process* in the sense that $\forall x \in \mathbb{B}$ $\langle x, F_v\rangle$ is a real random variable.

When the process F_v is adapted (resp. optional, resp. predictable, resp. weakly optional, ...) we say that v is adapted (resp. optional, resp. predictable, resp. weakly optional, ...).

When the process F_v is weakly adapted (resp. optional, resp. predictable, ... we say that v is weakly adapted (resp. optional, resp. predictable, ...).

When, with probability one, the measure $v(\omega,.)$ is of bounded variation on every bounded interval $[0, t]$, we say that v *is of finite variation* and that the process F_v is also of finite variation. As an abbreviation, we shall say that *such a process F_v is an F.V. process.*

When $\mathbb{B} = \mathbb{R}$ every right-continuous process F with paths of finite variation and with $F_0 = 0$ is an F. V. process.

15.2 Theorem *(Positive case).* (1) *Let μ be a positive admissible measure on $\mathscr{P}$, with $\mu([0, t] \times \Omega) < \infty$ for every $t \in \mathbb{R}^+$ and let $\tilde{\mu}$ be the dual predictable projection of μ. Then there exists a unique (up to indistinguability) right-continuous, measurable, increasing process V* with $V_0 = 0$ such that, for every $A \in \mathscr{B}_{\mathbb{R}^+} \otimes \mathfrak{A}$,

$$\tilde{\mu}(A) = E\left(\int_0^\infty 1_A(s,.)dV(s,.)\right).$$

(2) *This process V is predictable.*

(3) *If $\mu([T]) = 0$ for every predictable stopping time T, the process is continuous.*

(4) *If two positive admissible measures μ and v on $\mathscr{P}$ have the same restriction on $]0, \sigma] \cap \mathscr{P}$, where σ is a stopping time, the two corresponding predictable processes V^μ and V^v coincide on $]0, \sigma]$.*

Proof: (1) Since $\tilde{\mu}$ is admissible, for every $t \in \mathbb{R}^+$, the measure $F \leadsto \tilde{\mu}([0,t] \times F)$ on

$\mathfrak{A}$ is absolutely continuous with respect to P. We write $\tilde{V}(t,.)$ for the Radon-Nikodym density of this measure. Since $\tilde{\mu}$ is positive, we have

(15.2.1) $$\forall s, t \in \mathbb{R}^+,\ s \leqslant t,\ \tilde{V}_t \geqslant \tilde{V}_s \quad \text{a.s.},$$

and as $E(\tilde{V}_t - \tilde{V}_s) = \tilde{\mu}(]s,t] \times \Omega)$,

(15.2.2) $$\lim_{t \downarrow s} E(\tilde{V}_t - \tilde{V}_s) = 0.$$

Following (15.2.1), there exists a P-null set $\Omega_0 \subset \Omega$ such that

$$\forall \omega \notin \Omega_0,\ \forall t, s \in \mathbb{Q},\ s \leqslant t,\ \tilde{V}(t,\omega) \geqslant \tilde{V}(s,\omega).$$

It is therefore possible to define

(15.2.3) $$V(t,\omega) := \begin{cases} \lim\limits_{\substack{u \downarrow t \\ u \in \mathbb{Q}}} \tilde{V}(u,\omega) & \text{if} \quad (u,\omega) \in \mathbb{R}^+ \times (\Omega \backslash \Omega_0), \\ 0 & \text{if} \quad (u,\omega) \in \mathbb{R}^+ \times \Omega_0, \end{cases}$$

where V is an increasing, right-continuous process with the properties $V_t = \tilde{V}_t$ a.s. (see (15.2.2)) and $V_0 = 0$.

Therefore,

(15.2.4) $$\forall]s,t] \times F \in \mathscr{B}_{\mathbb{R}^+} \otimes \mathfrak{A}$$

$$\tilde{\mu}(]s,t] \times F) = E(1_F \cdot (V_t - V_s)) = E\left(\int_0^\infty 1_{]s,t] \times F}(u,.)\, dV(u,.)\right)$$

This expresses the equality of the measures $\tilde{\mu}$ and μ_V on $\mathscr{B}_{\mathbb{R}^+} \otimes \mathfrak{A}$ and therefore the formula in part 1) of the theorem.

The equality $\tilde{\mu}(]0,t] \times F) = E(1_F \cdot V_t)$ for every $t \in \mathbb{R}^+$ and $F \in \mathfrak{A}$ show immediately that any process V with the property of (1) must be such that $V_t = \tilde{V}_t$ a.s. for all t. The right-continuous process V is thus uniquely defined up to indistinguishability.

(2) The predictability of V follows from Theorem 14.9.

(3) For every predictable stopping time T, we have

$$\mu([T]) = \tilde{\mu}([T]) = E(V_T - V_{T^-}).$$

Since V is predictable, the equality $E(V_T - V_{T^-}) = 0$ for every predictable stopping time T implies the continuity of V as a consequence of the structure theorem, Proposition 7.7.

(4) Since for every measurable set $A \subset [0,\sigma]$, the process $1_A^{\mathscr{P}}$ is clearly null outside $[0,\sigma]$. This is an immediate consequence of the definition of $1_A^{\mathscr{P}}$. The dual predictable projections $\tilde{\mu}$ and $\tilde{\nu}$ therefore coincide on $[0,\sigma] \cap (\mathscr{B}_{\mathbb{R}^+} \otimes \mathfrak{A})$, which implies that

$$E(1_F \cdot V^{\mu}_{t \wedge \sigma}) = E(1_F \cdot V^{\nu}_{t \wedge \sigma})$$

for every t and $F \in \mathfrak{A}$. In others words, the random variables $V^{\mu}_{t \wedge \sigma}$ and $V^{\nu}_{t \wedge \sigma}$ are a.s. equal. The right continuity of the paths of V^{μ} and V^{ν} then implies the equality of the paths a.s on $[0,\sigma]$. □

15.3 Corollary 1. *Let μ be a positive admissible measure on $\mathscr{P}$ with $\mu([0, t] \times \Omega) < \infty$ for every $t \in \mathbb{R}^+$. There exists a unique, predictable, right-continuous, increasing process V with $V_0 = 0$ and $\mu = \mu_V$.*

If $\mu([T]) = 0$ for every predictable stopping time T, V is continuous.

Proof: Since the dual predictable projection $\tilde{\mu}$ of μ coincide with μ on $\mathscr{P}$, the existence of V and the second statement of the corollary follow from Theorem 15.2.

To prove the uniqueness, we prove the following implication.

"V predictable and $\mu_V = 0$ on $\mathscr{P}$" $\Rightarrow$ "V is evanescent".

The property $\mu_V = 0$ on $\mathscr{P}$ says that V is a martingale. Theorem 13.7 then says that V is a continuous martingale. For every m, we consider the beginning T_m of the predictable set $\{|V| \geq m\}$. The continuity of V and the properties of μ_V then imply

$$0 = \int_{[0, T_m \wedge t]} V \, d\mu_V = E\Big(\int_{[0, T_m \wedge t]} V_s dV_s\Big) = \tfrac{1}{2}(E|V_{T_m \wedge t}|^2)$$

This implies

$$E(|V_{T_m \wedge t}|^2) = 0$$

and, since $\lim_{m \to \infty} P\{T_m \geq t\} = 1$, we obtain

$$V_t = 0 \quad \text{a.s. for all } t.$$

15.4 Corollary 2 *(Doob-Meyer Decomposition Theorem). Let S be a submartingale of class $[L. D]$. There exists a unique, predictable, right-continuous, increasing process V with $V_0 = 0$ such that $S - V$ is a martingale.*

If $E(|\Delta S_T|) = 0$ for every predictable stopping time T, the process V is continuous.

Proof: Let μ be the (positive) Doleans measure of the submartingale S and V be the corresponding predictable increasing process of the measure μ as defined in Corollary 1. Since $\mu_V = \mu$ the Doleans measure of the process $S - V$ is zero. This says that $S - V$ is a martingale.

The second statement of the corollary follows immediately from the second statement of 15.3.

15.5 Corollary 3 *(real case). Let μ be a real admissible measure on $\bigcup_{\alpha \in \mathbb{R}^+} \mathscr{P} \cap ([0, \alpha] \times \Omega)$.*

There exists a unique, predictable (right-continuous) F.V. process with $V_0 = 0$ such that $\mu = \mu_V$.

If $\mu([T]) = 0$ for every predictable stopping time T, the process V is continuous.

Proof: Let μ^+ (resp. μ^-) be the positive part (resp. the negative part) of μ and let P^+ and P^- be two disjoint predictable sets in $\mathbb{R}^+ \times \Omega$ such that, for every $\alpha \in \mathbb{R}^+$,

$$\mu^+(P^+ \cap [0, \alpha] \times \Omega) = \mu(P^+ \cap [0, \alpha] \times \Omega)$$

and

$$\mu^-(P^- \cap [0, \alpha] \times \Omega) = -\mu(P^- \cap [0, \alpha] \times \Omega).$$

Let V^+ and V^- be increasing predictable right-continuous processes with the properties $\mu_{V^+} = \mu^+$, $\mu_{V^-} = \mu^-$. One has only to set $V = V^+ - V^-$ to get a predictable right-continuous F. V. Process such that $\mu_V = \mu$.

The same argument as in corollary 1 shows the uniqueness of V.

For every predictable stopping time T, we have

$$\mu^+([T]) = \mu([T] \cap P^+) = \mu(\text{[Beginning of the set } [T] \cap P^+]).$$

Since the beginning of $[T] \cap P^+$ is a predictable stopping time, we see that $\mu^+([T]) = 0$ for every predictable stopping time T as soon as $\mu([T]) = 0$.

The same holds for μ^-. Therefore if $\mu([T]) = 0$ for every predictable stopping time, the processes V^+ and V^- are continuous according to Corollary 2. V is then continuous.

15.6 Definition. Let X be a real quasimartingale of class [L. D] and λ_X its Doleans measure. The predictable process of the measure λ_X, as defined in Corollary 3, is called the *dual predictable projection* of X[1]).

To avoid any confusion with the predictable projection, we will denote by $\tilde{X}$ the dual predictable projection of X.

As an immediate consequence of the definition, a quasimartingale X of class [L. D] is a martingale if and only if $\tilde{X} = 0$.

15.7 Examples. (1) Let A be a Poisson process with parameter α. As $(A_t - \alpha t)_{t \in \mathbb{R}^+}$ is a martingale and the deterministic process $(\alpha t)_{t \in \mathbb{R}^+}$ is predictable, the uniqueness of the Doob-Meyer decomposition implies

$$\tilde{A}_t = \alpha t.$$

(2) Let A be a process with independent increments such that the function $t \rightsquigarrow E(A_t)$ has finite variation. It is immediately seen that $(A_t - E(A_t))_{t \in \mathbb{R}^+}$ is a martingale. Therefore

$$\tilde{A}_t = E(A_t).$$

(3) Let A be the increasing process: $A := 1_{[T, \infty[}$, where T is a stopping time.

If T is predictable, the process A is predictable. Therefore $A = \tilde{A}$.

If T is totally inaccessible, the process $A - \tilde{A}$ is an F. V.-process and a martingale. Since $E(|\Delta A_S|) = 0$ for every predictable stopping time S. Corollary 3 then implies the continuity of $\tilde{A}$. We thus find that *for T totally inaccessible, the process $\widetilde{1_{[T, \infty[}}$ is continuous.*

(4) Let T be a predictable stopping time and h an $\mathscr{F}_T$-measurable integrable random variable. Since $(h - E(h|\mathscr{F}_{T-})) \cdot 1_{[T, \infty[}$ is a right-continuous martingale,

---

[1]) In P. A. Meyer and others' works it is often called the *compensator* of X.

according to Corollary 13.8, and since $E(h|\mathscr{F}_{T-})\cdot 1_{[T,\infty[}$ is predictable (Proposition 6.8), the process $E(h|\mathscr{F}_{T-})\cdot 1_{[T,\infty[}$ is the dual predictable projection of $h\cdot 1_{[T,\infty[}$.

(5) *Markov processes.* Let $(X_t)_{t\in\mathbb{R}^+}$ be a Markov process with respect to the given filtration on the stochastic basis and suppose it has the locally compact state space E and infinitesimal operator L. For every bounded function φ in the domain of the operator L, the process $(M_t^\varphi)_{t\in\mathbb{R}^+}$, where

$$M_t^\varphi := \Big(\varphi(X_t)-\varphi(X_0)-\int_0^t L\varphi(X_s)\,ds\Big)_{t\in\mathbb{R}^+}$$

is a martingale with respect to the given filtration. Because of the adaptation and continuity and, therefore, the predictability of the process $\big(\int_0^t L\varphi(X_s)\,ds\big)_{t\in\mathbb{R}^+}$, this latter process is the dual predictable projection of $(\varphi(X_t))_{t\in\mathbb{R}^+}$.

15.8 Theorem* *(vector case). Let $\mathbb{B}'$ be a Banach space which is dual space of a separable Banach space $\mathbb{B}$. Let μ be an admissible, $\mathbb{B}'$-valued measure on $\bigcup_\alpha([0,\alpha]\times\Omega)\cap\mathscr{P}$ and $\tilde{\mu}$ the dual predictable projection of μ. We assume, moreover, that μ has a finite variation on $[0,\alpha]\times\Omega$ for every $\alpha>0$.*

(1) *There exists a unique, weak, $\mathbb{B}'$-valued, random measure ν such that, for every $\alpha>0$ and $A\in([0,\alpha]\times\Omega)\cap(\mathscr{B}_{\mathbb{R}^+}\otimes\mathfrak{A})$, the following equality holds.*

(15.8.1) $$\tilde{\mu}(A)=E\big(\textstyle\int 1_A(s,.)\,\nu(ds,.)\big).$$

The random measure ν is, moreover, weakly predictable and can be characterized as the unique, weakly predictable, $\mathbb{B}'$-valued, random measure ν, which, for every $\alpha>0$ and $A\in([0,\alpha]\times\Omega)\cap\mathscr{P}$, satisfies the relation (15.8.1).

The primitive process F_ν of ν is an F.V. process and, if $\mu([T])=0$ for every predictable stopping time T, is also continuous.

(2) *There exists a unique (up to indistinguishability) weakly predictable, $\mathbb{B}'$-valued process g, with $\|g\|=1$, such that, for every bounded set $A\in\mathscr{P}$ and $x\in\mathbb{B}$,*

(15.8.2) $$\langle x,\mu(A)\rangle=\int_A\langle x,g(s,\omega)\rangle\,|\mu|(ds,\omega).$$

For every bounded Borel subset B of $\mathbb{R}^+$ and P-almost, all $\omega\in\Omega$

(15.8.3) $$\langle x,\nu(B,\omega)\rangle=\int_B\langle x,g(s,\omega)\rangle\,dV(s,\omega),\quad \forall x\in\mathbb{B},$$

where V is the predictable increasing process of the positive measure $|\mu|$. If $\mathbb{B}'$ is separable, g is actually predictable and ν is a predictable $\mathbb{B}'$-valued random measure.

Proof: If the Banach space $\mathbb{B}'$ has the Radon-Nikodym property (see [21] or §11.4), in particular, if $\mathbb{B}'$ is separable, there exists a $\mathbb{B}'$-valued, $\mathscr{P}$-measurable function g on $\mathbb{R}^+\times\Omega$, such that, for every bounded $A\in\mathscr{P}$,

(15.8.4) $$\mu(A) = \int_A g(s,\omega)\,|\mu|(ds,d\omega),$$

where $|\mu|$, as always, denotes the variation of μ. Moreover $\|g(s,\omega)\| = 1$, $|\mu|$. a. e.

In any case, it is easy to show directly the existence of g such that $\forall x \in \mathbb{B}$, $\langle x, g\rangle$ is a predictable real function and (15.8.4) holds in the following sense for every $A \in \mathscr{P}$.

(15.8.5) $$\forall x \in \mathbb{B} \quad \langle x, \mu(A)\rangle = \int_A \langle x, g(s,\omega)\rangle\,|\mu|(ds,d\omega).$$

For every $x \in \mathbb{B}$ we may indeed consider the predictable density g^x of the real measure $A \rightsquigarrow \langle x, \mu(A)\rangle$ with respect to $|\mu|$. Clearly $\|g^x\|_\infty \leqslant \|x\|$ and $x \rightsquigarrow g^x$ is linear continuous from $\mathbb{B}$ into $L^\infty([0,\alpha] \times \mathbb{R}^+, \mathscr{P}, |\mu|)$. Using a lifting theorem of A. and C. Ionescu-Tulcea (cf. [IoT. 2]), one may, for each x, choose the function $g^x(s,\omega)$ for all s,ω in such a way that $\sup_{s,\omega \in [0,\alpha] \times \Omega} |g^x(s,\omega)| \leqslant \|x\|$, the mapping $x \rightsquigarrow g^x(s,\omega)$ being linear and therefore an element of $\mathbb{B}'$ with norm $\leqslant 1$. We denote this mapping by $g(s,\omega)$. This proves (15.8.5).

The separability of $\mathbb{B}$ shows that $\|g(s,\omega)\| = \sup_{\substack{x \in \mathbb{B} \\ \|x\| \leq 1}} |\langle x, g(s,\omega)\rangle|$ is predictable.

The inequality $|\mu|(A) \leqslant \int_A \|g(s,\omega)\|\,|\mu|(ds,d\omega)$, being an easy consequence of the definition of $|\mu|$, clearly implies $\|g(s,\omega)\| = 1$ $|\mu|$ a.e.

Let V be the predictable increasing process of $|\mu|$ (Corollary 15.3). Then for every $x \in \mathbb{B}$ and every bounded predictable set A, it follows from (15.8.5) that

(15.8.6) $$\langle x, \mu(A)\rangle = E\Big(\int_A \langle x, g(s,.)\rangle\, dV(s,.)\Big)$$

For every Borel set $G \subset [0,\alpha]$ and every $\omega \in \Omega$ there exists an element of $\mathbb{B}'$ (the so-called weak integral), denoted by $\int_G g(s,\omega)\, dV(s,\omega)$, such that $\forall x \in \mathbb{B}$,

$$\langle x, \int_G g(s,\omega)\, dV(s,\omega)\rangle = \int_G \langle x, g(s,\omega)\rangle\, dV(s,\omega).$$

Since the mapping $G \rightsquigarrow \int_G \langle x, g(s,\omega)\rangle\, dV(s,\omega)$ is σ-additive for each ω, it follows from a theorem of Pettis (cf. Dunford and Schwartz [DuS], ch. 4–10) that $G \rightsquigarrow \int_G g(s,\omega)\, dV(s,\omega)$ defines a $\mathbb{B}'$-valued measure, which we denote by $\nu(ds,\omega)$. We write this symbolically as

(15.8.7) $$\nu(ds,\omega) = g(s,\omega)\, dV(s,\omega).$$

Since, by definition, $\langle x, \tilde{\mu}(A)\rangle = \langle x, \int (1_A)^{\mathscr{P}}\, d\mu\rangle$ for every $A \in \mathscr{B}_{\mathbb{R}^+} \otimes \mathfrak{A}$, it follows from (15.8.6) that $\langle x, \tilde{\mu}(A)\rangle = E(\int \langle x, g(s,.)\rangle (1_A)^{\mathscr{P}}\, dV(s,.))$. Since $\langle x, g\rangle$ and V are predictable real processes, one has

$$\begin{aligned}\langle x, \tilde{\mu}(A)\rangle &= E\Big(\int \langle x, g(s,.)\rangle\, 1_A(s,.)\, dV(s,.)\Big)\\ &= \langle x, E\Big(\int 1_A(s,.)\, g(s,.)\, dV(s,.)\Big)\rangle\end{aligned}$$

This proves (15.8.1) and the uniqueness of ν follows from the separability of $\mathbb{B}$ and the uniqueness of $\langle x, \nu(ds,.)\rangle$, which is the unique, predictable, real random measure associated with $\langle x, \mu\rangle$.

Since $\|g\| \leqslant 1$, the measure $\nu(ds, \omega)$, defined by (15.8.7), is of bounded variation on $[0, \alpha]$ for every ω; so is the primitive process F_ν of ν.

If $\mu([T]) = 0$ for every predictable stopping time T, it follows from Corollary 15.5 and formula (15.8.5) that $E(\langle x, \Delta F_\nu(T, \omega)\rangle) = 0$ for every $x \in \mathbb{B}$ and every stopping time T (predictable or not). This implies $\Delta F_\nu(T, \omega) = 0$ a.s. because of the separability of $\mathbb{B}$. Therefore, F_ν has no jump and all the assertions of (1°) are proved.

The statements in (2°) are merely reformulations of the definition and properties of ν and g as established above.

15.9* Corollary. *Let S be a weak $\mathbb{B}'$-valued [L. D]-quasimartingale. There exists a unique, $\mathbb{B}'$-valued, weakly predictable, random measure ν such that $\langle x, S - F_\nu\rangle$ is a martingale for every $x \in \mathbb{B}$, where F_ν is the primitive process of ν. This process is moreover, an F.V process and is predictable when $\mathbb{B}'$ itself is separable. In this latter case, $S - F_\nu$ is a martingale.*

If $E(\|\Delta S_T\|) = 0$ for every predictable stopping time T, the process F_ν is continuous.

Proof. This corollary is an immediate consequence of Theorem 15.8 and of Corollary 15.4 of Theorem 15.2.

15.10* Definition. Let X be a weak $\mathbb{B}'$-valued [L. D]-quasimartingale, where $\mathbb{B}'$ is separable and the dual space of a Banach space $\mathbb{B}$. The $\mathbb{B}'$-valued weakly predictable process of the Doleans measure λ_X, the existence of which follows from 15.9, is called the *dual predictable projection* of X and denoted by $\tilde{X}$.

15.11 Dual predictable projection of a point process

Let E be an open subset of $\mathbb{R}^d$. The stochastic basis $(\Omega, (F_t)_{t \in \mathbb{R}^+}, \mathfrak{A}, P)$ assumed given (as always), we consider, for every $\omega \in \Omega$, a positive measure $\nu(\omega, ds, du)$ on the Borel σ-algebra $\mathscr{B}_{\mathbb{R}^+} \otimes \mathscr{B}_E$ of the following form.

$$\nu(\omega, ds, du) = \sum_{(t,x) \in I(\omega)} \varepsilon_{(t,x)}(ds, du),$$

where $I(\omega)$ is a denumerable subset of $\mathbb{R}^+ \times E$ and $\varepsilon_{t,x}$ denotes the measure of total mass one, concentrated at the point (t, x) ("Dirac measure" in (t, x)).

If, for every continuous function φ on $\mathbb{R}^+ \times E$ with compact support, the mapping $\omega \frown \int_{\mathbb{R}^+ \times E} \varphi(s, u)\, \nu(\omega, ds, du)$ is a random variable, we call ν a *point process*. It is easy to see from a classical extension argument in measure theory that, for every Borel set $B \in \mathscr{B}_{\mathbb{R}^+} \otimes \mathscr{B}_E$, the real mapping $\nu(\omega, B)$ is a random variable.

Let p be a strictly positive continuous function on E. We denote by $\mathfrak{M}^p(E)$ the

Banach space of real measures μ on $\mathscr{B}_E$ with the property $\int_E p(x)|\mu|dx < \infty$ and norm $\|\mu\|_p := \int_E p(x)|\mu|(dx)$.

The space $\mathfrak{M}^p(E)$ is the dual Banachspace of the Banachspace $\mathscr{C}^p(E)$, the elements of which are the real continuous functions φ on E such that $\frac{|\varphi(x)|}{p(x)} \in \mathscr{C}_0(E)$ and the norm of φ is $\|\varphi\|_p := \sup_{x\in E} \frac{|\varphi(x)|}{p(x)}$. ($\mathscr{C}_0(E)$ is the space of continuous functions "vanishing at infinity" (i.e.: $f \in \mathscr{C}_0(E)$ iff, for every $\varepsilon > 0$ there exists a compact $K_\varepsilon \subset E$ such that $\sup_{x \notin K_\varepsilon} |f(x)| \leqslant \varepsilon$).

We say that a point process ν is of order p if, for every $t \in \mathbb{R}$ and $\omega \in \Omega$, the measure $\nu(\omega,]0, t], .)$ on $\mathscr{B}_E$ is an element of $\mathfrak{M}^p(E)$. Every point process ν can therefore be viewed as an $\mathfrak{M}^p(E)$-valued random measure as defined in 15.1. From the positivity of the measure $\nu(\omega, ds, du)$ for every ω, it follows immediately that

$$\begin{aligned}\sum_{i=1}^{k} \|F_\nu(\omega, t_{i+1}) - F_\nu(\omega, t_i)\|_p &= \sum_{i=1}^{k} \|\nu(\omega]t_i, t_{i+1}], .)\|_p \\ &= \sum_{i=1}^{k} \int p(x)\, 1_{]t_i, t_{i+1}]}(s)\, \nu(\omega, ds, du) \\ &= \int p(x)\, 1_{]t_0, t_{k+1}]}(s)\, \nu(\omega, ds, du) \\ &\leqslant \|\nu(\omega,]0, \alpha], .\|_p\end{aligned}$$

for every subdivision $t_0 < \ldots < t_{k+1}$ in $[0, \alpha]$.

A point process ν of order p is therefore an $\mathfrak{M}^p$-valued random measure with finite variation.

The *dual predictable projection* of ν is, by definition, the dual predictable projection of the $\mathfrak{M}^p(E)$-valued process $(\nu(.,]0, t], .))_{t\in\mathbb{R}^+}$.

For example, let ν be a point process of order p with the following two properties.

(i) For every two disjoint "rectangles" $]s_1, t_1] \times B_1,]s_2, t_2] \times B_2$, with $B_1, B_2 \in \mathscr{B}_E$, the random variables $\nu(.,]s_1, t_1], B_1)$ and $\nu(.,]s_2, t_2], B_2)$ are independent.

(ii) There exists a measure $\alpha \in \mathfrak{M}^p(E)$ such that, for every "rectangle" $]s, t] \times B$, $B \in \mathscr{B}_E$,

$$E(\nu\ .,]s, t] \times B) = (t - s)\, \alpha(B).$$

Such a pointprocess is called a *time-homogeneous Poisson point process with Levy measure* α.

It follows immediately from properties (i) and (ii) that, for every function $\varphi \in \mathscr{C}^p(E)$, the process

$$Y := \Big(\int_{]0,t]\times E} \varphi(x)\, \nu(., ds, dx) - t \int_E \varphi(x)\, \alpha(dx)\Big)_{t\in\mathbb{R}^+}$$

is a process with increments $Y_{t+h} - Y_t$ independent of $\mathscr{F}_t$ for all $t, h > 0$.

This process is therefore a martingale with respect to the filtration $(\mathscr{F}_t)_{t\in\mathbb{R}^+}$.

This means that *the deterministic* $\mathcal{M}^p(E)$*-valued process* $(t\cdot\alpha)_{t\in\mathbb{R}^+}$ *is the dual predictable projection of the process* F_v.

15.12 Point process associated with an R. R. C. process. Case of processes with independent increments.

Let X be an $\mathbb{R}^d$-valued R.R.C. process. If D is the subset of $\mathbb{R}^+\times\Omega$ defined by $D:=\{(t,\omega):\|\Delta X_t(\omega)\|>0\}$, the regularity of the paths implies that, for every $t>0$, every compact subset $K\subset\mathbb{R}^d-\{0\}$ and every ω, there is only a finite number of $t_i<t$ such that $\Delta X_{t_i}(\omega)\in K$. We therefore define a point process μ if we set

(15.12.1) $$\mu(\omega, ds, dx) := \sum_{(t,\omega)\in D} \varepsilon_{(t,\,\Delta X_t(\omega))}(ds, dx).$$

This point process is called the *point process of jumps of* X.

We shall make clean later that for a wide range of processes X (the semimartingales) one has $\sum_{s\leqslant t}\|\Delta_s(\omega)\|^2<\infty$ a.s. In this case, the function $\dfrac{\|x\|^2}{1+\|x\|^2}$ is a.s. integrable with respect to all the measures $\mu(\omega,]0,t],dx)$ on $\mathbb{R}^d-\{0\}$.

If we set

(15.12.2) $$p(x)=\frac{\|x\|^2}{1+\|x\|^2},$$

we see that the point process of jumps of such an X is a point process of order 2 for the function p in (15.12.2).

Before arriving at this general situation (see chapter 5), we state the following.

Let X *be an* $\mathbb{R}^d$*-valued R. R. C. process with stationary independent increments (we recall that stationary means that the law of* $X_{t+r}-X_r$ *is the same for all* r*). We assume* X *to be stochastically continuous, i.e.,* $\lim_{h\to 0} P\{\|X_{t+h}-X_t\|>\eta\}=0$ *for every* $t\geqslant 0$ *and* $\eta>0$*, and that the jumps of* X *are bounded. Then the point process of jumps of* X *is a time homogeneous Poisson point process with Levy measure* α *such that* $t\,\alpha(B)=E(\sum_{s\leqslant t} 1_B(\Delta X_s))$.

The proof of this is given as an exercise in exercise E. 3.3. below.

Exercises and supplements

1. Uniform integrability of the random variables (X_t) – Processes of class [D] – Stopping Theorem

From the convergence theorem for martingales it follows that every martingale $(X_t)_{t\in\mathbb{R}^+}$, for which the random variables X_t, $t\in\mathbb{R}^+$ form a uniformly integrable family, converges in $L^1(\Omega,\mathcal{F}_\infty,P)$ towards a random variable X_∞. As a consequence of Proposition 8.8 and Theorem 10.1, this implies that X belongs to the class [D].

Exercise E. 1.1 shows that the analogous property holds for quasimartingales indexed by $\mathbb{N}$, while exercise E. 1.3 shows it is false even for positive supermartingales indexed by $\mathbb{R}^+$.

Exercise E. 1.2 shows an application of the stopping theorem to computation of exit times for Brownian motion.

E. 1.1. Let $I = \bar{\mathbb{N}}$ and let $(X_t)_{t \in I}$ be a quasimartingale with the property that X_n converges in $L^1(\Omega, \mathcal{F}_\infty, P)$ to X_∞ when n tends to infinity.

(1) Prove that X belongs to the class $[D]$ on I.

(*Hint:* The content $|\lambda_X|$, in this case, is σ-additive on every σ-algebra $(]n, n+1] \times \Omega) \cap \mathcal{P}$ and is therefore σ-additive on $\mathcal{A}_I$ if and only if $\lim_{n \to \infty} |\lambda_X|(]n, \infty] \times \Omega) = 0)$.

(2) Application: Every positive supermartingale (X_n) for which $\lim_{n \to \infty} X_n = 0$ a.s. and $\lim_{n \to \infty} E(X_n) = 0$ belongs to the class $[D]$.

E. 1.2. Let $I = \mathbb{R}^+$ and let $(X_t)_{t \in \mathbb{R}^+}$ be the canonical Brownian motion as defined in 2.2, having state-space $\mathbb{R}^3$.

For every closed set B in $\mathbb{R}^3$, we call T_B the stopping time

$$T_B := \inf \{t; X_t \notin B\}$$

and, for each $x \in \mathbb{R}^3$, we denote by $\Pi_B(x, dy)$ the measure in $\mathbb{R}^3$ defined by

$$\Pi_B(x, A) = P_x \{T_B \in A\},$$

where P_x is the law of the canonical Brownian motion in $\mathbb{R}^3$ starting at x. (The corresponding mathematical expectation is written E_x).

(1) The measure $\Pi_B(x,.)$ has total mass $\leqslant 1$. It is carried by the boundary δ_B of B (use the continuity of paths) for every $x \in B^C$ and $\Pi_B(x,.) - \varepsilon_x$ (Dirac measure in x) for every $x \in B$.

(2) We write $H_B \varphi(x)$ for $\int \varphi(y) \Pi_B(x, dy) = E_x(\varphi(X_{T_B}))$. Let $K(x, r)$ be the ball with center x and radius r and μ_r the uniform probability law on the surface of $K(x, r)$. Using the martingale stopping theorem, prove the following relation for every bounded Borel function φ.

$$H_B \varphi(x) = \int H_B \varphi(y) \mu_r(dy)$$

(3) Show that the function $H_B \varphi$ as defined in (2) is continuous and such that, in the open set B^C, $H_B \varphi$ is harmonic, that is to say, twice differentiable with

$$\sum_{i=1}^{3} \frac{\partial^2}{(\partial x^i)^2} H_B \varphi(x) = 0 \text{ for every } x \in B^C, \text{ and } H_B \varphi = \varphi \text{ on } B.$$

(*Hint:* To prove the differentiability of $H_B \varphi$ in B^C, consider for each $r > 0$ a function ϱ_r on $\mathbb{R}^+$ which is zero outside $[0, r]$ and such that $x \rightsquigarrow \varrho_r(\|x\|)$ is infinitely differentiable and $\int_{\mathbb{R}^3} \varrho_r(\|x\|) dx = 1$. Set $g_r(x) = \varrho_r(\|x\|)$ and $\tilde{h}_r(x) = h * g_r(x)$. Show that

$\tilde{h}_r(x) = h(x)$ as soon as $K(x,r) \subset B^C$. Next, use a Taylor expansion of h and formula of (2) to prove the harmonicity of $H_B\varphi)$.

(4) Let B be the set $\{\|x\| \leqslant \alpha\} \cup \{\|x\| \geqslant \beta\}$ with $0 < \alpha < \beta$. We set $\tau_\alpha := \inf\{t : \|X_t\| \leqslant \alpha\}$, $\sigma_\beta := \inf\{t : \|X_t\|_t \geqslant \beta\}$. Show that

$$P_x\{\tau_\alpha < \sigma_\beta\} = \begin{cases} 1 & \text{if } \|x\| \leqslant \alpha, \\ \dfrac{\alpha}{\beta - \alpha} \dfrac{\beta - \|x\|}{\|x\|} & \text{if } \alpha < \|x\| < \beta, \\ 0 & \text{if } \|x\| \geqslant \beta. \end{cases}$$

(*Hint:* Remark that $P_x\{\tau_\alpha < \sigma_\beta\} = \int 1_{\{\|x\| \leqslant \alpha\}}(y)\, \Pi_B(x, dy)$ and apply (3)).

(5) Use (4) to show that

$$P_x\{\tau_\alpha < \infty\} = \begin{cases} 1 & \text{if } \|x\| < \alpha, \\ \dfrac{\alpha}{\|x\|} & \text{if } x > \alpha. \end{cases}.$$

E. 1.3. Let X be the canonical Brownian motion in $\mathbb{R}^+$ with starting point $x_0 \neq 0$.

(1) Show that the process $Y := \left(\dfrac{1}{\|X_t\|}\right)_{t \in \mathbb{R}^+}$ is a positive supermartingale.

(2) Consider the stopping time

$$T_n := \tau_{1/n} := \inf\left\{t : \|X_t\| \leqslant \frac{1}{n}\right\}.$$

Using the result in E. 1.2. (5), show that

$$\int\limits_{\{Y_{T_n} \geqslant n\}} Y_n \, dP \geqslant \frac{1}{\|x_0\|}$$

for every $n \geqslant \dfrac{1}{\|x_0\|}$.

This proves that Y does not belong to the class [D] although the random variables $(Y_t)_{t \in \mathbb{R}^+}$ form a uniformly integrable family according to chapter II, exercise E. 2.2. (2). This counterexample is due to Johnson and Helms [HeJ].

2. Predictable and dual predictable projection

E. 2.1. Compute the predictable projection for all the processes considered in §15.7 and compare them with the dual predictable projection given in 15.7.

Answers:

(1) $A_t^{\mathscr{P}} = A_{t-}$.

(2) $A_t^{\mathscr{P}} = A_{t-} - \Delta_t E(A_t)$.

(3) When T is predictable $(1_{[T,\infty[})^{\mathscr{P}} = 1_{[T,\infty[}$. When T is totally inaccessible, $A_t^{\mathscr{P}} = A_{t-}$.

(4) $(h\,1_{[T,\infty[})^{\mathscr{P}} = h\,1_{[T,\infty]} + E(h|\mathscr{F}_{T-})\,1_{[T]}$.

E. 2.2. Let τ be a predictable stopping time, X a quasimartingale in the class [D] and λ_X its Doléans measure. Prove that, for every positive predictable process X,

$$\int_{\mathbb{R}^+\times\Omega} 1_{\{\tau\}} X(s,\omega)\,\lambda_X(ds,d\omega) = E(X_\tau \Delta X_\tau).$$

(*Hint:* First consider the case $X_\tau = 1_H$, where $H \in \mathscr{F}_{\tau_n}$ and (τ_n) is an increasing sequence of stopping times converging to τ with $\tau_n < \tau$ a.s.).

E. 2.3. As in exercise E. 1.1 of chapter II, consider a canonical Brownian motion β on its canonical basis $(\Omega_c, (\mathscr{C}_t)_{t\in\mathbb{R}^+}, P)$ and the extended filtration $(\mathscr{F}_t)_{t\in\mathbb{R}^+}$, where $\mathscr{F}_t$ is the σ-algebra generated by $\mathscr{C}_t$ and $\mathscr{C}_1$.

What is the Doob-Meyer decomposition of the quasimartingale $(\beta_t)_{t\in\mathbb{R}^+}$ with respect to the filtration $(\mathscr{F}_t)$? (The answer is immediately given by exercise E. 1.1 of chapter 2).

E. 2.4. Let $(\Omega, (\mathscr{F}_t)_{t\in\mathbb{R}^+}, \mathfrak{A}, P)$ be the stochastic basis defined in example 4.8 and the totally inaccessible stopping time $T(\omega) := \omega$. Consider the process $A := 1_{[T,\infty[}$.

What is the dual predictable projection $\tilde{A}$ of A?

Answer: $A_t(\omega) = \left(\int_0^{t\wedge\omega} \frac{1}{1-u}\,du\right)$.

E. 2.5. *Predictable and "natural processes"*. Let A be an adapted, increasing, right-continuous, real process with $E(A_t) < \infty$ for every $t \in \mathbb{R}^+$ and $A_0 = 0$. Following P. A. Meyer (cf. [Mey. 1]) the process A is called "natural" if, for every right-continuous, positive, bounded martingale $(Y_s)_{s\in\mathbb{R}^+}$ and every $t \in \mathbb{R}^+$, the following relation holds.

$$(N) \qquad E\left(\int_0^t Y_s\,dA_s\right) = E\left(\int_0^t Y_{s-}\,dA_s\right).$$

Show that an adapted, increasing, right-continuous, real process A with $E(A_t) < \infty$ for every $t \in \mathbb{R}^+$ and $A_0 = 0$ is natural if and only if it is predictable.

(*Hint:* The relation (N) above is equivalent to the predictability of the Doléans measure λ_A).

3. Point Processes

E.3.1. *Dual predictable projection of* $\sum_n \varepsilon_{(T_n,\beta_n)}(ds,dx)$. Let $(T_n)_{n\geq 0}$ be an increasing sequence of positive finite random variables on $(\Omega, \mathfrak{A}, P)$ with $T_0 = 0$, $T_{n+1} > T_n$ P.

a. s. and let $(\beta_n)_{n\geq 1}$ be a sequence of random variables with values in the open subset E of $\mathbb{R}^d$. For every t we call $\mathscr{F}_t$ the σ-algebra generated by the sets $\{T_n \leq s\} \cap \{\beta_n \in B\}$, $s \leq t$, $n \in \mathbb{N}$, $B \in \mathscr{B}_E$. We denote by $(\mathscr{F}_t^P)_{t<0}$ the smallest filtration with $\mathscr{F}_t^P \supset \mathscr{F}_t$ satisfying the "usual hypotheses" with respect to P.

(1°) If, for every $\omega \in \Omega$, $\mu(\omega, ds, dx)$ is the measure on $\mathbb{R}^+ \times E$ and $\mathscr{B}_{\mathbb{R}^+} \otimes \mathscr{B}_E$ is defined by $\mu(\omega, ds, dx) := \sum_{n\geq 1} \varepsilon_{(T_n, \beta_n)}(ds, dx)$, then μ is a point process with respect to $(\mathscr{F}^P)$.

(2°) For every n let $G_n(\cdot\, ds, dx)$ be the conditional law of (T_{n+1}, β_{n+1}) with respect to $\mathscr{F}^P_{T_n}$. Show that the dual predictable projection of μ is

$$v(\omega, ds, dx) = \sum_{n\geq 0} 1_{]T_n, T_{n+1}]}(s) \frac{G_n(\omega, ds, dx)}{G_n(\omega, [s, \infty[, E)}.$$

(*Hint:* For every bounded Borel function φ on E set

$$A_t^{\varphi}(\omega) := \sum_{n\geq 0} 1_{[T_n, \infty[}(t) \int_{T_n}^{T_{n+1}} \int_E \frac{1}{G_n(\omega, [s, \infty[, E)} G_n(\omega, ds, dx)\, \varphi(x),$$

$$X_t^{\varphi}(\omega) = \sum_{n\geq 1} 1_{[T_n, \infty[}(t)\, \varphi(\beta_n).$$

A^{φ} is easily seen to be predictable and $E\, 1_F(X_u^{\varphi} - A_u^{\varphi} + X_t^{\varphi} - A_t^{\varphi}) = 0$ for every $t < u$ and $F = \{T_n \leq s\} \cap \{\beta_n \in B\}$ with $s \leq t$).

E. 3.2. *An example in queuing theory.* We consider one queue, the evolution of which is entirely described by the arrival process $\sum_{n\geq 1} 1_{[\tau_n^1, \infty[}(t)$ and the departure process $\sum_{n\geq 1} 1_{[\tau_n^2, \infty[}(t)$, where τ_n^1 (resp. τ_n^2) are the arrival times (resp. departure times). In this model customers arrive one by one and only one customer is served at a time. We introduce the random measure

$$\mu(\omega, ds, dx) = \sum_n \varepsilon_{(\tau_n^1(\omega), 1)}(ds, dx) + \varepsilon_{(\tau_n^2(\omega), -1)}(ds, dx).$$

The state of the queue (number of customers in the system at time t) is given by

(1) $$N_t(\omega) = \int_{]0,t] \times \mathbb{R}} x\, \mu(\omega, ds, dx).$$

(Remark that the sequences $(\tau_n^1(\omega))$ and $(\tau_n^2(\omega))$ are such that $N_t(\omega) \geq 0$ for all t).

If $(T_n)_{n\geq 0}$ is the sequence of random times defined by $T_0 = 0$ $T_{n+1} := \min\{\tau_k^i;\ i = 1, 2\ \tau_k^i > T_n\}$, then μ is a random measure as in exercise E. 3.1. Let us assume that "the inter-arrival times and service times of the customers are all independent", the common law of the inter-arrival times (resp. service times) being the probability on $\mathbb{R}^+$ with density $\alpha^1(s)$) (resp. $\alpha^2(s)$).

Call Θ_k^2 the start of the service of the kth customer. ($\Theta_k^2 = \tau_{k-1}^2$ if $N_{\tau_{k-1}^2} \neq 0$). To simplify the formulas below, we write $\Theta_k^1 = \tau_{k-1}^1$ for all k.

(1°) The hypothesis on the law of the inter-arrival times and service times can be expressed using the terminology of E. 3.1. by

(2) $$G_n(\omega, dt, \{1\}) = (\alpha^1(t-\tilde{\tau}_n^1)\int_t^\infty \alpha^2(u-\tilde{\tau}_n^2)\,du)\,1_{\{N_{T_n}\neq 0\}}dt + 1_{\{N_t=0\}}\alpha^1(t-\tilde{\tau}_n^1)dt,$$

where $\tilde{\tau}_n^1$ is the time of the last arrival before T_n and $\tilde{\tau}_n^2$ is the time of the start of the service of the last customer beeing served at T_n.

(3) $$\tilde{\tau}_n^i := \max\{\Theta_h^i : \Theta_k^i \leqslant T_n\}.$$

Analogously,

(4) $$G_n(\omega, dt, \{-1\}) = (\alpha^2(t-\tilde{\tau}_n^2)\int_t^\infty \alpha^1(u-\tilde{\tau}_n^1)\,du)\,1_{\{N_{T_n}\neq 0\}}\,dt.$$

(2°) Let ν be the dual predictable projection of μ with respect to the filtration generated by μ (the one introduced in E. 3.1.) Prove.

$$\nu(.,dt,\{1\}) = \sum_{k\geqslant 0} 1_{[\Theta_k^1,\tau_{k+1}^1[}(t)\frac{\alpha^1(t-\Theta_k^1)}{\int_t^\infty \alpha^1(u-\Theta_k^1)\,du}dt$$

$$\nu(.,dt,\{-1\}) = \sum_{k\geqslant 0} 1_{[\Theta_k^2,\tau_{k+1}^2[}(t)\frac{\alpha^2(t-\Theta_k^2)}{\int_t^\infty \alpha^2(u-\Theta_k^1)\,du}dt$$

(3°) The laws of inter-arrivals and serving times are exponential with parameters λ and μ respectively (in which case the process (N_t) is Markovian) if and only if

$$\nu(.,dt,\{1\}) = \sum_{k\geqslant 0} 1_{[\Theta_k^1,\tau_{k+1}^1[}(t)\,\lambda\,dt$$

$$\nu(.,dt,\{-1\}) = \sum_{k\geqslant 0} 1_{[\Theta_k^2,\tau_{k+1}^2[}(t)\,\mu\,dt$$

E. 3.3. *The Poisson point process of a process with stationary independent increments.* We prove the statement on the point process of jumps of a process with stationary independent increments made at the end of section 15.12.

(1°) If I and J are disjoint intervals and B_1 and B_2 are Borel sets in $\mathbb{R}^d - \{0\}$, show that the $\bar{\mathbb{N}}$-valued random variables $\mu(., I\times B_1)$ and $\mu(., J\times B_2)$ are independent.

(2°) Let I be a bounded interval $[s,t]\subset\mathbb{R}^+$ and B a bounded Borel set with $\bar{B}\subset\mathbb{R}^d-\{0\}$ (The random variable $\mu(., I\times B)$ is then finite). For every n, $\{I_{n,k}: k=1..n\}$ is a family of disjoint subintervals of I with equal length $|I_{n,k}| = \frac{|I|}{n}$.

Set $p := P\{\mu(., I\times B)>0\}$ and $p_n := P\{\mu(., I_{n,k}\times B)>0\}$.

Show that $\sum_{k\leqslant n} P\{\mu(I_{n,k}\times B)>0\}$ is bounded.

(Remark that $(1-p_n)^n = 1-p$ and therefore $p_n \sim \frac{1}{n}|\log(1-p)|$).

(3°) Prove that the random variable $\mu(.,I\times B)$ of (2°) obeys a Poisson law. As a consequence $E(\mu(.,I\times B))<\infty$.

Hints: Set $\tilde{N}_n := \sum_{k\leqslant n} 1_{\{\mu(.,I_{n,k}\times B)>0\}}$ and note that $\mu(.,I\times B)=\lim_n \tilde{N}_n$ a.s. and that

$$e^{i\lambda\mu(.,I\times B)}=\lim_n \prod_{k\leqslant n}((1-p_n)+p_n e^{i\lambda})^n.$$

(4°) The point process $\mu(.,ds,dx)$ is a time homogeneous Poisson point process.

Hints: The problem is reduced by (1°) to proving that if B_1 and B_2 are disjoint, bounded Borel sets with $\bar{B}_1$ and $\bar{B}_2\subset\mathbb{R}^d-\{0\}$, then the random variables $\mu(I\times B_1)$ and $\mu(I\times B_2)$ are independent. Let $I_{n,k}$ be as in (2°) and set $N^1=\mu(I\times B_1)$ $N^2=\mu(I\times B_2)$, $N^i_{n,k}=\mu(I_{n,k}\times B)$, $i=1,2$. Use the independence of pairs $(N^1_{n,k}, N^2_{n,k})$, k $=1..n$ to evaluate

$$|E e^{i\lambda(N^1+N^2)}-E(e^{i\lambda N^1})\cdot E(e^{i\lambda N^2})| \doteq$$
$$=\prod_{k\leqslant n} E(e^{i\lambda(N^1_{n,k}+N^2_{n,k})})-\prod_{k\leqslant n} E(e^{i\lambda N^1_{n,k}})\,E(e^{i\lambda N^2_{n,k}})$$

Now use the elementary inequality $|\prod_{k\leqslant n} a_k-\prod_{k\leqslant n} b_k|\leqslant\sum_{k\leqslant n}|a_k-b_k|$, which is valid as soon as $|a_k|\leqslant 1$ and $|b_k|\leqslant 1$ for all k and the relation $N^1_{n,k}N^2_{n,k}=0$, which implies $e^{i\lambda(N^1_{n,k}+N^2_{n,k})}=e^{i\lambda N^1_{n,k}}+e^{i\lambda N^2_{n,k}}-1$, to get $E e^{i\lambda(N^1+N^2)}-(e^{i\lambda N^1})\,E(e^{i\lambda N^2}\leqslant$ $\leqslant 2\sup_k|E(e^{i\lambda N^1_{n,k}}-1|(\sum_{k\leqslant n}P\{N^2_{n,k}>0\})\big|$ The stochastic continuity of the process and the result in (2°) above imply the independence of N^1 and N^2.

The time homogeneity of μ is a straightforward consequence of the increments of X being stationary.

E. 3.4. *The Poisson point process of a process with independent increments.* Let X be a $\mathbb{R}^d$-valued R.R.C. stochastically continuous process with independent increments. Show that the point process of jumps of X is a Poisson point process μ with Levy measure $\nu(I\times B)=E(\sum_{s\in I}1_B(\Delta X_s))$. This means

(i) For every pair $]s_1,t_1]\times B_1,]s_2,t_2]\times B_2$ of disjoint rectangles with $B_1,B_2\in\mathscr{B}_E$, $\mathbb{E}=\mathbb{R}^d-\{0\}$, the random variables $\mu(.,]s_1,t_1],B_1)$ and $\mu(.,]s_2,t_2],B_2)$ are independent.

(ii) There exists a measure $\nu(ds,dx)$ on $\mathbb{R}^+\times E$ such that, for every $]s,t]\times B$,

$$E(\mu(.,]s,t],B))=\nu(]s,t]\times B).$$

Hints: To establish this property of processes with independent increments do the same as in E. 3.3. The only difference comes from the fact that, in the non-stationary case, the probabilities $p_{n,k}:=P(\{\mu(I_{n,k}\times B>0\})$ are not equal for $k=1..n$. Use the stochastic continuity to show that $\lim_n\sup_k p_{n,k}=0$.

4. Martingales and Markov processes

E. 4.1. *The Dynkin operator.* Let $(T_t)_{t\geqslant 0}$ be a Feller semi-group on $C_0(\mathbb{R}^d)$ and $\mathcal{G}$ its generator as defined in chapter 2, exercise E. 4.1. We assume that, for every $x \in \mathbb{R}^d$, there exists a Markov process X^x associated with $(T_t)_{t\geqslant 0}$ which is defined on a stochastic basis $(\Omega^x, (\mathcal{F}_t^x)_{t>0}, P_x)$ with right-continuous paths such that $P_x\{X_0^x = x\} = 1$. Prove the following.

(a) For every finite stopping time τ and $\varphi \in D_{\mathcal{G}}$,

$$(1) \qquad E_x(\varphi(X_\tau^x) - \varphi(x)) = E_x\left(\int_0^\tau \mathcal{G}\varphi(X_s^x)\,ds\right).$$

(Remember the Dynkin formula of 2–E. 4.1.)

(b) If (G_n) is a decreasing sequence of neighborhoods of x with $\bigcap_n G_n = \{x\}$, if $\tau_n := \inf\{t : t > 0\; X_t^x \notin G_n\}$, one has

$$(1) \qquad \mathcal{G}\varphi(x) = \lim_n \frac{1}{E(\tau_{G_n})} E_x[\varphi(X_{\tau_{G_n}}^x) - \varphi(x)\, 1_{\{\tau_{G_n} < \infty\}}]$$

for every $\varphi \in D_{\mathcal{G}}$.

(Write $E_x(\tau_{G_n} \wedge T)(\mathcal{G}\varphi(x) + \varepsilon_n) = E_x[\varphi(X_{\tau_{G_n}\wedge T}^x) - \varphi(x)]$ for every $T > 0$).

The *Dynkin operator* is defined by formula (1) for all the functions φ for which the limit in (1) exists at every point x. It is therefore an extension of the infinitesimal generator.

E. 4.2. *The infinitesimal generator of a Markov pure jump process.* By a Markov pure jump process we mean a Markov process with right-continuous paths which are constant between isolated jumps.

If, as usual, we call P_x the law of the process starting from x and $\tau_x := \inf\{t : X_t \neq x\}$ the exit time of x, we know from exercise E. 19, chapter 1, that τ_x obeys an exponential law with parameter $\lambda(x)$ if $0 < \lambda(x)$ and $P_x\{\tau_x = +\infty\} = 1$ if $\lambda(x) = 0$.

Show that if f belongs to the domain $\mathcal{D}_{\mathcal{G}}$ of the infinitesimal generator, then

$$\mathcal{G}f(x) = \lambda(x) \int \nu(x, dy)\,(f(y) - f(x)),$$

where $\nu(x,.)$ is called the *law of jumps from* x and is defined by $\nu(x, B) = P_x\{X_{\tau_x} \in B\}$ for $\lambda(x) \neq 0$.

Hint: Write the Dynkin formula as

$$E_x(f(X_{\tau_x}) - f(x)) - E_x\left(\int_0^{\tau_x} \mathcal{G}f(X_s)\,ds\right) = 0\,.$$

Historical and bibliographical notes

What we have called the Doléans measure was introduced by C. Doléans in [Dol. 1]. The proof of necessary and sufficient conditions for its existence, which is given here, comes from [MeP. 2]. If the "usual hypotheses" on the filtration ($\mathscr{F}$) are dropped, the 'necessary' part of Theorem 13.3 is no longer true. H. Föllmer gave a theorem where conditions on Ω and ($\mathscr{F}$) imply the σ-additivity of λ_X for every quasimartingale X. [Föl]

The central theorem of this chapter is the Doob-Meyer theorem, Theorem 15.4, which was proved by P.A. Meyer [Mey. 1, 4, 6]. This type of decomposition was first stated by Doob in the case of a process indexed by $\mathbb{N}$, where it is easy to obtain (see [Doo. 1] and [Nev. 6]). The first proof by P.A. Meyer, which was based on discrete approximation called "approximate potentials" was quite complicated. A simpler proof was given by K. M. Rao [Rao. 1]. But the introduction of the "Doléans measure" and Dellacherie's predictable projection theorem, as in [Dol. 1], leads back to a transparent proof and gives the theorem its full meaning. We have adopted this approach here.

There is a decomposition theorem for quasimartingales not of class [L. D], due essentially to Ito-Wanabe [ItW] (see also Dellacherie-Meyer [DeM], chap. 7), the martingale in the decomposition being replaced by a local martingale. The reader will find a proof of this with many complements in [DeM], chap. 7. A proof of this theorem (valid for Banach-valued processes), starting with a decomposition of λ_X into its σ-additive part and its purely simply additive part, may be found in [MeP. 2].

The dual predictable projection of point processes was introduced by J. Jacod [Jac. 3] (see also [Ore. 2]). The reader will find in [BrJ] a simple introduction to the theory of point processes and their applications. The systematic treatment in [Jac. 2] will be more easily read if one is familiar with the main topics in the first seven chapters of the present book.

Chapter 4

Square integrable martingales and semimartingales

Among the spaces of martingales which may be considered, the space of square integrable martingales is certainly the simplest because of its Hilbert space structure and it plays a fundamental role.

To illustrate this, one could say that, traditionally, most of the classical types of stochastic integrals have been introduced (at least implicitly!) as isomorphic transformations of some particular space of square integrable martingales.

This chapter gives a fairly extensive account of the structure properties of this fundamental space. Sections 16 to 18 expound in some detail the material contained in P.A. Meyer [Mey. 3] (see also [Met. 4]). This concerns the notion of Meyer process $\langle M \rangle$ and of strong orthogonality. The structure theorem (Th. 17–7) is given, saying that a square integrable martingale is the orthogonal sum of its continuous and purely discontinuous parts. Then one studies the isometric stochastic integral of a real square integrable martingale and the concept of quadratic variation.

Section 19 provides the reader with inequalities in constant use giving upper bounds of expressions such as $E(\sup_{t \leq \tau} |M_t|^2)$, where τ is a stopping time. The inequality of Theorem 19.1 is essentially a Doob inequality, while 19.4 due to M. Metivier and J. Pellaumail [MeP. 3] is much more elaborate and very useful in the theory of stochastic differential equations.

Sections 20 to 22 are devoted to Hilbert-valued square integrable martingales. We consider finite and infinite-dimensional Hilbert spaces. Except for those properties of §20 which are trivial extensions of the real case, we are led to introduce new notions as the second Meyer process $\langle\langle M \rangle\rangle$ and the tensor quadratic variation $[\![M]\!]$. This makes it necessary, for the sake of the reader, to recall a few basic facts about tensor products, Hilbert-Schmidt and nuclear operators. In the finite-dimensional case the consideration of matrices with suitable topologies would be enough. It should, however, be emphasized that *even in the finite-dimensional case* there are phenomena which could not be dealt with by using only the 1-dimensional theory. Section 22, which is devoted to the isometric stochastic integral and completed by exercise E. 9 (dealing with a two-dimensional Gaussian martingale), illustrates this fact.

Section 23 considers the processes which are obtained by localization of the concepts of Martingales and processes with bounded variation. The semimartingales are actually defined as the sum of a local martingale and a process with finite variation, as in P. A. Meyer [Mey. 3]. They are also characterized as being the processes admitting a "control process" (cf. section 23.9 and 23.11). The importance of this control process has been underlined in the book by M. Metivier and J. Pellaumail (MeP. 1] (where a process admitting a control process is called a Π^*-process) and plays, in particular, a major role in the study of strong solutions of stochastic equations in a very general context.

In the whole chapter we consider implicitly a stochastic Basis which is *assumed to fulfil the usual hypotheses* (right continuity and completness, see section 1.1).

16. Spaces of real L^2-martingales

16.1 Definitions Martingales have been defined in § 8. We say that the martingale $(M_t)_{t \in \mathbb{R}^+}$ (with respect to a given filtration $(\mathscr{F}_t)_{t \in \mathbb{R}^+}$ on Ω) is an *L^2-Martingale* if, for every $t \in \mathbb{R}^+$, one has $E(|M_t|^2 < \infty$.

It is said to be *square integrable* if $\sup_{t \in \mathbb{R}^+} E|M_t|^2 < \infty$. This implies that the random variables $\{M_t : t \in \mathbb{R}^+\}$ are a uniformly integrable family, so by Theorem 10.4 there exists an $\mathscr{F}_\infty$-measurable random variable M_∞, such that

(16.1.1) $$M_t = \mathrm{E}(M_\infty | \mathscr{F}_t) \quad \text{a. s.}$$

According to Theorem 10.6, M_∞ is in $L(\Omega, \mathscr{F}_\infty, P)$.

Conversely, to every $M_\infty \in L^2(\Omega, \mathscr{F}_\infty, P)$ there corresponds a martingale defined by (16.1.1) and, the filtration $(\mathscr{F}_t)_{t \in \mathbb{R}^+}$ being assumed to be right-continuous, Corollary 10–10 shows that there is a unique (up to indistinguishability) right-continuous integrable martingale M (relative to $(\mathscr{F}_t)_{t \in \mathbb{R}^+}$) satisfying (16.1.1). As $(|M_t|^2)_{t \in \mathbb{R}^+}$ is a submartingale one also has

(16.1.2) $$\sup_{t \in \mathbb{R}^+} E(|M_t|^2) = E(|M_\infty|^2) = 0 \, .$$

M is evanescent if and only if $E(|M_\infty|^2)$.

If we call $\mathscr{M}^2_\infty(\mathscr{F})$ (or simply $\mathscr{M}^2_\infty$ if $\mathscr{F}$ is clear) the vector space of right-continuous square integrable martingales (defined up to indistinguability), this space is algebraically isomorphic to $L^2(\Omega, \mathscr{F}_\infty, P$ and can therefore be endowed with a Hilbert space structure by considering the scalar product $(M, N) := E(M_\infty N_\infty)$ which makes it isomorphic to $L^2(\Omega, \mathscr{F}_\infty, P)$.

We call $\mathscr{M}^{2,c}_\infty$ the vector subspace of $\mathscr{M}^2_\infty$ which consists of continuous square integrable martingales. It will be proved below that it is a Hilbert subspace of $\mathscr{M}^2_\infty$. We denote by $\mathscr{M}^2$ (resp., $\mathscr{M}^{2,c}$) the vector space of right-continuous L^2-martingales (resp., continuous L^2-martingales) with the locally convex structure defined by the seminorms $M \rightsquigarrow E|M_t|^2$, $t \in \mathbb{R}^+$.

If $M \in \mathscr{M}^2_\infty$, then since

$$E(|M_t|^2) = E(|E(M_\infty|\mathscr{F}_t)|^2) \leqslant E(|M_\infty|^2),$$

the Hilbert space $\mathscr{M}^2_\infty$ is clearly a topological subspace of $\mathscr{M}^2$.

It is also very easy to check that $\mathscr{M}^2$ is a complete metrizable space, that is, a Fréchet space.

As mentioned above for $\mathscr{M}^2_\infty$, when we want to specify the filtration $\mathscr{F}$ with respect to which martingales are considered, we write $\mathscr{M}^2(\mathscr{F}), \mathscr{M}^{2,c}(\mathscr{F}), \mathscr{M}^{2,c}_\infty(\mathscr{F}),\ldots$ instead of simply $\mathscr{M}^2, \mathscr{M}^{2,c}, \mathscr{M}^{2,c}_\infty \ldots$

16.2 Examples. (1) The canonical Brownian motion as defined in §2.2 is an element of $\mathscr{M}^{2,c}$ $(\mathscr{C})$. (For the martingale property of Brownian motion we again draw the attention of the reader to the remark in 2.5).

(2) Let Π be a Poisson process with respect to the filtration $\mathscr{F}$ as defined in §2.3. We have already seen that $(\Pi_t - \alpha t)_{t\in\mathbb{R}^+}$, where $\alpha = E(\Pi_1 - \Pi_0)$, is a martingale. Since $E((\Pi_t - \Pi_0 - \alpha t)^2) = \alpha t$, if we assume that Π is a right-continuous version of the process, we have $\Pi \in \mathscr{M}^2$.

(3) Let T be a stopping time, h an element of $L^2_+(\Omega, \mathscr{F}_T, P)$ and A the increasing process $h \cdot 1_{[T,\infty[}$. Let $\tilde{A}$ denote the dual predictable projection of A. This means, in particular, that $N := \mathrm{A} - \tilde{A}$ is a martingale (see §15). We show that N belongs to $\mathscr{M}^2_\infty$. To this end we prove $E(|\tilde{A}_\infty|^2) < \infty$.

Let T_e be the accessible part of T (see §7) and let (T_n) be a sequence of predictable stopping times with disjoint graphs such that $[T] \subset \bigcup_n [T_n]$ (see Theorem 7.3).

Following example 4 in 15.7 we may write

$$(16.2.1) \qquad \tilde{A} = \tilde{A}_0 + \sum_n E(h\,1_{\{T_n = T < \infty\}}|\mathscr{F}_{T_n-})\,1_{[T_n\infty[},$$

where $\tilde{A}_0$ is the (continuous) dual predictable projection of $h\,1_{[T=T_e]}\,1_{[T,\infty[}$.

$$\text{As} \quad \sum_n h\,1_{\{T_n = T\}} \leqslant h \quad \text{is an element of} \quad L^2_+(\Omega, \mathscr{F}_T, P)$$

the sum on the right-hand side of (16.2.1) converges a.s. in $L^2_+(\Omega, \mathscr{F}_T, P)$.

According to a classical formula for Stieltjes integrals,

$$\forall \omega \in \Omega,\ |\tilde{A}(\infty,\omega)|^2 = 2\int_0^\infty \tilde{A}(s,\omega)\,d\tilde{A}(s,\omega) - \sum_n (E(h1_{\{T_n = T<\infty\}}|\mathscr{F}_T-)|^2\,1_{\{T_n<\infty\}}$$

From the definition of $\tilde{A}$ and its predictability,

$$\begin{aligned} E(|\tilde{A}(\infty,.)|^2) &\leqslant 2E\left(\int_0^\infty \tilde{A}(s,.)\,dA(s,.)\right) \\ &\leqslant 2E(\tilde{A}_T\,h\,1_{\{T<\infty\}}) \leqslant 2E(\tilde{A}_\infty\,h\,1_{\{T<\infty\}}) \end{aligned}$$

The Schwarz inequality gives

$$E(|\tilde{A}(\infty)|^2) \leqslant 2\sqrt{E(h^2)}\sqrt{E(|\tilde{A}_\infty|^2)}$$

and therefore

$$E(|\tilde{A}(\infty)|^2) \leqslant 4\,E(h^2)\,.$$

This proves $E(|N_\infty|^2) < \infty$.

16.3 Lemma. *Let $(M^n)_{n\in\mathbb{N}}$ be a sequence of martingales in $\mathcal{M}^2_\infty$ (resp., $\mathcal{M}^2$), which converge to N in $\mathcal{M}^2_\infty$ (resp., $\mathcal{M}^2$). Then there exists a subsequence $(M^{n_k})_{k\in\mathbb{N}}$ with the following property. For P-almost all $\omega\in\Omega$ the sequence $(M^{n_k}(.,\omega))_{k\in\mathbb{N}}$ converges to $N(.,\omega)$ uniformly on $\mathbb{R}^+$ (resp., on every compact interval).*

Proof. Let $(M^n)_{n\in\mathbb{N}}$ be a sequence in $\mathcal{M}^2_\infty$ and $N\in\mathcal{M}^2_\infty$ with

$$\lim_{n\to\infty} E(|M^n_\infty - N_\infty|^2) = 0 \tag{16.3.1}$$

We take a subsequence $(M^{n_k})_{k\in\mathbb{N}}$ with the property

$$E(|M^{n_k}_\infty - N_\infty|^2) \leqslant \frac{1}{2^k} \quad \text{for every} \quad k\,.$$

According to the Doob-inequality of §9 (inequality 9.1.3),

$$P\left(\left\{\sup_{t\in\mathbb{R}^+} |M^{n_k}_t - N_t|^2 > \frac{1}{k}\right\}\right) \leqslant \frac{k}{2^k}\,.$$

The Borel-Cantelli lemma gives

$$P\left(\bigcap_p \bigcup_{k\geqslant p} \left\{\sup_{t\in\mathbb{R}^+} |M^{n_k}_t - N_t|^2 > \frac{1}{p}\right\}\right) = 0 \tag{16.3.2}$$

Moreover, for every $\omega\in\bigcup_p\bigcap_{k\geqslant p}\left\{\sup_{t\in\mathbb{R}^+}|M^{n_k}_t(\omega) - N_t(\omega|^2 \leqslant \frac{1}{p}\right\}$ the sequence $(M^{n_k}(.,\omega))_{k\in\mathbb{N}}$ of functions converges uniformly on $\mathbb{R}^+$ to $N(.,\omega)$. The relation (16.3.2) therefore proves the lemma for a sequence in $\mathcal{M}^2_\infty$. For a sequence $(M^n)_{n\in\mathbb{N}}$ in $\mathcal{M}^2$ we consider, for every $T\in\mathbb{R}^+$, the sequence $(M^{n,T})_{n\in\mathbb{N}}$ and the martingale N^T defined respectively by $M^{n,T}(t,\omega) := N^n(t\wedge T,\omega)$ and $N^T(t,\omega) = N(t\wedge T,\omega)$.

$$\text{Since} \quad E(|M^{nT}_\infty - N^T_\infty|^2 = E(E(|M^{nT}_\infty - N^T_\infty|^2|\mathcal{F}_T)) = E(|M^n_T - N_T|^2)$$

we may apply the result just proved to the sequence $(M^{T,n})$. For each T we get a subsequence which converges a.s. uniformly on $[0,T]$ and by a diagonalization procedure we get a subsequence which a.s. converges uniformly on every $[0,T]$. □

16.4 Proposition. *The vector space $\mathcal{M}^{2,c}_\infty$ is a Hilbert subspace of $\mathcal{M}^2_\infty$. The vector space $\mathcal{M}^{2,c}$ is a Fréchet subspace of $\mathcal{M}^2$.*

Proof. This proposition is a very easy consequence of Lemma 16.3. □

17. The first increasing process and orthogonality of L^2-martingales

If $M \in \mathcal{M}^2$, then M^2 is a submartingale which, according to Theorem 10.1 is an [L. D]-submartingale. Theorem 13.3 then implies the existence of the Doleans measure of M^2. If M and N are two L^2-martingales, then $MN = \frac{1}{4}[(M+N)^2 - -(M-N)^2]$ is clearly an [L. D]-quasimartingale and MN therefore admits a Doleans measure λ_{MN}.

17.1 Definition. Let $M \in \mathcal{M}^2$. The Doleans measure of $|M|^2$ will be called the *admissible measure* of M and denoted by $\lambda_{|M|^2}$ or, for the sake of simplicity, just μ_M.

Two L^2-martingales M and N will be called *strongly orthogonal* when $\lambda_{MN} = 0$.

17.2 Proposition. *Let M and N be two L^2-martingales. There exists a unique (up to P-equality) predictable F.V. process V with the property that $MN - V$ is a martingale and $V_0 = 0$.*

When $M = N, V$ is increasing. If $(\Delta MN)_T = 0$ a.s. for every predictable stopping time T, in particular if M or N is continuous, V is continuous.

Proof. This follows immediately from Corollary 15.4. □

17.3 Definition. If M and N are two L^2-martingales, we write $\langle M, N\rangle$ for the process V of Proposition 17.2.

To abbreviate we shall often write $\langle M\rangle$ instead of $\langle M, M\rangle$ and this process $\langle M, M\rangle$ will be called the *first increasing process* or the *Meyer process of M*.

For the Brownian motion W with variance σ^2 it is immediately checked that $\langle W\rangle_t = \sigma^2 t$.

17.4 Proposition. *Let M and N be two L^2-martingales. The following statements are equivalent.*

(a) *M and N are strongly orthogonal*

(b) *MN is a martingale*

(c) $\langle M, N\rangle = 0$

(d) *For every bounded stopping time T: $E(M_T N_T) = E(M_0 N_0)$*

(e) *Let $\mathcal{T}$ be a family of bounded stopping times such that the stochastic intervals $\{[0, T] : T \in \mathcal{T}\}$ generate $\mathcal{P}$. For every $T \in \mathcal{T}$ one has $E(M_T N_T) = E(M_0 N_0)$.*

If $M \in \mathcal{M}^2_\infty$, $N \in \mathcal{M}^2_\infty$ with $N_0 = 0$ and if M and N are strongly orthogonal they are orthogonal in the Hilbert space $\mathcal{M}^2_\infty$.

Proof. Definitions 17.1 and 17.3 and Theorem 17.2 clearly imply the equivalence of (a), (b) and (c).

Since for every stopping time T, the property $\lambda_{MN}(]0, T]) = 0$ is necessary and sufficient for $\lambda_{MN} = 0$, (a), (d) and (e) are equivalent.

The last statement of the proposition follows immediately from the equation $\lambda_{MN}(\mathbb{R}^+ \times \Omega) = E(M_\infty N_\infty)$. □

17.5 Examples. (1) Let A be a regular right-continuous, increasing process A with $E(A_t) < \infty$ for every t, let $\tilde{A}$ be the dual predictable projection of A and $N := A - \tilde{A}$. According to Lemma 14.8 we may write

$$(17.5.1) \qquad E(M_T \cdot N_T) = E\left(\int_0^T M_s \, dN_s\right)$$

for every bounded martingale M and every bounded stopping time T.

The definition of $\tilde{A}$ implies

$$(17.5.2) \qquad E\left(\int_0^T X \, dN_s\right) = 0$$

for every bounded predictable process X. Let us consider $M \in \mathscr{M}^{2,c}$ and $T_n := \inf\{t: M_t \geqslant n\}$. When T runs through the family of all stopping times and $n \in \mathbb{N}$ the stochastic intervals $]0, T \wedge T_n]$, generate the σ-algebra $\mathscr{P}$. The martingales $(M_{t \wedge T_n})_{t \in \mathbb{R}^+}$ are continuous and bounded. Therefore (17.5.1) gives $E(M_{T \wedge T_n} N_{T \wedge T_n}) = 0$. According to Proposition 17.4, the above N is *strongly orthogonal to every* $M \in \mathscr{M}^{2,c}$.

(2) Let T be a stopping time which is assumed to be either totally inaccessible or predictable. We write $A := h \, 1_{[T,\infty[}$ where $h \in L^2(\Omega, \mathscr{F}_T, P)$. Denoting by $\tilde{A}$ the dual predictable projection of A, we know (cf. 16.2, example 3) that $N := A - \tilde{A}$ is an element of $\mathscr{M}^2_\infty$.

We prove that N is strongly orthogonal to every $M \in \mathscr{M}^2$ such that $|\Delta M_T| = 0$ a.s. We know that N has no jump outside $[T]$. Indeed, if T is totally inaccessible, $\tilde{A}$ is continuous and, if T is predictable, $\tilde{A} = E(h | \mathscr{F}_{T^-}) \, 1_{[T,\infty[}$ (cf. §15.7, Ex. 4).

We remark that every martingale M in $\mathscr{M}^2_\infty$ can be written as the difference of two positive martingales in $\mathscr{M}^2_\infty$ (write $M_\infty = M^+_\infty - M^-_\infty$ and consider the two positive martingales generated by M^+_∞ and M^-_∞). Using then Lemma 14.8 applied to positive martingales and increasing processes A and $\tilde{A}$, we may write, for every martingale M in $\mathscr{M}^2$ and every bounded stopping time τ,

$$(17.5.3) \qquad E(M_\tau N_\tau) = E\left(\int_0^\tau M_{s^-} \, dN_s\right) + E\left(\int_0^\tau (\Delta M_s) \, dN_s\right).$$

Since $(M_{s^-})_{s \in \mathbb{R}^+}$ is predictable, $E\left(\int_0^\tau M_{s^-} \, dN_s\right) = \int_{]0,\tau]} M_{s^-} \, d\lambda_N = 0$

and $\qquad E(M_\tau N_\tau) = E(\Delta M_T \Delta N_T \, 1_{\{\tau > T\}}) = 0$.

This proves the strong orthogonality of M and N.

17.6 Proposition. *Let M be an L^2-martingale, μ_M its admissible measure defined on $\mathscr{P}$ and $\tilde{\mu}_M$ the dual predictable projection of μ_M defined on $B_{\mathbb{R}^+} \times \mathscr{F}_\infty$.*

(1°) *If M is continuous the σ-algebra $\mathscr{G}$ is contained in the $\tilde{\mu}_M$-completion of $\mathscr{P}$.*

(2°) *If μ_M is absolutely continuous with respect to $\lambda^1 \otimes P$, restricted to $\mathscr{P}$, where λ^1 is the Lebesgue measure on $(\mathbb{R}^+, \mathscr{B}_{\mathbb{R}^+})$, then $\tilde{\mu}_M$ is also absolutely continuous with respect to $\lambda^1 \otimes P$ and the $\tilde{\mu}_M$-completion of $\mathscr{P}$ contains the σ-algebra $\mathscr{M}^1$ of progressively measurable sets.*

Proof. (1°) From the definition of $\tilde{\mu}_M$ and property 14.5, it follows that $\tilde{\mu}_M([S]) = 0$ for every totally inaccessible stopping time S. For every predictable stopping time T we have $\mu_M([T]) = \tilde{\mu}_M([T])$ and the continuity of M then implies $\tilde{\mu}_M([T]) = E(|\Delta M|_T^2) = 0$. The decomposition theorem, Theorem 7.3, shows that $\tilde{\mu}_M([T]) = 0$ for every stopping time and therefore $\tilde{\mu}_M(]0,T[) = \tilde{\mu}_M(]0,T])$. All the stochastic intervals $]0,T[$ are therefore in the $\tilde{\mu}_M$-completion of $\mathscr{P}$, since $]0,T]$ is predictable. Since they generate $\mathscr{G}$, a classical extension argument shows that $\mathscr{G}$ itself lies in the $\tilde{\mu}_M$-completion of $\mathscr{P}$.

(2°) Let f be the predictable density of μ_M with respect to $\lambda^1 \otimes P$ on $\mathscr{P}$. The increasing process A defined by

$$A(t,\omega) = \int_{]0,t]} f(s,\omega)\,ds$$

is continuous and therefore predictable. Theorem 14.9 shows that the measure μ with density f with respect to $\lambda^1 \otimes P$ on $\mathscr{B}_{\mathbb{R}^+} \otimes \mathscr{F}_\infty$ is also predictable. Since its restriction to $\mathscr{P}$ is μ_M, we have $\mu = \tilde{\mu}_M$. Let X be a bounded progressively measurable process. We set

$$\forall t \in \mathbb{R}^+_{,} \forall h > 0, \forall \omega \in \Omega, \;\; X_t^h(\omega) := \frac{1}{h} \int_{t-h}^{t} X_s(\omega)\,ds\,.$$

A classical result in differentiation theory gives

$$\forall \omega \in \Omega, \lim_{h \downarrow 0} X_t^h(\omega) = X_t(\omega) \quad \text{for} \quad \lambda^1\text{-almost all} \quad t \in \mathbb{R}^+\,.$$

Since X^h is clearly adapted and bounded with continuous paths, it is predictable. For every $\alpha > 0$ we also have

$$\lim_{h\downarrow 0} \int_{\{|f| \leqslant \alpha\}} |X^h - X|\,d\tilde{\mu}_M = \lim_{h\downarrow 0} \int_{\{|f| \leqslant \alpha\}} |X_t^h(\omega) - X_t(\omega)|\,f(t\,\omega)\,\lambda^1(dt)\,P(d\omega) = 0\,.$$

The process X therefore coincides on $\{|f| \leqslant \alpha\}$ $\tilde{\mu}_M$- a.s. with a predictable process. Since $\mathbb{R}^+ \times \Omega = \bigcup_n \{|f| \leqslant n\}$, we have proved the proposition. □

17.7 Theorem (Structure of L^2-martingales). *For every $M \in \mathscr{M}^2$ (resp., $\mathscr{M}^2_\infty$) there exists a sequence (T_n) of stopping times with disjoint graphs and a unique (up to indistinguability) $M^c \in \mathscr{M}^{2,c}$ (resp., $\mathscr{M}^{2,c}_\infty$) with the following properties.*

(1) *If $A_n := \Delta M_{T_n} 1_{[T_n,\infty[}$ and if we denote by $\tilde{A}_n$ the dual predictable projection of A_n, then the martingales $M^c, A_1 - \tilde{A}_1, \dots A_n - \tilde{A}_n, \dots$ are strongly orthogonal elements of $\mathscr{M}^2$ (resp., $\mathscr{M}^2_\infty$), for every $t \in \mathbb{R}^+$*

(17.7.1) $$E(|M_t|^2) = E(|M_t^c|^2) + \sum_n E(|A_n(t) - \tilde{A}_n(t)|^2)$$

and M is the sum in $\mathcal{M}^2$ (resp., $\mathcal{M}^2_\infty$) of the family $\{M^c, A_1 - \tilde{A}_1, \ldots, A_n - \tilde{A}_n, \ldots\}$

(2) *When* $M \in \mathcal{M}^2_\infty$, $E(|A_n(\infty) - \tilde{A}_n(\infty)|^2) = E(|\Delta M_{T_n}|^2)$ for every n and

$$\sum_n E(|\Delta M_{T_n}|^2 \, 1_{\{T_n < \infty\}}) = \|\sum_n (A_n - \tilde{A}_n)\|^2_{\mathcal{M}^2_\infty} < \infty$$

Proof. It is sufficient to consider only the case $M \in \mathcal{M}^2_\infty$. The case $M \in \mathcal{M}^2$ is easily reduced to the previous one by considering the martingales $(M_{t \wedge u})_{t \in \mathbb{R}^+}$ for fixed $u \in \mathbb{R}^+$.

Let (T_n) be a sequence of stopping times with disjoint graphs which are either totally inaccessible or predictable, and such that $\{(t,\omega): |\Delta M(t,\omega)| > 0\} \subset \bigcup_n [T_n]$ (Theorem 7.3).

From the inequalities $|\Delta M_{T_n}| \leqslant 2 \sup_{t \leqslant T_n} |M_t|$ and $E(\sup_{t<\infty} |M_t|^2) \leqslant 4E(|M_\infty|^2)$ (Theorem 10.12) it follows that

$$E(|\Delta M|^2_{T_n}) < \infty . ^{1)}$$

We proved in § 16.2 (example 3) that $V_n := A_n - \tilde{A}_n$ is an element of $\mathcal{M}^2_\infty$ and, in § 17.5 (example 2), that the martingale $M - \sum_{i \leqslant k} V_i$ is, for every $n \leqslant k$, strongly orthogonal to V_n.

Since $V_n(0) = 0$, all the martingales V_n are mutually orthogonal in $\mathcal{M}^2_\infty$ (Prop. 17.4) and therefore

$$E(|M_\infty|^2) = E(|M_\infty - \sum_{i \leqslant k} V_i(\infty)|^2) + \sum_{i \leqslant k} E|V_i(\infty)|^2$$

This proves, in particular, that

$$\sum_{i \geqslant 0} E(|V_i(\infty)|^2) \leqslant E(|M_\infty|)^2$$

and that the family $(V_i)_{i \geqslant 0}$ is summable in $\mathcal{M}^2_\infty$. We may therefore define

$$M^c := \lim_{n \to \infty} (M - \sum_{i \leqslant n} V_i) \text{ (lim in } \mathcal{M}^2_\infty).$$

The martingales $(M - \sum_{i \leqslant n} V_i)_{n>k}$ have no jump outside $\bigcup_{i \leqslant k} [T_i]$ and, in view of Lemma (16.3), this therefore holds for the limit M^c, which has thus no jump outside $\bigcap_n (\bigcup_{k \geqslant n} [T_k]) = \emptyset$ and, as a consequence, is continuous. We have thus proved

$$M^c := M - \sum_{n \geqslant 0} V_n \in \mathcal{M}^c_\infty, \text{ (sum in } \mathcal{M}^2_\infty). \tag{17.7.2}$$

The strong orthogonality of M^c and all the V_n's follows from § 17.5 (example 1). This implies the orthogonality in $\mathcal{M}^2_\infty$ of all the martingales $M^c, V_1, \ldots, V_n, \ldots$ (since $V_n(0) = 0$ and by virtue of Prop. 17.4). This with formula (17.7.2) implies the first statement of the theorem.

---

1) Since $\Delta M_\infty = 0$ there is no ambiguity in writing $|\Delta M_{T_n}|$ $|\Delta M_{T_n}| 1_{\{T_n \quad \infty\}}$.

To prove the second statement we first remark that each V_n has only one jump. This is equal to ΔM_{T_n} if T_n is totally inaccessible (in this case $\tilde{A}_n$ is continuous according to Corollary 15.4) and to $\Delta M_{T_n} - E(\Delta M_{T_n^-})$ (see 15.7, example 4) if T_n is predictable. But, since M is a martingale, in this latter case we have $E(\Delta M_{T_n} | \mathscr{F}_{T_n^-}) = 0$ (see Theorem 13.6). In any case ΔM_{T_n} is therefore the only jump of V_n and a simple formula for Stieltjes integrals gives for P-almost all ω

$$|V_n(\infty, \omega)|^2 = 2 \int_0^\infty V_n(s, \omega)\, dV_n(s, \omega) - |\Delta M_{T_n}(\omega)|^2$$

$$= 2 \int_0^\infty V_n(s-, \omega)\, dV_n(s, \omega) + |\Delta M_{T_n}(\omega)|^2 .$$

Since V_n is a martingale its Doléans measure is null, so

$$E\left(\int_0^\infty V_n(s-, \omega)\, dV_n(s, \omega)\right) = 0$$

and

$$E|V_n(\infty, \omega)|^2 = E(|\Delta M_{T_n}|^2) .$$

This, and the orthogonality of the V_n's in $\mathscr{M}^2_\infty$, prove both equalities in the second statement of the theorem. □

17.8 Definition. In the decomposition formula (17.7.1) the process M^c will be called the *continuous part* of the square integrable martingale M and $M^d = M - M^c$ will be called the *purely discontinuous part* of M.

We denote by $\mathscr{M}^{2,d}$ (resp., $\mathscr{M}^{2,d}_\infty$) the subspace of $\mathscr{M}^2$ (resp., $\mathscr{M}^2_\infty$) consisting of those M such that $M^c = 0$.
It is clear from Theorem 17.7 that $\mathscr{M}^{2,d}_\infty$ is the subspace orthogonal to $\mathscr{M}^{2,c}_\infty$ in $\mathscr{M}^2_\infty$.
It is therefore a Hilbert subspace of $\mathscr{M}^2_\infty$.

17.9 Spaces $\mathscr{M}^p_\infty$ of martingales $p \geqslant 1$

It is clearly possible for $p \geqslant 1$ to consider the space $\mathscr{M}^p_\infty$ of R.R.C uniformly integrable martingales M with the property $E(|M_\infty|^p) < \infty$ and the norm

$$\|M\|_p := [E(|M_\infty|^p)]^{1/p}$$

The above considerations in the case $p = 2$ can, in part, be easily extended. For example, for every Cauchy sequence $(M_n)_{n \geqslant 0}$ in $\mathscr{M}^p_\infty$ it is possible to find a subsequence $(M_{n_k})_{k \geqslant 0}$, the paths of which converge a.s. uniformly on $\mathbb{R}^+$. This follows, as in the proof of Lemma 16.3, from the Doob-Inequality 9.1.3 and the Borel-Cantelli lemma. As in Proposition 16.4, one can therefore show quite easily that $\mathscr{M}^p_\infty$ is a Banach space for every $p \geqslant 1$ and also that the continuous martingales in $\mathscr{M}^p_\infty$ form a closed subspace of $\mathscr{M}^p_\infty$.

18. The L^2-stochastic integral and the quadratic variation of an L^2-martingale

L^2-stochastic integral

18.1 Notation. Let $M \in \mathcal{M}^2$ and μ_M be the Doleans measure of M^2 as defined in § 17.1. We will also denote by μ_M the dual predictable projection of μ_M defined on the σ-algebra $\mathcal{B}_{\mathbb{R}^+} \otimes \mathcal{F}_\infty$ (see 14.7).

For any σ-algebra $\mathfrak{U}$ with $\mathcal{P} \subset \mathfrak{U} \subset \mathcal{B}_+ \otimes \mathcal{F}_\infty$, we define the following spaces

$$L^2(\mathfrak{U}, M) := L^2(\mathbb{R}^+ \times \Omega, \mathfrak{U}, \mu_M) \quad \text{and}$$
$$\Lambda^2(\mathfrak{U}, M) := \{X : X 1_{[0,t]} \in L^2(\mathfrak{U}, M) \, \forall t \in \mathbb{R}^+\}$$

The space $\Lambda^2(\mathfrak{U}, M)$ will be endowed with the topological structure defined by the semi-norms $p_t : X \curvearrowright \sqrt{\int X^2 1_{[0,t]} d\mu_M}$ $t \in \mathbb{R}^+$. This clearly turns $\Lambda^2(\mathfrak{U}, M)$ into a Fréchet space.

The spaces we shall mostly consider are $\mathcal{L}^2(\mathcal{P}, M)$ and $\Lambda^2(\mathcal{P}, M)$ and, when M is continuous (see, in this case, Proposition 17.6), $\mathcal{L}^2(\mathcal{G}, M)$ and $\Lambda^2(\mathcal{G}, M)$.

18.2 Theorem. *Let $M \in \mathcal{M}^2$ be given. There exists a linear isometry from $\mathcal{L}^2(\mathcal{P}, M)$ into $\mathcal{M}^2_\infty$ (resp. $\Lambda^2(\mathcal{P}, M)$ into $\mathcal{M}^2$), which is uniquely defined by the following condition. For every indicator function $X := 1_{]u,v] \times F}$ of a predictable rectangle $]u,v] \times F$ the image of X is the process $(1_F \cdot (M_{v \wedge t} - M_{u \wedge t}))_{t \geq 0}$.*

When $M \in \mathcal{M}^{2,c}$, this isometry has a unique extension into a linear isometry from $\mathcal{L}^2(\mathcal{G}, M)$, resp. $(\Lambda^2(\mathcal{G}, M))$, into $\mathcal{M}^{2,c}_\infty$ (resp. $\mathcal{M}^2_\infty$).

Proof. Let $X := \sum_{i=1}^{n} \alpha_i 1_{]s_i,t_i] \times F_i}$ be a linear combination of indicator functions of predictable rectangles with $t_i \leq T \in \mathbb{R}^+$. We may assume that these rectangles are disjoint. One can see immediately that the process N, where

$$N_t := \sum_{i=1}^{n} \alpha_i 1_{F_i}(M_{t_i \wedge t} - M_{s_i \wedge t})$$

is a martingale. Further, since the martingale property for M implies

$$E(1_{F_i}(M_{t_i \wedge t} - M_{s_i \wedge t}) 1_{F_j}(M_{t_j \wedge t} - M_{s_j \wedge t})) = 0 \text{ when } j \neq i, \text{ we see that}$$

$$\begin{aligned} E(|N_T|^2) &= \sum_{i=1}^{n} \alpha_i^2 E(1_{F_i}|M_{t_i} - M_{s_i}|^2) \\ &= \sum_{i=1}^{n} \alpha_i^2 \mu_M(]s_i, t_i] \times F_i) \\ &= \int_{]0,T] \times \Omega} |X|^2 d\mu_M . \end{aligned}$$

It is also easy to check that the process N does not depend on the chosen representation of X as sum of indicators of predictable rectangles. This, with the above formula, shows that $X \to N$ is a linear isometry of a dense linear subspace of $\Lambda^2(\mathscr{P}, M)$ into $\mathscr{M}^2$ for the seminorms $(\int_{]0,T]\times\Omega} |X|^2 d\mu_M)^{\frac{1}{2}}$ and $(E(|N_T|^2))^{\frac{1}{2}}$.

This proves the first statement of the theorem. The second follows from Proposition 17.6, which says that, when M is continuous, $L^2(\mathscr{P}, M)$ is dense in $\mathscr{L}^2(\mathscr{G}, M)$. □

18.3 Definition and notation. The martingale image of the process by the isometry of Theorem 18.2 will be denoted by $\int X dM$ or $(\int_{]0,T[} X dM)_{t\in}$.

In "Strasbourg notation" it is often written as $X o M$. The random variable of the process $\int X dM$ at time τ (where τ is a finite stopping time) will be denoted by $\int_{]0,\tau]} X dM$ or $(X o M)_\tau$.[1])

The mapping $X \rightsquigarrow \int X dM$ is called the *L^2-stochastic integral* with respect to M.

18.4 Theorem (Uniform Approximation Theorem). *Let (X^n) be a sequence of processes which converge in $\mathscr{L}^2(\mathscr{P}, M)$ (resp. in $\Lambda^2(\mathscr{P}, M)$) to a process X. Then there exists a subsequence $(X^{nk})_{k\in\mathbb{N}}$ such that, for P-almost all ω, the paths $(X^{nk} o M)(\omega)$ converge uniformly on $\mathbb{R}^+$ (resp. uniformly on every compact interval in $\mathbb{R}^+$) to the path $X o M(\omega)$.*

Proof. This follows immediately from the convergence in $\mathscr{M}^2_\infty$ (resp. in $\mathscr{M}^2$) of the sequence $X^n oM$ to $X oM$ and from Lemma 16.3. □

18.5 Classical examples of Brownian motion and Poisson process

(1) If, for the martingale M, we take the Brownian motion, the Doléans measure of β^2 is the measure $\lambda \otimes P$, where λ is the Lebesgue measure on $\mathscr{B}_{\mathbb{R}^+}$. In this case, the space $\mathscr{L}^2(\mathscr{P}, M)$ is $\mathscr{L}^2(\mathbb{R}^+, \times \Omega, \mathscr{B}_{\mathbb{R}^+} \times \mathscr{F}_\infty, \lambda \otimes P)$. Note in this case that, since the paths $t \rightsquigarrow \beta_t(\omega)$ of Brownian motion have no finite variation, the random variable $\int_{]0,t]} X_s d\beta_s$ *cannot be defined* as a Stieltjes integral on P-almost all paths of the function $s \rightsquigarrow X(s,\omega)$ with respect to $d\beta(s,\omega)$.

(2) If $(N_t)_{t\in\mathbb{R}^+}$ is a right-continuous Poisson process of parameter α, the process $M_t := N_t - \alpha t$ is a martingale and the computation of $\mu_M(]s,t] \times F) = E(1_F(M_t - M_s))$ for a predictable rectangle shows immediately that $\mu_M = \lambda \otimes P$, as in the case of Brownian motion. It is immediately seen from the definition that, when X is a linear combination of indicators of predictable reetangles, the process $\int X dM$ coincides for P-almost all paths with the process $(\int_0^t X_s(dN_s - \alpha ds))_{t\in\mathbb{R}^+}$, where the latter

[1]) For reasons explained in section 26, we shall seldom use this notation.

integral is taken in the sense of the Stieltjes integral with respect to the function with finite variation $s \rightsquigarrow N_s - \alpha s$. Since this Stieltjes integral is defined P. a.s. for every $X \in \mathscr{L}^2(\mathbb{R}^+ \times \Omega, \mathscr{B}_{\mathbb{R}^+} \otimes \mathscr{F}_\infty, \lambda \otimes P)$ and the mapping $X \rightsquigarrow \int_0^t X_s(dN_s - \alpha ds)$ in $L^2(\Omega, \mathscr{F}_t, P)$ is continuous, the random variables $(X o M)_t$ and $\int_0^t X_s(dN_s - \alpha ds)$ are a.s. equal. This shows that, in this case, where the Stieltjes integral has a meaning on each path ω for any $X \in \Lambda^2(\mathscr{P}, M)$, it coincides with the above L^2-stochastic integral.

Quadratic variation

18.6 Theorem (Quadratic variation of a martingale). *Let $M \in \mathscr{M}^2$ (resp. $\mathscr{M}^2_\infty$). There exists an increasing R. R. C. process which is uniquely defined up to P-equality denoted by $[M, M]$ and called the quadratic variation of M with the following properties.*

(1) *For every increasing sequence (Π_n) of increasing subsequences of $\mathbb{R}^+$: $\Pi_n := \{0 < t_0 < t_1 \ldots < t_n < \ldots\}$ such that $\lim_{k \uparrow \infty} t_k = +\infty$ and $\lim_{n \to \infty} \delta(\Pi_n) = 0$, where $\delta(\Pi_n)$ is the "mesh" of Π_n defined by $\delta(\Pi_n) := \sup_{t_i \in \Pi_n} (t_{i+1} - t_i)$, one has*

$$(18.6.1) \qquad [M, M]_t = \lim_n (L^1) \sum_{t_i \in \Pi_n} (M_{t_{i+1} \wedge t} - M_{t_i \wedge t})^2$$

(2) *$|M|^2 - [M, M]$ is a martingale.*

(3) *The processes $[M, M]$, $\langle M, M \rangle$ and $|M|^2$ admit the same Doléans measure. If M is continuous, $[M, M] = \langle M, M \rangle$.*

(4) *For every t,*

$$[M, M]_t = \langle M^c, M^c \rangle_t + \sum_{s \leqslant t} |\Delta M_s|^2 = [M^c, M^c]_t + \sum_{s \leqslant t} |\Delta M_s|^2 \quad \text{a.s.},$$

where M^c is the continuous part of M as defined in 17.8.

Proof. We consider only the case $M \in \mathscr{M}^2_\infty$, the case $M \in \mathscr{M}^2$ being easily reduced to this one by considering the martingales $(M_{t \wedge T})_{t \in \mathbb{R}^+}$, $T \in \mathbb{R}^+$.

We write

$$(18.6.2) \qquad \begin{aligned} M_t^2 - M_0^2 &= \sum_{t_i \in \Pi_n} M^2_{t_{i+1} \wedge t} - M^2_{t_i \wedge t} \\ &= \sum_{t_i \in \Pi_n} (M_{t_{i+1} \wedge t} - M_{t_i \wedge t})^2 + 2 M_{t_i \wedge t}(M_{t_{i+1} \wedge t} - M_{t_i \wedge t}) \end{aligned}$$

and set

$$v_n(M, t) := \sum_{t_i \in \Pi_n} (M_{t_{i+1} \wedge t} - M_{t_i \wedge t})^2$$

and

$$X^n(t,\omega) := \sum_{t_i \in \Pi_n} M_{t_i}(\omega)\, 1_{]t_i, t_{i+1}]}.$$

Next we observe that (18.6.2) can be written

(18.6.3) $$M_t - M_0 = v_n(M,t) + 2 \int_{]0,t[} X^n dM.$$

For every $K > 0$, we define $X^{K,n} := (-K) \vee (X^n \wedge K)$ and $M_{t^-}^K(\omega) := (-K) \vee (M_{t^-}(\omega) \wedge K)$. The sequence $(X^{K,n}(t,\omega))_{n\in\mathbb{N}}$ converges to $(M^K(t^-,\omega))$ for every (t,ω), so the processes converge in $\mathscr{L}^2(\mathscr{P}, M)$. The random variables $\int_{]0,t]} X_s^{K,n}\, dM_s$ then converge in $L^2(\Omega, \mathscr{F}_t, P)$ to $\int_{]0,t]} M_{s^-}^K\, dM_s$ for every $t \in \mathbb{R}^+$

Since on the event $F_K := \{\sup_{t\in\mathbb{R}^+} |M_t| \leqslant K\}$

(18.6.4) $$M_t^2 - M_0^2 = v_n(M,t) + 2 \int_{]0,t[} X^{K,n} dM,$$

the sequence $(1_{F_K} v_n(M,t))_{n\in\mathbb{N}}$ converges to

$$1_{F_K}[(M_t^2 - M_0^2) - 2 \int_{]0,t[} M_{s^-}^K\, dM_s] \quad \text{in} \quad L^2(\Omega, \mathscr{F}_t, P).$$

According to Theorem 18.4, it is even possible to extract a subsequence, the paths of which converge uniformly on $\mathbb{R}^+$ to the paths of a regular right-continuous process $[M,M]$ (and are continuous if M is continuous). Since $(v_n(M,t))_{t\in}$ is increasing, t>e same holds for $[M,M]$.

For every $\varepsilon > 0$ it is possible to choose $K > 0$ in such a way that $P(F_K) < \varepsilon$. This shows the stochastic convergence of $v_n(M,t)$ to $M_t^2 - M_0^2 - [M,M]_t$ for every t.

The random variables $[M,M]_t$ are therefore defined uniquely up to P-equivalence. The process $[M,M]$, being R.R.C., is therefore defined uniquely up to indistinguishability.

Following formula (18.6.3) and the martingale property of $\int X^n\, dM$, one has

(18.6.5) $$E(v_n(M,t)) = E(M_t^2) - E(M_0^2)$$

Passing to the limit when $K \uparrow \infty$ we get

(18.6.6) $$E(v_n(M,t)) = E[M,M]_t = E(M_t^2 - M_0^2) \quad \text{for every} \quad t \quad \text{and} \quad n.$$

The positivity of n, the convergence in probability of $v_n(M,t)$ to $[M,M]_t$ and equality (18.6.6) together imply the convergence of $v_n(M,t)$ to $[M,M]_t$ in $L^1(\Omega, \mathscr{F}_t, P)$ and of $(\int_{]0,t} X^{K,n}\, dM)_{t\in\mathbb{R}^+}$ in $\mathscr{M}_\infty^1$ to a martingale according to § 12.9. We have thus proved statements (1) and (2) of the theorem.

Statement (3) expresses that $|M|^2 - [M,M]$ and $|M^2| - \langle M,M\rangle$ are martingales, which has now been proved. If, moreover, M is continuous, then $[M,M]$ is continuous, therefore predictable and coincides with $\langle M,M\rangle$.

Proof of statement (4). We remark that

$$E(v_n(M,t)) = E(M_t^2) - E(M_0^2) \leqslant \|M\|^2_{\mathcal{M}^2_\infty}.$$

Let $(M^k)_{k>0}$ be the sequence

$$M^k := M^c + \sum_{n \leqslant k} A_n - \tilde{A}_n$$

with the same notation as in Theorem 17.7. Let us define $N^k := M - M^k$. This sequence converges to M in $\mathcal{M}^2_\infty$.

As a consequence of

$$v_n(M,t) = v_n(M^k,t) + v_n(M - M^k,t) + 2 \sum_{t_i \in \Pi_n} (M^k_{t_{i+1}\wedge t} - M^k_{t_i \wedge t})(N^k_{t_{i+1}\wedge t} - N^k_{t_i \wedge t})$$

and the Schwarz inequality

$$|v_n(M,t) - v_n(M^k,t)| \leqslant v_n(M^k,t) + 2\sqrt{v_n(M^k,t)}\sqrt{v_n(N^k,t)},$$

we have

(18.6.7) $$E|v_n(M,t) - v_n(M^k,t)| \leqslant 3\|N^k\|\|M\|.$$

This proves the convergence in $L^1(\Omega, \mathcal{F}, P)$ of $[M^k, M^k]_t$ to $[M,M]_t$.

Statement (4) will then be proved if we show

(18.6.8) $$[M^k, M^k]_t = \langle M^c, M^c \rangle_t + \sum_{n \leqslant k} |\Delta M_{T_n \wedge t}|^2$$

but this is a straightfoward consequence of the following lemma. □

18.7 Lemma. *Let $M = H + V$, where $H \in \mathcal{M}^{2,c}$ and V is a martingale with paths of finite variation having finitely many jumps. Then*

$$[M]_t = [H]_t + \sum_{s \leqslant t} |\Delta V_s|^2.$$

Proof. As seen above,

$$v_n(M,t) = v_n(H,t) + v_n(V,t) + 2 \sum_{t_i \in \Pi_n} (M^c_{t_{i+1}\wedge t} - M^c_{t_i \wedge t})(V_{t_{i+1}\wedge t} - V_{t_i \wedge t}).$$

The third term on the right-hand side is smaller than $\sup_{t_i \leqslant t} |M^c_{t_{i+1}\wedge t} - M^c_{t_i \wedge t}| \operatorname{var}(V,t)$, if $\operatorname{var}(V,t)$ denotes the variation of V on $[0,t]$. From the continuity of M^c we see that

(18.7.1) $$\lim_{n\to\infty} 2\left|\sum_{t_i \in \Pi_n} (M^c_{t_{i+1}\wedge t} - M^c_{t_i \wedge t})(V_{t_{i+1}\wedge t} - V_{t_i \wedge t})\right| = 0.$$

Now it is a known property of functions with finite variation that

(18.7.2) $$\lim_{n\to\infty} v_n(V,t) = \sum_{s \leqslant t} |\Delta V_s|^2,$$

so the formula of the lemma follows from (18.7.1). □

18.8 Corollary 1. *If $M \in \mathscr{M}$ and M has paths with finite variation, then $M^c = 0$.*

Proof. We have already mentioned that, for every path $t \curvearrowright M(t, \omega)$ with finite variation, $\lim\limits_{n\to\infty} v_n(M, t)(\omega) = \sum\limits_{s\leqslant t} |\Delta M_s(\omega|^2$. Statement (4) of Theorem 18.6 shows that, in this case, $\langle M^c, M^c \rangle$ is a P-null process, so $E|M_t^c|^2 = 0$ for every t and $M_t^c = 0$ a.s. for all t.

18.9 Corollary 2. *If M and N are two martingales in $\mathscr{M}^2$ there exists an R.R.C. process with paths of finite variation, denoted by $[M, N]$, which is uniquely defined up to P-equality with the following properties.*
(1°) *For every increasing sequence (Π_n) as in Theorem 18.6,*

(18.9.1) $$[M, N]_t = \lim_n (L^1) \sum_{t_i \in \Pi_n} (M_{t_{i+1}\wedge t} - M_{t_i \wedge t})(N_{t_{i+1}\wedge t} - N_{t_i \wedge t})$$

(2°) *$MN - [M, N]$ is a martingale.*
(3°) *The processes MN, $\langle M, N \rangle$ and $[M, N]$ admit the same Doleans measure.*
(4°) For every t

$$[M, N]_t = \langle M^c, M^c \rangle_t + \sum_{s\leqslant t} \Delta M_s \Delta N_s \quad \text{a.s.}$$

the series on the right-hand side being a.s. summable.

Proof. One may write $MN = \frac{1}{4}((M+N)^2 - (M-N)^2)$ and observe that the process $\frac{1}{4}([M+N, M+N] - [M-N, M-N])$ satisfies the properties required for $[M, N]$. This proves the existence of $[M, N]$. The uniqueness is guaranteed by the equality (18.9.1): two different sequences (Π_n) and (Π'_n) give rise to the same limit because the limit along each of them is necessarily the same as the limit relative to $(\Pi_n \vee \Pi'_n)$, where $\Pi_n \equiv \Pi'_n$ means the increasing subsequence of $\mathbb{R}^+$, built by taking all points in Π_n and all points in Π'_n.

18.10 Remark. In formulas (18.6.1) and (18.9.1) one may replace the sequences Π_n of fixed times by sequences $\{0 < \tau_0^n < \tau_1^n < \ldots < \tau_k^n < \ldots\}$ of stopping times with the properties

(i) $$\forall n, \lim_k \tau_k^n = +\infty,$$

(ii) $$\lim_{n\to\infty} \sup_{\tau_i \in \Pi_n} (\tau_{i+1}(\omega) - \tau_i(\omega)) = 0 \quad \text{a.s.}$$

The reader can check for himself that the proof of Theorem 18.6 can be reproduced in this case without any change.

18.11 Notation. For brevity we will write $\langle M \rangle$ instead of $\langle M, M \rangle$ and $[M]$ instead of $[M, M]$ when there is no risk of confusion.

Further properties of stochastic integrals

18.12 Proposition. *Let* $M \in \mathscr{M}^2$, $N \in \mathscr{M}^2$ $X \in \Lambda^2(\mathscr{P}, M)$ *and* $Y \in \Lambda^2(\mathscr{P}, N)$. *Then* $\lambda_{(X \circ M)(Y \circ N)} = XY \lambda_{MN}$.

Proof. We consider simple processes X and Y of the form

$$X = 1_{]u,\infty] \times G}, \quad G \in \mathscr{F}_u$$
$$Y = 1_{]v,\infty] \times H}, \quad H \in \mathscr{F}_v.$$

Let us assume $u < v$ and show that for every predictable rectangle $]s,t] \times F$ one has

(18.12.1) $$\lambda_{(X \circ M)(Y \circ N)}(]s,t] \times F) = \int_{]s,t] \times F} XY \, d\lambda_{MN}.$$

The left-hand side of (18.12.1) can be written (using $u < v$)

$$E\{((M_{u \vee t} - M_u)(N_{v \vee t} - N_v) - (M_{u \vee s} - M_u)(N_{v \vee s} - N_v)) 1_{G \cap H \cap F}\} =$$
$$= E\{((M_{v \vee t} - M_u)(N_{v \vee t} - N_v) - (M_{v \vee s} - M_u)(N_{v \vee s} - N_v)) 1_{G \cap H \cap F}\}$$

Using the martingale property of M and N, this simplifies to

$$E\{(M_{v \vee t} N_{v \vee t} - M_{v \vee s} N_{v \vee s}) 1_{G \cap H \cap F}\} = \int_{]s,t] \times F} 1_{[v,\infty[\times (G \cap H)} d\lambda_{MN},$$

which proves (18.12.1). □

The mappings

$$(X,Y) \rightsquigarrow \lambda_{(X \circ M)(Y \circ N)}(]s,t]) \times F = E(1_F \int_{]s,t]} X dM \int_{]s,t]} Y dN)$$

and

$$(X,Y) \rightsquigarrow \int_{]s,t] \times F} XY \, d\lambda_{MN}$$

are bilinear continuous mappings from $\Lambda^2(\mathscr{P}, M) \times \Lambda^2(\mathscr{P}, N)$ into $L^1(\Omega, \mathscr{F}, P)$ which, according to the beginning of the proof, coincide on a set of processes whose linear combinations are dense in $\Lambda^2(\mathscr{P}, M)$ and $\Lambda^2(\mathscr{P}, N)$. These two mappings are therefore identical.

18.13 Proposition. *Let* M *and* N *be martingales in* $\mathscr{M}^2$, $X \in \Lambda^2(\mathscr{P}, M)$ *and* $Y \in \Lambda^2(\mathscr{P}, N)$. *Then*

(18.13.1) $$\langle \int X dM, \int Y dN \rangle_t = \int_{]0,t]} X_s Y_s \, d\langle M, N \rangle_s$$

and

(18.13.2) $$[\int X dM, \int Y dN]_t = \int_{]0,t]} X_s Y_s \, d[M,N]_s,$$

where the integrals on the right-hand side of these equalities are Stieltjes integrals taken for every ω.

Proof. Proposition 18.12 says that the processes $\int X dM$, $\int Y dN$ and $(\int_{]0,t]} XYd\langle M,N\rangle)_{t\geq 0}$ possess the same Doléans measure. Since the latter is clearly predictable, this shows (18.13.1).

To prove (18.13.2), we use Corollary 18.9.4°. We leave it to the reader to prove (using 18.13.1) that $(\int X dM)^c = \int X dM^c$ and $(\int X dN)^c = \int Y dN^c$ (see E. 2 at the end of the chapter).

We may then write

$$[\int X dM, \int Y dN]_t = \int_{]0,t]} X_s Y_s d\langle M^c, N^c\rangle_s + \sum_{s\leq t} X_s Y_s \Delta M_s \Delta N_s$$

Again using 18.9.4°, we get (18.13.2). □

18.14 Proposition. *Let $M\in\mathcal{M}^2$, $X, Y\in\Lambda^2(\mathcal{P}, M)$ and τ be a stopping time. Then the equality $X 1_{]0,\tau]} = Y 1_{]0,\tau]}$ implies $1_{]0,\tau]}\int X dM = 1_{]0,\tau]}\int Y dM$.*

Proof. We clearly only have to prove the implication

$$X 1_{]0,\tau]} = 0 \Rightarrow 1_{]0,\tau]}\int X dM = 0 .$$

Proposition 18.12 gives

$$\forall t\in\mathbb{R}^+,\ \lambda_{(\int X dM)^2}(]0,\tau\wedge t]) = \int 1_{]0,\tau\wedge t]} X^2 d\lambda_{M^2} = 0 .$$

Therefore

$$\forall t\in\mathbb{R}^+,\ E(\int_{]0,t\wedge\tau]} X dM)^2 = 0 ,$$

which implies

$$\forall t\in\mathbb{R}^+,\quad \int_{]0,t\wedge\tau]} X dM = 0 \quad \text{a. s.}$$

We conclude this section with a representation theorem (see [Mey. 3]). □

18.15 Proposition. *Let $M\in\mathcal{M}^2_\infty$, $\mathcal{H}_1 := \{\int X dM : X\in L^2(\mathcal{P}, M)\}\subset\mathcal{M}^2_\infty$ and $\mathcal{H}_2$ be the subspace of $\mathcal{M}^2_\infty$ orthogonal to $\mathcal{H}_1$. Then every element of $\mathcal{H}_1$ is strongly orthogonal to every element of $\mathcal{H}_2$ and every martingale $L\in\mathcal{M}^2_\infty$ has a unique decomposition of the form*

$$L = \int X dM + N ,$$

where $X\in L^2(\mathcal{P}, M)$, $N\in\mathcal{H}_2$.

Proof. The isometry property of the L^2-stochastic integral shows that $\mathcal{H}_1$ is a Hilbert subspace of $\mathcal{M}^2_\infty$. The existence and uniqueness of the decomposition of L is therefore immediate.

Let us consider $N\in\mathcal{H}_2$. For every finite stopping time τ we may, on account of the martingale property, write

$$E(N_\tau \int_{]0,\infty]} X\,dM = E(N_\tau \int_{]0,\tau]} X\,dM) = E(N_\infty \int_{]0,\tau]} X\,dM)$$

but since $\int_{]0,\tau]} X\,dM = \int_{]0,\infty]} 1_{]0,\tau]} X\,dM$, we have

$$E(N_\infty \int_{]0,\tau]} X\,dM) = 0$$

and therefore

$$E(N_\tau \int_{]0,\infty]} X\,dM) = 0\,,$$

which shows that the martingale $(N_{\tau\wedge t})_{t\in\mathbb{R}^+}$ is in $\mathscr{H}_2$. The subspaces $\mathscr{H}_1$ and $\mathscr{H}_2$ therefore have the property called *stability* in [Mey. 3], which is defined as follows. If $N \in \mathscr{H}_2$ (resp. $\mathscr{H}_1$), then $(N_{\tau\wedge t})_{t\in\mathbb{R}^+}$ belongs to $\mathscr{H}_2$ (resp. $\mathscr{H}_1$) for every stopping time τ. It follows at once that $E(N_\tau H_\tau) = 0$ for every stopping time τ if $N \in \mathscr{H}_1$ and $H \in \mathscr{H}_2$. This expresses the strong orthogonality of $\mathscr{H}_1$ and $\mathscr{H}_2$. □

19. Stopped martingales Inequalities

Let $M \in \mathscr{M}^2(\mathscr{F})$ and let τ be a stopping time with respect to $\mathscr{F}$. According to the Stopping Theorem, Theorem 13.5, the process $(M_{t\wedge\tau})_{t\geq 0}$ is still an element of $\mathscr{M}^2(\mathscr{F})$ with $M_{\tau\wedge\infty} = M_\tau$. Let us assume $E(|M_\tau|^2) < \infty$. Then the Doob inequality 10.12 applied to the martingale $(M_{t\wedge\tau})_{t\geq 0}$ gives

$$E(\sup_{t\leq\tau} |M_t|^2) \leq 4E(|M_\tau|^2)$$

We then obtain the following property.

19.1 Theorem. *For every $M \in \mathscr{M}^2$ with $M_0 = 0$ and any stopping time τ we may write*

$$E(\sup_{t\leq\tau} |M_t|^2) \leq 4E(|M_\tau|^2) = 4E(\langle M\rangle_\tau) = 4E([M]_\tau)\,.$$

Proof. Theorem 18.6.3° says that, for every stopping time, $E(|M_\tau|^2) = E(\langle M\rangle_\tau) = E([M]_\tau)$. These three non-negative numbers are therefore finite together, if at all. When they are finite, the above reasonning shows the validity of the formula and when they are infinite the inequality is trivial. □

19.2 Application and remark. Let $M \in \mathscr{M}^{2,c}$ and $K > 0$. We define the following stopping time.

$$\tau := \inf\{t\colon \langle M\rangle_t > K\}$$

From the continuity of $\langle M\rangle$, it follows that $\langle M\rangle_\tau \leq K$ (with equality if $\tau < \infty$).

The above theorem then gives

$$E(\sup_{t\leqslant\tau}|M_t - M_0|^2) \leqslant 4K.$$

In many instances one has to answer questions of the following type. Let σ be a finite stopping time and $\varepsilon > 0$. Does there exist a stopping time τ, strictly greater than σ such that, in the stochastic inverval $[\sigma, \tau[$, the following inequality holds?

(19.2.1) $$E(\sup_{\sigma\leqslant t<\tau}|M_t - M_\sigma|^2) \leqslant \varepsilon$$

If $M \in \mathscr{M}^{2,c}$ and if σ is a stopping time such that $E(|M_\sigma|^2) < \infty$, we may say, after observing that $([M]_{t\vee\sigma} - [M]_\sigma)_{t\in\mathbb{R}^+}$ is the quadratic variation of the martingale $(M_{t\vee\sigma} - M_\sigma)_{t\in\mathbb{R}^+}$, that the inequality (19.2.1) is satisfied when τ is defined as follows.

$$\tau := \inf\{t\colon t > \sigma, [M]_{t\vee\sigma} - [M_\sigma] \geqslant \varepsilon\}$$

The right continuity of $[M]$ implies, moreover, that $\tau > \sigma$. Now, if M is not continuous, it is not possible to use the same reasoning because, with the same definition of τ, one has not, in general, $[M]_\tau - [M]_\sigma \leqslant \varepsilon$ but only $[M]_{\tau^-} - [M]_\sigma \leqslant \varepsilon$. This suggests the following question.

Let $M \in \mathscr{M}^2$. Do there exist inequalities of the type

(19.2.2) $$E(\sup_{t<\tau}|M_t - M_0|^2) \leqslant KE([M]_{\tau^-})$$

or

(19.2.3) $$E(\sup_{t<\tau}|M_t - M_0|^2) \leqslant KE(\langle M\rangle_{\tau^-})$$

for a suitable constant K and all stopping times?

It is immediately clear, as a consequence of 19.1, that, when M is continuous, (19.2.2) and (19.2.3) hold with $K = 4$, because in this case $[M]$ and $\langle M\rangle$ are continuous.

If $M \in \mathscr{M}^2$ has jumps only at totally inaccessible stopping times, we know that $\langle M\rangle$ is continuous (Corollary 15.2). Therefore, (19.2.3) still holds in this case with $K = 4$ but there may exist no constant K such that (19.2.2) holds, as shown by exercise E. 6 at the end of the chapter.

Let τ now be a finite predictable stopping time. Then there exists an increasing family (τ_n) of stopping times such that $\tau_n < \tau$ a. s. and $\tau = \lim_n \tau_n$ a. s. We may therefore write

$$\begin{aligned}
E(\sup_{t<\tau}|M_t|^2) &\leqslant \sup_n E(\sup_{t\leqslant\tau_n}|M_t|^2)\\
&\leqslant 4\sup_n E(\langle M\rangle_{\tau_n})\\
&= 4\sup_n E([M]_{\tau_n})\\
&= 4E(\langle M\rangle_{\tau^-}) = 4E([M]_{\tau^-}).
\end{aligned}$$

We then see that (19.2.2) and (19.2.3) hold for any $M \in \mathcal{M}^2$ and any predictable τ with $K = 4$ but, unfortunately, as shown by exercises E. 5 and E. 6 at the end of the chapter, when M has jumps at stopping times which are accessible but non-predictable, then one may have neither (19.2.2) nor (19.2.3) for any value of K.

We may, however, state inequalities which give a useful bound to the first members of (19.2.2) and (19.2.3) and the end of this section is concerned with doing so.

First we give a definition.

19.3 Definition. *The pure jump-martingale part of* $M \in \mathcal{M}^2$. Let $M \in \mathcal{M}^2$ and let (T_n) be a sequence of predictable stopping times with mutually disjoint graphs such that, for every predictable stopping time σ with the property $\sum_n P\{T_n = \sigma < \infty\} = 0$, one has $|\Delta M_\sigma| = 0$ a. s. (such a sequence exists according to section 7.6 and the process $\sum_n |\Delta M_{T_n}| 1_{[T_n, \infty]}$ is uniquely defined, independently of the particular sequence (T_n) considered). We set

$$\check{M} := \sum_n \Delta M_{T_n} 1_{[T_n, \infty[}$$

We know that $\check{M}$ is a martingale in $\mathcal{M}^2$ (see Theorem 17.7, where if T_n is predictable, one has $\tilde{A}_n = 0$) and we shall call $\check{M}$ the *pure jump-martingale* part of M.

This $\check{M}$ is to be distinguished from the purely discontinuous part M^d of M as defined in 17.8.

19.4 Theorem. *Let* $M \in \mathcal{M}^2$ *with* $M_0 = 0$. *For every stopping time* τ *one has*

$$E(\sup_{t<\tau} |M|^2) \leqslant 4E(\langle M \rangle_{\tau-} + [\check{M}]_{\tau-})$$

Proof. We write $N = M - \check{M}$. Since N has jumps only at totally inaccessible stopping times, $\langle N \rangle$ is continuous. Moreover, $\langle M \rangle = \langle N \rangle + \langle \check{M} \rangle$ because of the strong orthogonality of N and $\check{M}$ (see section 17).

Our proof will consist in showing, for every stopping time τ, the existence of a martingale $W \in \mathcal{M}^2$ for which

(i) $\quad 1_{[0,\tau[} W = 1_{[0,\tau[} \check{M}$,

(ii) $\quad E([W]_\tau) = E(\langle W \rangle_\tau) \leqslant E([\check{M}]_{\tau-} + \langle \check{M} \rangle_{\tau-})$,

(iii) $\quad W$ is trongly orthogonal to N.

If such a martingale W exists, the latter property will imply $\langle N + W \rangle = \langle N \rangle + \langle W \rangle$ and, on account of (i) and the Doob inequality 19.2,

$$E(\sup_{t<\tau} |M_t|^2) = E(\sup_{t<\tau} |N_t + W_t|^2)$$
$$\leqslant E(\sup_{t\leqslant\tau} |N_t + W_t|^2) \leqslant 4E(\langle N \rangle_\tau + \langle W \rangle_\tau).$$

Property (ii) will then establish the inequality of the theorem, which will therefore follow immediately from the next proposition.

19.5 Proposition. *Let $M \in \mathcal{M}^2$ and let (T_n) be a sequence of predictable stopping times as in 19.3. We set $M^n := \Delta M_{T_n} 1_{[T_n, \infty[}$. Then, given a stopping time τ, there exists a sequence (W^n) of martingales in $\mathcal{M}^2$, which are mutually strongly orthogonal, orthogonal to $N := M - \check{M}$ and possess the following properties.*

(a) *For every n one has $1_{[0,\tau[} W^n = 1_{[0,\tau[} M^n$.*

(b) *$E(\langle W^n \rangle_\tau) = E([W^n]_\tau) \leqslant E([M^n]_{\tau-} + \langle M^n \rangle_{\tau-})$.*

(c) *The sum $\sum_n W^n$ converges in $\mathcal{M}^2$ to a martingale W, satisfying* (i), (ii) and (iii) above.

Proof. Let $A := \{T_n < \tau\}$ and $B := \Omega \backslash A$. We define $\mathcal{G} := \mathcal{F}_{T_n^-}$ and write $\mathcal{G}^*$ for the the σ-algebra generated by $\mathcal{G}$ and A.

We set

$$\begin{aligned}(19.5.1)\qquad h &:= \Delta M_{T_n} 1_B - E(\Delta M_{T_n} 1_B | \mathcal{G}^*)\\ &= 1_B(\Delta M_{T_n} - E(\Delta M_{T_n} | \mathcal{G}^*))\end{aligned}$$

Since $E(h|\mathcal{G}^*) = 0$ one has, a fortiori, $E(h|\mathcal{G}) = 0$. As a consequence, the process $h 1_{[T_n, \infty[}$ is a square integrable martingale. (cf. Corollary 13.8 and 16.2.3).

We set

$$(19.5.2)\qquad W^n := (\Delta M_{T_n} - h)\, 1_{[T_n, \infty[}$$

and show that this martingale has the properties (i) to (iii). By virtue of (19.5.1) we may write

$$(19.5.3)\qquad W^n := 1_A \Delta M_{T_n} 1_{[T_n, \infty[} + 1_B E(\Delta M_{T_n} 1_B | \mathcal{G}^*)\, 1_{[T_n, \infty[}.$$

The validity of (i) is immediate from this formula and the definition of B.

The Meyer process of M^n is (cf. 15.7.4°)

$$(19.5.4)\qquad \langle M^n \rangle = E(|\Delta M_{T_n}|^2 | \mathcal{F}_{T_n^-})\, 1_{[T_n, \infty[}.$$

From the definition of W^n we derive

$$E([W^n]_\tau) = E(1_A |\Delta M_{T_n}|^2 + 1_{\{T_n = \tau\}} |E(\Delta M_{T_n} | \mathcal{G}^*)|^2)$$

and then

$$(19.5.5)\qquad E([W^n]_\tau) \leqslant E(1_A |\Delta M_{T_n}|^2 + 1_B |E(\Delta M_{T_n} | \mathcal{G}^*)|^2).$$

Let us write for convenience $Z := E(\Delta M_{T_n} | \mathcal{G}^*)$, $a := E(1_A | \mathcal{G})$, $b := E(1_B | \mathcal{G})$. The definition of $\mathcal{G}^*$ shows we can write $Z = 1_A X + 1_B Y$, where X and Y are $\mathcal{G}$-measurable random variables. The relation $E(Z|\mathcal{G}) = E(\Delta M_{T_n} | \mathcal{G}) = 0$ gives $aX + bY = 0$. Since $Z^2 = X^2 1_A + Y^2 1_B$ one has

$$E(1_A E(Z^2 | \mathcal{G})) = E(1_A(aX^2 + bY^2)) = E(a^2 X^2 + abY^2)$$

and from $aX = -bY$ we deduce (note that $a+b=1$)

(19.5.6) $$E(1_A E(Z^2|\mathcal{G})) = E(Y^2(b^2+ab)) = E(Y^2 b).$$

Analogously we write

(19.5.7) $$E(Z^2 1_B) = E(Y^2 1_B) = E(Y^2 b).$$

The three inequalities (19.5.5) to (19.5.7) together give

(19.5.8) $$E([W^n]_\tau) \leqslant E(1_A|\Delta M_{T_n}|^2 + 1_A E(Z^2|\mathcal{G}))$$

In view of (19.5.4) and Jensen's inequality,

$$1_A E(Z^2|\mathcal{G}) \leqslant 1_A E(|\Delta M|^2_{T_n}|\mathcal{G}) = \langle M^n\rangle_{\tau-}$$

The inequality (19.5.8) is therefore nothing but

(19.5.9) $$E([W^n]_\tau) \leqslant E([M^n]_{\tau-} + \langle M^n\rangle_{\tau-})$$

The strong orthogonality of all the W^n's and N is trivial from their definition and the convergence of the sequence (W^n) to a W follows from

$$\sum_n E(|W^n_\infty|^2) = \sum_n E(|W^n_\tau|^2) \leqslant \sum_n E(|\Delta M_{T_n}|^2) < \infty .$$

From the strong orthogonality of the W^n's it follows that

$$\langle W\rangle = \sum_n \langle W^n\rangle ,$$

and, since the jumps of W^n occur at stopping times with disjoint graphs, $[W] = \sum [W^n]$.

The property (a) of the martingales W^n then implies (i) for W while (b) implies (ii). This completes the proof of the proposition and therefore of Theorem 19.4. □

20. Spaces of Hilbert valued martingales

$\mathbb{H}$ denotes a Hilbert space with scalar product denoted by $x \cdot y$.

20.1 Immediate extensions of the real case

It is clear that if we define $\mathcal{M}^2_\infty(\mathbb{H})$ (or more specifically $\mathcal{M}^2_\infty(\mathbb{H};\mathcal{F})$, if we want to recall the filtration in question) as the vector space of $\mathbb{H}$-valued martingales M such that $E\|M_\infty\|^2 < \infty$ (defined up to P-equality), exactly the same considerations as in the real case show that $\mathcal{M}^2_\infty(\mathbb{H})$, endowed with scalar product $(M|N) := E(M_\infty \cdot N_\infty)$, is a Hilbert space.

One may define in exactly the same way the spaces $\mathcal{M}^2(\mathbb{H};\mathcal{F})$, $\mathcal{M}^{2,c}_\infty(\mathbb{H},\mathcal{F})$, $\mathcal{M}^{2,c}(\mathbb{H};\mathcal{F})$, these latter two spaces being the complete subspaces of $\mathcal{M}^2_\infty(\mathbb{H};\mathcal{F})$ and $\mathcal{M}^2(\mathbb{H};\mathcal{F})$ respectively consisting of continuous martingales.

Lemma 16.3 holds in exactly the same way for sequences $(M^n)_{n\in\mathbb{N}}$ *in* $\mathcal{M}^2_\infty(\mathbb{H})$ (resp. $\mathcal{M}^2(\mathbb{H})$).

The *admissible real measure* of $M \in \mathcal{M}^2(\mathbb{H})$ is the (real) Doléans measure $\lambda_{\|M\|^2}$ of the real submartingale $\|M\|^2$. We shall continue to denote it by μ_M.

Two martingales M and N in $\mathcal{M}^2(\mathbb{H})$ will be called *strongly orthogonal* if the Doleans measure $\lambda_{M\cdot N}$ of the real process $M\cdot N = \frac{1}{4}[(M+N)\cdot(M+N) - (M-N)\cdot(M-N)]$ is zero.

We denote by $\langle M, N\rangle$ the predictable, increasing process of $\lambda_{M\cdot N}$ resulting from Corollary 15.4. It is the unique (up to P-equivalence) predictable, real process V with paths of finite variation such that $V_0 = 0$ and $M\cdot N - V$ is a real martingale.

If M or N is continuous or, more generally, if $\Delta(M\cdot N)_T = 0$ a.s. for every predictable stopping time, $\langle M, N\rangle$ is continuous. The process $\langle M, M\rangle$ also denoted by $\langle M\rangle$) is still called the *first increasing process of* M (or *Meyer process of* M).

Proposition 17.4 holds without change except that the ordinary product $M_t N_t$ of real variables is replaced by the scalar product $M_t\cdot N_t$ when they are Hilbert-valued.

Finally, as in the real case, we have the following structure theorem.

20.2 Theorem*. *For every* $M\in\mathcal{M}^2(\mathbb{H})$ *(resp.* $\mathcal{M}^2_\infty(\mathbb{H})$*) there exists a sequence* $(T_n)_{n\geq 0}$ *of stopping times with disjoint graphs and a unique (up to P-equivalence)* $M^c\in\mathcal{M}^{2,c}(\mathbb{H})$ *(resp.* $M^{2,c}_\infty(\mathbb{H})$*) with the following properties.*

(1) *If* $A_n := \Delta M_{T_n} 1_{[T_n,\infty[}$, *if* $\tilde{A}_n$ *is the dual predictable projection of* A_n *and the martingales* $M^c, A_1 - \tilde{A}_1, \ldots A_n - \tilde{A}_n, \ldots$ *are mutually strongly orthogonal*[1]), *then for every* $t\in\mathbb{R}^+$ we have

$$E(\|M_t\|^2) = E(\|M_t^c\|^2) + \sum_n E(\|A_n(t) - \tilde{A}_n(t)\|^2)$$

and M *is the sum in* $\mathcal{M}^2(\mathbb{H})$ *(resp.* $\mathcal{M}^2_\infty(\mathbb{H})$*) of the family* $\{M^c, A_1 - \tilde{A}_1, \ldots, A_n\ \tilde{A}_n, \ldots\}$.

(2) *When* $M\in\mathcal{M}^2_\infty(\mathbb{H})$ *one has, moreover,* $E(\|A_n(\infty) - \tilde{A}_n(\infty)\|^2) = E(\|\Delta M_{T_n}\|^2)$ *for every n and*

$$\sum_n E(\|\Delta M_{T_n}\|^2\, 1_{\{T_n<\infty\}} = \|\sum_n (A_n - \tilde{A}_n)\|^2_{\mathcal{M}^2_\infty(\mathbb{H})} < \infty\,.$$

Proof. Entirely analogous to the proof of 17.7.

The martingale $M - M^c$ will still be called the purely discontinuous part of M and be denoted by M^d. □

20.3 Example. Let $Y(x,t)$ describe the value of a variable depending on x, for x belonging to some domain G, and time t. (For example, a physical parameter in a given volume at time t). Random perturbations at time t produce, during the interval of time Δt, a perturbation of Y given by

---

[1]) and even very strongly orthogonal (see E. 7).

$$Y(x, t + \Delta t) = Y(x, t) + f(x, t)\Delta_t,$$

where Δ_t is a centered scalar random variable and $f(x, t)$ a given function.

If the perturbation for the interval $[t, t + \Delta t]$ is produced by a number of independent small perturbations, an idealized modeling of the phenomenon is obtained by writing

$$Y(x, t) = Y(x, 0) + \int_0^t f(x, s)\, dW_s,$$

where $\int_0^t f(x, s)\, dW_s$ is to be understood for each x as the stochastic integral with respect to the real Brownian motion W.

If $f(., s)$, as a function of x, belongs to some Hilbert space $\mathbb{H}$ (for example $L^2(G)$ or a Sobolev space of functions, as used in the theory of partial differential equations) and if, for example, $\sup \|f(., s)\|_{\mathbb{H}}$ is bounded, it is easy to see that $(Y(., t))_{t \geq 0}$ is an $\mathbb{H}$-valued process and actually an element of $\mathcal{M}^{2,c}_\infty(\mathbb{H}, \mathcal{F})$, where $\mathcal{F}$ is the filtration generated by W.

It is, moreover, a very easy extension of § 18 to see that, for every $\mathbb{H}$-valued process f such that $E\left(\int_0^\infty \|f(., t)\|^2_{\mathbb{H}}\, dt\right) < \infty$, the stochastic integral with respect to W is defined for this latter norm as the continuous extension of the mapping $f \to \int f(., s) dN_s$ (with values in $\mathcal{M}^2_\infty(\mathbb{H}, \mathcal{F})$) when $\int f dW$ is given by

$$\int_0^t f(., s)\, dW_s := \sum_i \varphi_i(.)\, 1_{F_i}(W_{t_i \wedge t} - W_{s_i \wedge t}),$$

f having the form

$$f(x, t) = \sum_i \varphi_i(x)\, 1_{]s_i, t_i] \times F_i}, \quad \varphi_i \in \mathbb{H}, \quad F_i \in \mathcal{F}_{s_i}.$$

20.4 L^2-stochastic Integral

This example suggests that the L^2-stochastic integral with respect to Hilbert-valued martingales can be defined exactly as in the real case. In fact, if μ_M is the admissible measure of $M \in \mathcal{M}^2_\infty(\mathbb{H})$ and $X := \sum_{i=1}^n a_i\, 1_{]s_i t_i] \times F_i}$, $F_i \in \mathcal{F}_i$, where a_i is a linear bounded operator from $\mathbb{H}$ into some other Hilbert space $\mathbb{G}$, it is easy to see that the process

$$N_t := \sum_{i=1}^n 1_{F_i}(a_i(M_{t_i \wedge t}) - a_i(M_{s_i \wedge t}))$$

is a $\mathbb{G}$-valued martingale with

$$E(\|N_\infty\|_{\mathbb{G}}^2) := \sum_{i=1}^{n} E(1_{F_i} \cdot \|a_i(M_{t_i}) - a_i(M_{s_i})\|_{\mathbb{G}}^2)$$
$$\leqslant \sum_{i=1}^{n} \|a_i\|^2 \mu_M(]s_i, t_i] \times F_i) = \int_{\mathbb{R}^+ \times \Omega} \|X\|^2 \, d\mu_M$$

There is therefore a continuous extension of the mapping $X \to N$ from $\mathscr{L}^2(\mathbb{R}^+_{\mathbb{H}} \times \Omega, \mathscr{P}, \mu_M)$ into $\mathscr{M}^2_\infty(\mathbb{H})$ which is no longer an isometry except if $\mathbb{H} = \mathbb{G} = \mathbb{R}$. This is a contraction.

In some sense, therefore, this definition of the integral is too crude and uses too strong a norm on the space of processes to be integrated.

In § 21 below we define a more appropriate integral (which naturally coincides with the one just defined when this latter exists). However, since it is more elaborate we first mention some useful results which do not need it.

Quadratic variation and mutual quadratic variation

20.5 Theorem. *Let M and N be two martingales in $\mathscr{M}^2(\mathbb{H})$. There exists a real R. R. C. process with paths of finite variation, denoted by $[M, N]$, which is uniquely defined up to P-equality with the following properties.*

(1) *For every increasing sequence (Π_n) of subdivisions of $\mathbb{R}^+$ as in Theorem 18.6,*

(20.5.1) $$[M, N]_t = \lim_n (L^1) \sum_{t_i \in \Pi_n} (M_{t_{i+1} \wedge t} - M_{t_i \wedge t}) \cdot (N_{t_{i+1} \wedge t} - N_{t_i})$$

(2) *$M \cdot N - [M, N]$ is a martingale.*

(3) *The processes $M \cdot N$, $\langle M, N\rangle$, $[M, N]$ admit the same Doleans measure.*

(4) *For every t*

$$[M, N]_t = \langle M^c, N^c\rangle_t + \sum_{s \leqslant t} \Delta M_s \cdot \Delta N_s \quad \text{a.s.,}$$

the real series on the right-hand side being a. s. summable.

Proof. The proof is exactly as in 18.6 and 18.9: one has only to replace the ordinary product xy of real numbers by the scalar product $x \cdot y$ in $\mathbb{H}$. We therefore omit it. $[M, N]$ is the *mutual quadratic variation of M and N* while $[M, M]$ is the *quadratic variation of M*.

Inequalities for stopped martingales

The definitions and arguments in § 19 apply to martingales in $\mathscr{M}^2(\mathbb{H})$. Just as in 19.3 we define the *pure jump martingale part* of $M \in \mathscr{M}(\mathbb{H})$ and obtain the following theorem.

20.6 Theorem. *Let $M \in \mathcal{M}(\mathbb{H})$ with $M_0 = 0$. For every stopping time τ one has*

(1) (Doob Inequality) $E(\sup_{t \leq \tau} \|M_t\|^2) \leq 4E(\langle M \rangle_\tau) = 4E([M]_\tau) = 4E(\|M_\tau\|^2)$

(2) *(cf. Metivier-Pellaumail* [MeP.3]*)* $E(\sup_{t < \tau} \|M_t\|^2) \leq 4E(\langle M \rangle_{\tau^-} + [\check{M}]_{\tau^-})$

Proof. As we mentioned, the proof is exactly the same as in § 19. □

21. The process $\langle\langle M \rangle\rangle$ of a square integrable Hilbert-valued martingale

21.1 Introduction

To have some feeling for what is to be done, consider finite dimensional Hilbert spaces $\mathbb{H}$ and $\mathbb{G}$ with orthonormal basis in which the coordinates of M are written M^k and a_i is represented by a matrix $a_i^{k\ell}$. Considering the Doleans measure $\alpha_M^{\ell,r}$ of $M^\ell M^r$ we easily find from the definition of N above that

$$\begin{aligned} E(\|N_\infty\|_{\mathbb{G}}^2) &= \sum_{k=1}^{n} \sum_{k\,l,r} 1_{F_i} a_i^{kl} E((M_{t_i}^l - M_{s_i}^l)(M_{t_i}^r - M_{s_i}^r))\, a_i^{kr} \\ &= \sum_i \sum_{l,r} (a_i \circ a_i^*)^{l,r}\, \alpha_M^{l,r}(]s_i, t_i] \times F_i) \\ &= \sum_{l,r} \int (X_i \circ X_i^*)^{lr}\, d(\alpha_M^{l,r})\,. \end{aligned}$$

This equality illustrates, in particular, the purpose of introducing the matrix-valued Doleans measure (with values in $\mathbb{H} \otimes \mathbb{H}$) associated with the quasimartingale $(M_t \otimes M_t)_{t \geq 0}$, where $x \otimes y$ denotes the matrix $[x^i x^j]$.

This is what we are going to do now but first we recall some facts about tensor products of Hilbert and Banach spaces. For details on tensor-product spaces, the reader is referred to the books by A. H. Schaeffer [Sch], F. Treves [Tre] and W. Rudin [Rud].

A review of operators and Tensor products for Hilbert and Banach spaces

21.2 Tensor products

(1) $\mathbb{H} \hat{\otimes}_2 \mathbb{H}$: *the Hilbert-Schmidt tensor product.*

$\mathbb{H}$ being a real Hilbert space, we write $x \cdot y$ for the scalar product of x and y in $\mathbb{H}$. If several Hilbert spaces are being considered, then we write $\langle x, y \rangle_{\mathbb{H}}$ instead of $x \cdot y$ to prevent confusion.

We shall write $\mathbb{H} \otimes \mathbb{H}$ for the tensor product of $\mathbb{H}$ with itself, denoting by $x \otimes y$ the tensor product of $x \in \mathbb{H}$ and $y \in \mathbb{H}$. If $x = y$, we shall write $x \otimes x = x^{\otimes 2}$.

Let $(x_i, y_i)_{i \in I}$ be a finite family of pairs of elements of $\mathbb{H}$ and let $z = \sum_{i \in I} x_i \otimes y_i$

be the element of $\mathbb{H} \otimes \mathbb{H}$ associated with this family. We likewise consider $z' = \sum_{j \in J} x'_j \otimes y'_j$.

By putting
$$\langle z, z' \rangle = \sum_{i \in I} \sum_{j \in J} \langle x_i, x'_j \rangle \cdot \langle y_i, y'_j \rangle ,$$

we define a scalar product on $\mathbb{H} \otimes \mathbb{H}$.

We shall denote by $\mathbb{H} \hat{\otimes}_2 \mathbb{H}$ the space $\mathbb{H} \otimes \mathbb{H}$ completed for the norm associated with this scalar product. The corresponding norm on $\mathbb{H} \hat{\otimes}_2 \mathbb{H}$ is called the Hilbert-Schmidt norm and will be denoted by $\|.\|_2$. With this canonical extension of the scalar product defined above, $\mathbb{H} \hat{\otimes}_2 \mathbb{H}$ is a separable Hilbert space. If $(k_n)_{n \geq 0}$ is an orthonormal basis of $\mathbb{H}$, $(k_n \otimes k_m)_{n,m \geq 0}$ is an orthonormal basis of $\mathbb{H} \hat{\otimes}_2 \mathbb{H}$. If x and y are two elements of $\mathbb{H}$, we have

$$\|x \otimes y\| = \|x\|_{\mathbb{H}} \cdot \|y\|_{\mathbb{H}} .$$

The mapping $(x, y) \rightsquigarrow x \otimes y$ from $(\mathbb{H} \times \mathbb{H})$ into $(\mathbb{H} \hat{\otimes}_2 \mathbb{H})$ is a continuous bilinear mapping.

All the previous properties are well known and easy to prove. We recall also that, if $\mathbb{H}$ is d-dimensional, $\mathbb{H} \otimes \mathbb{H} = \mathbb{H} \hat{\otimes}_2 \mathbb{H}$ is isomorphic to the space of all $d \times d$ matrices. More precisely, let $(h_n)_{1 \leq n \leq d}$ be an orthonormal basis of $\mathbb{H}$, $(x_i, y_i)_{i \in I}$ be a finite family of pairs of elements of $\mathbb{H}$, with $x_i = \sum_{n=1}^{d} x_{i,n} h_n$ and $y_i = \sum_{n=1}^{d} y_{i,n} h_n$, and (X^i, Y^i) be the pair of matrices defined by $X^i = \begin{pmatrix} x_{i,1} \\ \vdots \\ x_{i,d} \end{pmatrix}$ and $Y^i = \begin{pmatrix} y_{i,1} \\ \vdots \\ y_{i,d} \end{pmatrix}$. Then the one-to-one mapping which assigns the $d \times d$ matrix $((\sum_{i \in I} x_{i,j} y_{i,k}))_{j,k} = \sum_{i \in I} X^i (Y^i)^{tr}$ to the element $(\sum_{i \in I} x_i \otimes y_i)$ of $\mathbb{H} \otimes \mathbb{H}$, is an isomorphism from $\mathbb{H} \otimes_2 \mathbb{H}$ into the vector space of all $d \times d$ matrices, with the Hilbert norm $|\sum_{i \in I} \text{trace } (X^i (X^i)^{tr})|^{1/2}$.

(2) $\mathbb{B} \hat{\otimes}_1 \mathbb{B}$ *for a Banach space* $\mathbb{B}$.

$\mathbb{B} \hat{\otimes}_1 \mathbb{B}$ can be defined for every Banach space $\mathbb{B}$ as the completion of $\mathbb{B} \otimes \mathbb{B}$ for a norm (written $\|.\|_1$) with the following property.

For every Banach space $\mathbb{G}$ and every continuous bilinear mapping b from $\mathbb{B} \times \mathbb{B}$ into $\mathbb{G}$, there exists a unique, continuous (for the norm $\|.\|_1$) linear mapping $\bar{b}$: $\mathbb{B} \hat{\otimes} \mathbb{B} \to \mathbb{G}$ such that $b(x, y) = \bar{b}(x \otimes y)$.

In other words, every bilinear mapping b can be factored as $\bar{b} \circ \pi$, where $\pi(x, y) = x \otimes y$, and $\bar{b}$ linear is continuous on $\mathbb{B} \hat{\otimes}_1 \mathbb{B}$.

Taking $\mathbb{G} = \mathbb{R}$, we see in particular that the dual $(\mathbb{B} \hat{\otimes}_1 \mathbb{B})'$ of $\mathbb{B} \hat{\otimes}_1 \mathbb{B}$ is isomorphic to the Banach space of continuous bilinear forms, with the usual norm.

For a Hilbert space $\mathbb{H}$, there is a continuous linear injection from $\mathbb{H} \hat{\otimes}_1 \mathbb{H}$ into $\mathbb{H} \hat{\otimes}_2 \mathbb{H}$, which extends the identical mapping $h \otimes g \rightsquigarrow h \otimes g$. In other words, $\|b\|_2 \leq \|b\|_1$ for every $b \in \mathbb{H} \hat{\otimes}_1 \mathbb{H}$, with the equality holding for every b which can be written $b = h \otimes g$. In this case, $\|h \otimes g\|_1 = \|h \otimes g\|_2 = \|h\|_{\mathbb{H}} \|g\|_{\mathbb{H}}$.

We recall also that the linear form trace on $\mathbb{H} \hat{\otimes}_1 \mathbb{H}$ is defined as the unique continuous linear extension to $\mathbb{H} \hat{\otimes}_1 \mathbb{H}$ of the mapping $h \otimes g \rightsquigarrow h \cdot g$.

Let us write $\langle x, x' \rangle$ for the canonical bilinear form on $\mathbb{B} \times \mathbb{B}'$. From the definition of $\mathbb{B} \hat{\otimes}_1 \mathbb{B}$ and $\mathbb{B}' \hat{\otimes}_1 \mathbb{B}'$, it is immediately seen that the mapping $(x \otimes y, x' \otimes y') \rightsquigarrow \langle x, x' \rangle \cdot \langle y, y' \rangle$ extends uniquely to a continuous bilinear form $\langle .,. \rangle$ on $(\mathbb{B} \hat{\otimes}_1 \mathbb{B}) \times (\mathbb{B}' \hat{\otimes}_1 \mathbb{B}')$.

Thus, $\mathbb{B} \hat{\otimes}_1 \mathbb{B}$ appears isomorphic to a subspace of the dual space $(\mathbb{B}' \hat{\otimes}_1 \mathbb{B}')'$.

This gives a meaning to the notion of positive and symmetric elements of $\mathbb{B} \hat{\otimes}_1 \mathbb{B}$: the element $b \in \mathbb{B} \hat{\otimes}_1 \mathbb{B}$ is said to be *positive* if $\langle b, x \otimes x \rangle \geqslant 0$ for all $x \in \mathbb{B}'$, and *symmetric* if $\langle b, x \otimes y \rangle = \langle b, y \otimes x \rangle$ for all $x, y \in \mathbb{B}'$.

21.3 Hilbert-Schmidt operators in a Hilbert space

With every $b \in \mathbb{H} \hat{\otimes}_2 \mathbb{H}$, it is possible to associate the bilinear form $(h, g) \rightsquigarrow \langle b, h \otimes g \rangle$. The tensor b is said to be *positive* if this bilinear form is such that $\langle b, h \otimes h \rangle \geqslant 0$ for all $h \in \mathbb{H}$, and it is called *symmetric* if $\langle b, h \otimes g \rangle = \langle b, g \otimes h \rangle$ for all g and h in $\mathbb{H}$.

$\mathbb{H} \hat{\otimes}_2 \mathbb{H}$ can thus be considered as a particular set of continuous bilinear forms on $\mathbb{H} \times \mathbb{H}$. With every bilinear form b on $\mathbb{H} \times \mathbb{H}$, there is a one-to-one association with a continuous linear operator $\tilde{b}$ through the formula

$$b(h, g) = \tilde{b}(h) \cdot g, \quad \text{all} \quad h, g \in \mathbb{H}.$$

The linear operator $\tilde{b}$ is called a *Hilbert-Schmidt operator* if $b \in \mathbb{H} \hat{\otimes}_2 \mathbb{H}$ and $\|b\|_2$ is its Hilbert-Schmidt norm. It is called self-adjoint if b is symmetric.

If we consider an orthonormal basis (h_n) of $\mathbb{H}$, we have the following useful characterization of Hilbert-Schmidt operators: *$\tilde{b}$ is a Hilbert-Schmidt operator iff $\sum_n \|\tilde{b}(h_n)\|^2 < \infty$ and, moreover, $\|\tilde{b}\|_2^2 = \sum_n \|b(h_n)\|$* (see, for example, Treves [Tre] or [GeW]).

For every self-adjoint Hilbert-Schmidt operator b in $\mathbb{H}$ there exists an orthogonal basis (h_n) in $\mathbb{H}$ such that

$$b(h) = \sum_n \lambda_n (h \cdot h_n) h_n, \quad \forall h \in \mathbb{H}, \tag{21.3.1}$$

where the λ_n are real numbers called the eigenvalues of $\tilde{b}$ and are such that

$$\sum_n \lambda_n^2 = \|\tilde{b}\|_2^2 < \infty.$$

These eigenvalues are positive if $\tilde{b}$ is positive.

The conjugate $\tilde{b}^*$ of a Hilbert-Schmidt operator, i.e., the operator $\tilde{b}^*$ defined by $\tilde{b}(h) \cdot g = h \cdot \tilde{b}^*(g)$, is equally Hilbert-Schmidt, with $\|\tilde{b}^*\|_2 = \|\tilde{b}\|_2$.

We denote by $\mathscr{L}_2(\mathbb{H}; \mathbb{H})$ the Hilbert space of Hilbert-Schmidt operators. From the above, the scalar product in $\mathscr{L}_2(\mathbb{H}; \mathbb{H})$ is expressed by

$$\langle \tilde{b}_1, \tilde{b}_2 \rangle = \operatorname{trace} \tilde{b}_1 \circ \tilde{b}_2^*$$

Starting form the formula $\langle h \otimes g, h' \otimes g' \rangle = (h \cdot h')(g \cdot g')$, it is possible, in an entirely analogous way, to define the space $\mathbb{H} \hat{\otimes}_2 \mathbb{G}$ and hence the space $\mathscr{L}_2(\mathbb{H}; \mathbb{G})$ of Hilbert-Schmidt operators from $\mathbb{H}$ into $\mathbb{G}$. Those operators $\tilde{b}$ are still characterized by the property

$$\sum_n \|\tilde{b}(h_n)\|^2 = \|\tilde{b}\|_2^2 < \infty ,$$

for every orthogonal basis (h_n) in $\mathbb{H}$.

21.4 Nuclear Operators

Let us now consider the tensor product $\mathbb{B} \hat{\otimes}_1 \mathbb{B}$, where $\mathbb{B}$ is a Banach space. Using a remark at the end of 21.2 according to which the elements of $\mathbb{B} \hat{\otimes}_1 \mathbb{B}$ are continuous linear forms on $\mathbb{B}' \hat{\otimes}_1 \mathbb{B}'$, we can associate with every element b of $\mathbb{B} \hat{\otimes}_1 \mathbb{B}$ a continuous linear operator $\tilde{b}$ from $\mathbb{B}'$ into $\mathbb{B}$, uniquely defined by

$$\langle \tilde{b}(h), g \rangle = \langle b, h \otimes g \rangle, \quad \text{when} \quad \mathbb{B} \quad \text{is reflexive.}$$

Such an operator is called *nuclear*.

Let us recall also that, if q_1 and q_2 are elements of $\mathscr{L}(\mathbb{B}; \mathbb{K})$, $q_1 \otimes q_2$ denotes the element of $\mathscr{L}(\mathbb{B} \hat{\otimes}_1 \mathbb{B}; \mathbb{K} \hat{\otimes}_1 \mathbb{K})$ which is the unique continuous linear extension of $x \otimes y \rightsquigarrow q_1(x) \otimes q_2(y)$.

Case of Hilbert spaces: Identifying $\mathbb{H}'$ with $\mathbb{H}$ as usual when $\mathbb{H}$ is a Hilbert space, we see that $\tilde{b} \in \mathscr{L}(\mathbb{H}'; \mathbb{H})$ is a *nuclear operator* iff the linear form $h \otimes g \rightsquigarrow \tilde{b}\, h \cdot g$ can be identified with an element b of $\mathbb{H} \hat{\otimes}_1 \mathbb{H}$.

Self-adjoint nuclear operators are characterized by the following property. $\tilde{b}$ is a self-adjoint nuclear operator iff there exists an orthonormal basis (h_n) such that

(21.4.1) $$\tilde{b}(h) = \sum_n \lambda_n (h \cdot h_n)\, h_n, \; \forall h \in \mathbb{H},$$

where the λ_n's are real numbers, called the eigenvalues, such that $\sum |\lambda_n| < \infty$.

One can check that, if b is the element of $\mathbb{H} \hat{\otimes}_1 \mathbb{H}$ associated with $\tilde{b}$, we have

(21.4.2) $$\operatorname{Trace}(\tilde{b}) = \sum_n \lambda_n$$

and

(21.4.3) $$\|\tilde{b}\|_1 = \sum_n |\lambda_n| .$$

The formulas (21.4.1) and (21.4.3) show that every positive self-adjoint operator $\tilde{b}$ can be written

$$\tilde{b} = \tilde{q} \circ \tilde{q}, \; \tilde{q} \in \mathscr{L}_2(\mathbb{H}; \mathbb{H}),$$

the eigenvalues of $\tilde{q}$ being the square roots of those of $\tilde{b}$. In this case, we write

(21.4.4) $$\tilde{q} = \tilde{b}^{\frac{1}{2}} .$$

We shall write $\mathscr{L}_1(\mathbb{H};\mathbb{H})$ for the space of nuclear operators in $\mathbb{H}$.

Finally, we mention the following properties.

(a) If $q \in \mathscr{L}_2(\mathbb{H};\mathbb{H})$, then $q \circ q \in \mathscr{L}_1(\mathbb{H};\mathbb{H})$ and $\|q\|_1 = \|q\|_2^2$.

(b) If $q \in \mathscr{L}_2(\mathbb{H};\mathbb{H})$ and $u \in \mathscr{L}(\mathbb{H};\mathbb{H})$, then $q \circ u \in \mathscr{L}_2(\mathbb{H};\mathbb{H})$ and $\|q \circ u\|_2 \leqslant \|q\|_2 \|u\|$.

(c) If $q \in \mathscr{L}_2(\mathbb{H};\mathbb{H})$ (resp. $q \in \mathscr{L}_1(\mathbb{H};\mathbb{H})$, then the adjoint q^* is an element of $\mathscr{L}_2(\mathbb{H};\mathbb{H})$ (resp. $\mathscr{L}_1(\mathbb{H};\mathbb{H})$) with the same norm.

21.5 The processes Q_M and $\langle\langle M \rangle\rangle$ of a square integrable martingale

With an $\mathbb{H}$-valued square integrable martingale, we have associated the Doléans measure α_M of $\|M\|^2$ (see § 17).

With M, we also associate the Doleans function of $M \otimes M$, that is

(21.5.1) $$d_{M \otimes M}(]s,t] \times F) := E\{1_F(M_t^{\otimes 2} - M_s^{\otimes 2})\} \in \mathbb{H} \hat{\otimes}_1 \mathbb{H} \subset \mathbb{H} \hat{\otimes}_2 \mathbb{H}$$

for every predictable rectangle $]s,t] \times F$ or, using the martingale property which gives $E(M_s \otimes (M_t - M_s)) = 0$, we may write

(21.5.2) $$d_{M \otimes M}(]s,t] \times F) = E\{1_F \cdot (M_t - M_s)^{\otimes 2}\} \in \mathbb{H} \hat{\otimes}_1 \mathbb{H} \subset \mathbb{H} \hat{\otimes}_2 \mathbb{H}.$$

Since the linear form "Trace" on $\mathbb{H} \hat{\otimes}_2 \mathbb{H}$ is the linear continuous extension of the mapping $(x,y) \rightsquigarrow \text{Trace}\,(x \otimes y) = x \cdot y$, it is clear from the above formula that

$$\alpha_M = \text{Trace}\; d_{M \otimes M}.$$

The existence of a σ-additive extension of $d_{M \otimes M}$ is therefore trivial. If μ_M denotes this extension, we have

$$\alpha_M = \text{Trace}\; \mu_M.$$

From the inequality

$$\|\mu_M(]s,t] \times F)\|_{\mathbb{H} \hat{\otimes}_2 \mathbb{H}} \leqslant E(\|1_F\ (M_t - M_s)^{\otimes 2}\|_{\mathbb{H} \hat{\otimes}_2 \mathbb{H}})$$

it follows that

(21.5.3) $$\|\mu_M(]s,t] \times F)\|_{\mathbb{H} \hat{\otimes}_2 \mathbb{H}} \leqslant \alpha_M(]s,t] \times F).$$

We easily deduce that the variation of the vector-valued measure μ_M is smaller than α_M.

As in § 15.8 we may use a Radon-Nikodym theorem for Hilbert-valued measures to immediately obtain the existence of an $\mathbb{H} \hat{\otimes}_2 \mathbb{H}$-valued predictable process Q_M such that, for every predictable G,

$$\mu_M(G) = \int_G Q_M \; d\alpha_M,$$

with $\|Q_M\| \leqslant 1$ in view of (21.5.3).

More in fact can be said: since

$$\|(M_t - M_s)^{\otimes 2}\|_{\mathbb{H} \hat{\otimes}_2 \mathbb{H}} = \|(M_t - M_s)^{\otimes 2}\|_{\mathbb{H} \hat{\otimes}_1 \mathbb{H}} = \|M_t - M_s\|^2_{\mathbb{H}},$$

the inequality (21.5.3) holds for the nuclear norm, which is a stronger inequality. In this case, it is in fact an equality because, from the definition, $\mu_M(]s, t] \times F)$ is clearly a positive element in $\mathbb{H} \hat{\otimes}_1 \mathbb{H}$ and

$$\|\mu_M(]s, t] \times F)\|_{\mathbb{H} \hat{\otimes}_1 \mathbb{H}} = \text{Trace } \|\mu_M(]s, t] \times F)\|.$$

It can, moreover, be proved that the Radon-Nikodym theorem holds for $\mathbb{H} \hat{\otimes}_1 \mathbb{H}$-valued measures with bounded variation, as a consequence of the Shatten theorem [Tre], which says that $\mathbb{H} \hat{\otimes}_1 \mathbb{H}$ is a separable dual space of a Banach space. Using this result from the theory of topological vector spaces we obtain the first part of the following theorem.

21.6 Theorem. (1) *There is one predictable $\mathbb{H} \hat{\otimes}_1 \mathbb{H}$-valued process Q_M, defined up to α_M-equivalence, such that, for every $G \in \mathcal{P}$,*

$$\mu_M(G) = \int_G Q_M \, d\alpha_M$$

Moreover, Q_M takes its values in the set of positive symmetric elements of $\mathbb{H} \hat{\otimes}_1 \mathbb{H}$ and

(21.6.1) $\qquad \text{Trace } Q_M(\omega, s) = \|Q_M(\omega, s)\|_{\mathbb{H} \hat{\otimes}_1 \mathbb{H}} = 1, \quad \alpha_M \quad \text{a.e.}$

(2) *The process*

$$\langle\langle M \rangle\rangle_t := \int_{[0,t]} Q_M \, d\langle M \rangle$$

has finite variation, is predictable, admits μ_M as its Doleans measure and is such that $M^{\otimes 2} - \langle\langle M \rangle\rangle$ is a martingale.

Proof: (1) The existence of Q_M has been proved above. Since μ_M takes its values in the set of positive symmetric elements of $\mathbb{H} \hat{\otimes}_1 \mathbb{H}$, the same holds for Q_M. The equality (21.6.1) follows immediately from this and the fact that $\alpha_M = \text{Trace } \mu_M$.

(2) The second part of the theorem follows immediately from § 15.8. □

21.7 Example. Brownian process with covariance Q, where $Q \in \mathbb{H} \hat{\otimes}_1 \mathbb{H}$

An $\mathbb{H}$-valued process W is called a Brownian motion adapted to the given stochastic basis with covariance Q if

(i) for every $s < t$, $W_t - W_s$ is independant of $\mathcal{F}_s$,

(ii) for every $s < t$ and $h \in \mathbb{H}$, the real random variable $\langle W_t - W_s, h\rangle$ is Gaussian centered with variance $(t - s)\langle Q, h \otimes h\rangle$.

Such a process is a square integrable martingale.

The reader can easily check that, in this case,

$$\langle\langle W \rangle\rangle_t = t\,Q,$$

while

$$\langle W \rangle_t = t \text{ Trace } Q.$$

22. The isometric stochastic integral with respect to Hilbert-valued martingales

22.1 The space $L^*(\mathbb{H}; \mathbb{G}; \mathscr{P}, M)$

With the $\mathbb{H} \hat{\otimes}_1 \mathbb{H}$-valued process Q_M we associate the $\mathscr{L}^1(\mathbb{H}; \mathbb{H})$-valued process $\tilde{Q}_M$ related to Q_M by

$$\tilde{Q}_M h \cdot g = Q_M(h \otimes g), \ (h, g) \in \mathbb{H} \times \mathbb{H}.$$

We may speak of the square root $\tilde{Q}_M^{\frac{1}{2}}$ of $\tilde{Q}_M$, which is a Hilbert-Schmidt operator-valued process. (See (21.4.4)).

Let $\mathbb{G}$ be another separable Hilbert space.

We call $L^*(\mathbb{H}; \mathbb{G}; \mathscr{P}, M)$ the space of processes Y, the values of which are (possibly non-continuous) linear operators from $\mathbb{H}$ into $\mathbb{G}$, with the following properties.

(i) The domain $\mathscr{D}X(t, \omega)$ of $X(t, \omega)$ contains $\tilde{Q}_M^{\frac{1}{2}}(\omega, t)(\mathbb{H})$ for every (t, ω).

(ii) For every $h \in \mathbb{H}$, the $\mathbb{G}$-valued process $X \circ \tilde{Q}_M^{\frac{1}{2}}(h)$ is predictable.

(iii) For every $(t, \omega) \in \mathbb{R}^+ \times \Omega$, $X(t, \omega) \circ \tilde{Q}_M^{\frac{1}{2}}(\omega, t)$ is a Hilbert-Schmidt operator and $\int \|X \circ \tilde{Q}_M^{\frac{1}{2}}\|_2^2 \, d\alpha_M < \infty$.

We then have the following proposition.

22.2 Proposition. *For every $X, Y \in L^*(\mathbb{H}; \mathbb{G}; \mathscr{P}, M)$, the process $X \circ \tilde{Q}_M \circ Y^*$ is an $\mathscr{L}_1(\mathbb{G}; \mathbb{G})$-valued predictable process, with $\int_\Omega$ Trace $(X \circ \tilde{Q} \circ Y^*) \, d\alpha_M < \infty$*[1]*). The bilinear form $(X, Y) \rightsquigarrow \int_\Omega$ Trace $(X \circ \tilde{Q}_M \circ Y^*) \, d\alpha_M$ is a scalar product on $L^*(\mathbb{H}; \mathbb{G}; \mathscr{P}, M)$, and for this scalar product, the space is complete.*

Proof. We first prove that Trace $(X \circ \tilde{Q}_M \circ Y^*)$ is a predictable process. Since

$$\begin{aligned}\text{Trace } (X \circ \tilde{Q}_M \circ Y^*) &= \text{Trace } (Y \circ \tilde{Q}_M \circ X^*) \\ &= \tfrac{1}{4}\{\text{Trace } [(X+Y) \circ \tilde{Q}_M \circ (X^* + Y^*)] \\ &\quad - \text{Trace } [(Y-X) \circ \tilde{Q}_M \circ (Y^* - X^*)]\}\end{aligned}$$

we have only to prove that, for every $X \in L^*(\mathbb{H}; \mathbb{G}; \mathscr{P}, M)$, the process Trace $(X \circ \tilde{Q}_M \circ X^*)$ is predictable.

Let (h_n) be an orthonormal basis of $\mathbb{H}$. Since

$$\text{Trace } (X \circ \tilde{Q}_M \circ X^*) = \|X \circ \tilde{Q}_M^{\frac{1}{2}}\|_2^2 = \sum_n \|X \circ \tilde{Q}_M^{\frac{1}{2}}(h_n)\|_{\mathbb{G}}^2,$$

---

[1]) $\circ$ denotes the composition of operators in vector spaces.

the hypothesis (ii) shows that this process is predictable. Since

$$\text{Trace}\,(X \circ \tilde{Q}_M \circ Y^*) \leqslant \|X \circ \tilde{Q}_M^{\frac{1}{2}}\|_2 \cdot \|Y \circ \tilde{Q}_M^{\frac{1}{2}}\|_2 ,$$

and in view of the Schwarz inequality, it is immediate that

$$(X,Y) \rightsquigarrow \int_\Omega \text{Trace}\,(X \circ \tilde{Q}_M \circ Y^*)\, d\alpha_M$$

is a positive continuous bilinear form on $L^*(\mathbb{H};\mathbb{G};\mathscr{P},M)$.

We now show that every Cauchy sequence for this scalar product has a limit in $L^*(\mathbb{H};\mathbb{G};\mathscr{P},M)$. Let us then consider (X_n) with

$$\text{(22.2.1)} \qquad \lim_{n,m\to\infty} \int \|(X_n - X_m) \circ \tilde{Q}_M^{\frac{1}{2}}\|_2^2\, d\alpha_M = 0 .$$

In the space $L^2_{\mathscr{L}_2(\mathbb{H};\mathbb{G})}(\mathbb{R}^+ \times \Omega, \mathscr{P}, \alpha_M)$, the sequence $X_n \circ \tilde{Q}_M$ converges to some Y and we can extract a subsequence $(X_{n_k})_{k\geqslant 0}$ such that

$$\lim_k X_{n_k}(t,\omega) \circ \tilde{Q}_M^{\frac{1}{2}}(t,\omega) = Y(\omega,t)\ \alpha_M \quad \text{a.e.}$$

Since $\tilde{Q}_M^{\frac{1}{2}}(t,\omega)f = 0$ implies $Y(t,\omega)f = 0$, it is possible to write Y as

$$Y(t,\omega) = X(t,\omega) \circ \tilde{Q}_M^{\frac{1}{2}}(t,\omega)$$

for some $X(t,\omega)$ which linearly maps $\tilde{Q}_M^{\frac{1}{2}}(t,\omega)\,\mathbb{H}$ into $\mathbb{G}$ and which clearly meets the conditions (i) to (iii). This completes the proof. □

22.3 The space $\Lambda^2(\mathbb{H};\mathbb{G};\mathscr{P},M)$

We call $\mathscr{E}(\mathscr{L}(\mathbb{H};\mathbb{G}))$ the space of $\mathscr{L}(\mathbb{H};\mathbb{G})$-valued, $\mathscr{A}$-simple processes, and $\Lambda^2(\mathbb{H};\mathbb{G};\mathscr{P},M)$ the closure of $\mathscr{E}(\mathscr{L}(\mathbb{H};\mathbb{G}))$ in $L^*(\mathbb{H};\mathbb{G};\mathscr{P},M)$. We thus obtain a Hilbert subspace of $L^*(\mathbb{H};\mathbb{G};\mathscr{P},M)$.

22.4 Proposition. *Every process X with the properties*
(i) *for every $(t,\omega) \in \mathbb{R}^+ \times \Omega$, $X(t,\omega) \in \mathscr{L}(\mathbb{H};\mathbb{G})$;*
(ii) *for every $h \in \mathbb{H}$, $X(h)$ is a predictable $\mathbb{G}$-valued process;*
(iii) $\int \text{Trace}\,(X \circ \tilde{Q}_M \circ X^*)\, d\alpha_M < \infty$,
belongs to $\Lambda^2(\mathbb{H};\mathbb{G};\mathscr{P},M)$.

Proof. Let us assume first that X is a measurable mapping from $\Omega' := \mathbb{R}^+ \times \Omega$ into the Banach space $\mathscr{L}(\mathbb{H};\mathbb{G})$ with $\sup_{\omega,t} \|X(t,\omega)\| \leqslant K < \infty$. There exists a uniformly bounded sequence (X_n) in $\mathscr{E}(\mathscr{L}(\mathbb{H};\mathbb{H}))$, which α_M- a.e. converges to $X(\omega,t)$ in $\mathscr{L}(\mathbb{H};\mathbb{G})$. For such a sequence, one has

$$\lim_{n\to\infty} \int_\Omega \|(X - X_n) \circ \tilde{Q}_M^{\frac{1}{2}}\|_2^2\, d\alpha_M = 0 .$$

If X is a process for which (i), (ii) and (iii) hold, then for a dense subset $\{h_n\}$ in the unit ball of $\mathbb{H}$, we have

$$\|X(t,\omega)\| = \sup_n \|X(t,\omega)(h^n)\|_{\mathbb{G}}$$

and $\|X\|$ is therefore predictable. For every $(s,\omega) \in \mathbb{R}^+ \times \Omega$,

$$\lim_{n\to\infty} \|1_{\{\|X\|<n\}} X(s,\omega)\| = O,$$

and therefore,

$$\lim_{n\to\infty} \|(1_{\{\|X\|\leqslant n\}} X(s,\omega) - X(s,\omega)) \circ \tilde{Q}_M^{\frac{1}{2}}(s,\omega)\|_2 = 0\,.$$

Since

$$\begin{aligned}\|(1_{\{\|X\|\leqslant n\}} X(s,\omega) - X(s,\omega)) \circ \tilde{Q}_M^{\frac{1}{2}}(s,\omega)\| &= 1_{\{\|X\|>n\}} \|X(s,\omega) \circ \tilde{Q}_M^{\frac{1}{2}}(s,\omega)\|_2^2 \\ &\leqslant \|X(s,\omega) \circ \tilde{Q}_M(s,\omega)\|_2^2\,,\end{aligned}$$

the following equation holds

$$\lim_{n\to\infty} \int \|(1_{\{\|X\|\leqslant n\}} X - X) \circ \tilde{Q}_M^{\frac{1}{2}}\|_2^2\, d\alpha_M = 0\,.$$

We have thus reduced the proof of the proposition to show that every process X with the properties (i), (ii) and (iii) and which is in norm bounded by some constant K is in $\Lambda^2(\mathbb{H}; \mathbb{G}; \mathscr{P}, M)$.

Let (h_i) (resp. (g_i)) be an orthonormal basis in $\mathbb{H}$ (resp. in $\mathbb{G}$). We call Π_1^n (resp. Π_2^n) the orthogonal projections from $\mathbb{H}$ (resp. from $\mathbb{G}$) into the subspaces generated by $\{h_1 \ldots . h_n\}$ (resp. $\{g_1 \ldots g_n\}$). We set

$$X_n := \Pi_2^n \circ X \circ \Pi_1^n\,.$$

For every i, we may write simultaneously

$$\lim_n \|(\Pi_2^n \circ X \circ \Pi_1^n - X) \circ \tilde{Q}_M^{\frac{1}{2}}(h_i)\|_{\mathbb{G}}^2 = 0$$

$$\|(\Pi_2^n \circ X \circ \Pi_1^n - X) \circ \tilde{Q}_M^{\frac{1}{2}}(h_i)\|_{\mathbb{G}}^2 \leqslant 4K^2 \|\tilde{Q}_M^{\frac{1}{2}}(h_i)\|_{\mathbb{H}}^2$$

and

$$\sum_i \|\tilde{Q}_M^{\frac{1}{2}}(h_i)\|_{\mathbb{H}}^2 = \|\tilde{Q}_M^{\frac{1}{2}}\|_2^2 < \infty\,.$$

From these three relations, we derive

$$\lim_n \sum_i \|(\Pi_2^n \circ X \circ \Pi_1^n - X) \circ \tilde{Q}_M^{\frac{1}{2}}(h_i)\|_{\mathbb{G}}^2 = \lim_n \|(X_n - X) \circ \tilde{Q}_M^{\frac{1}{2}}\|_2^2 = 0$$

with

$$\|(X_n - X) \circ \tilde{Q}_M^{\frac{1}{2}}\|_2^2 \leqslant 4K^2 \|X \circ \tilde{Q}_M^{\frac{1}{2}}\|_2^2\,.$$

Therefore,

$$\lim_{n\to\infty} \int \|(X_n - X)\circ \tilde{Q}_M^{\frac{1}{2}}\|_2^2 \, d\alpha_M = 0 .$$

However, the processes X_n are in $\Lambda^2(\mathbb{H}; \mathbb{G}; \mathscr{P}, M)$, according to the beginning of the proof. We have thus proved that X also belongs to this space. □

22.5 Remark and example. We should remark that $\Lambda^2(\mathbb{H}; \mathbb{G}; \mathscr{P}, M)$ contains, in general, other processes than those satisfying (i), (ii), (iii); in particular, processes whose values are unbounded operators. This is the case in the following very simple example. Let (h_i) and (g_i) be orthonormal basis in $\mathbb{H}$ and $\mathbb{G}$ as above. It is easy to define a process M with independent increments such that Q is non-random. Thus,

$$Q(\omega, s) = \sum_{i=1}^{\infty} \frac{1}{2^i} h_i \otimes h_i \quad \text{for every} \quad (\omega, s)$$

(for example, the Brownian motion associated with Q: see § 19.10).

Let us then consider the deterministic processes X_n, where

$$X_n(h) := \sum_{i=1}^{n} \sqrt{i}\,(h \otimes h_i)\, g_i .$$

Clearly,

$$\|(X_n - X_{n+k})\circ \tilde{Q}_M^{\frac{1}{2}}\|_2^2 = \sum_{i=n+1}^{n+k} \frac{i}{2^i} .$$

The sequence (X_n) is therefore a Cauchy sequence in $\Lambda^2(\mathbb{H}; \mathbb{G}; \mathscr{P}, M)$. Considering then a subsequence (X_{n_k}) such that $X_{n_k} \circ \tilde{Q}_M^{\frac{1}{2}}$ converges to $X \circ \tilde{Q}_M^{\frac{1}{2}}$ in $\mathscr{L}_2(\mathbb{H}; \mathbb{G})$, we obtain

$$\lim_k \frac{1}{2^i} X_{n_k}(h_i) = \frac{1}{2^i} X(h_i) ,$$

and, as a consequence,

$$X(h_i) = \sqrt{i}\, g_i .$$

The operator X is certainly not bounded!

22.6 Theorem (The isometric stochastic integral). *Let M be an element of $\mathscr{M}^2_\infty(\mathbb{H})$. There exists a unique, isometric, linear mapping from $\Lambda^2(\mathbb{H}; \mathbb{G}; \mathscr{P}, M)$ into $\mathscr{M}^2_\infty(\mathbb{G})$ such that the image of $X := 1_{F\times]r,s]} u$, for every predictable rectangle $F \times]r,s]$ and $u \in \mathscr{L}(\mathbb{H}; \mathbb{G})$ is the martingale $(1_F\,[u(M_{s\wedge t}) - u(M_{r\wedge t})])_{t\in\mathbb{R}^+}$.*

Proof. We have clearly only to prove that the mapping

$$X := \sum_{i=1}^{n} 1_{F_i \times]r_i, s_i]}\, u_i \rightsquigarrow \Big(\sum_{i=1}^{n} 1_{F_i}\big[u_i(M_{s_i\wedge t}) - u_i(M_{r_i\wedge t})\big]\Big)_{t\in\mathbb{R}^+}$$

is an isometric mapping from $\mathscr{E}(\mathscr{L}(\mathbb{H};\mathbb{G})$ into $\mathscr{M}^2_\infty(\mathbb{G})$. It is always possible to assume that the predictable rectangles $F_i \times]r_i, s_i]$ are disjoint. We can then write

$$E\,(\|\sum_{i=1}^n 1_{F_i}[u_i(M_{s_i}) - u_i(M_{r_i})]\|^2_{\mathbb{G}}) = E\,(\sum_{i=1}^n \|1_{F_i} u_i\,(M_{s_i} - M_{r_i})\|^2_{\mathbb{G}})$$

$$= E\,(\sum_{i=1}^n 1_{F_i}\,\text{Trace}\,[u_i \otimes u_i(M_{s_i} - M_{r_i})^{\otimes 2}])$$

$$= \text{Trace}\sum_{i=1}^n u_i \otimes u_i\,E\,[1_{F_i}(M_{s_i} - M_{r_i})^{\otimes 2}]$$

$$= \sum_{i=1}^n \text{Trace}\,[u_i \otimes u_i\,(\int_{F_i \times]r_i, s_i]} \tilde{Q}_M\,d\alpha_M)]$$

$$= \sum_{i=1}^n \int_{F_i \times]r_i, s_i]} \text{Trace}\,(u_i \circ \tilde{Q}_M \circ u_i^*)\,d\alpha_M$$

$$= \int \text{Trace}\,(X \circ \tilde{Q}_M \circ X^*)\,d\alpha_M\,.$$

This proves the theorem.

The image of X in $\mathscr{M}^2_\infty(\mathbb{G})$ under the previous mapping is called the *stochastic integral process of X with respect to M* and is denoted by $(\int X\,dM)$. If we want to be specific about the linear operation and if it is denoled by $\circ$, we write $(\int X \circ dM)$.

22.7 Properties of the stochastic integral

Proposition. *We consider* $M \in \mathscr{M}^2_\infty(\mathbb{H})$, $X \in \Lambda^2(\mathbb{H};\mathbb{G};\mathscr{P},M)$, $N = (\int X\,dM) \in \mathscr{M}^2_\infty(\mathbb{G})$. *Then, the following formulas hold.*

(22.7.1) $$\alpha_N = \text{Trace}\,(X \circ \tilde{Q}_M \circ X^*)\,\alpha_M\,;$$

(22.7.2) $$\tilde{Q}_N = \frac{1}{\text{Trace}\,(X \circ \tilde{Q}_M \circ X^*)}\,X \circ \tilde{Q}_M \circ X^*\,;$$

(22.7.3) $$\mu_N = (X \circ \tilde{Q}_M \circ X^*)^M{}_M = (X \otimes X)\,\mu_M\,;$$

(22.7.4) $$\langle N \rangle_t = \int_{]0,t]} \text{Trace}\,(X \circ \tilde{Q}_M \circ X^*)\,d\langle M \rangle\,;$$

(22.7.5) $$\langle\langle N \rangle\rangle_t = \int_{]0,t]} (X \circ \tilde{Q}_M \circ X^*)\,d\langle M \rangle = \int_{]0,t]} X \otimes X\,d\langle\langle M \rangle\rangle$$

(In the two latter formulas the integrals on the right-hand side are to be taken on each path.)

Proof. From the definition and the isometry property we derive, for every predictable rectangle $F \times]s, t]$,

$$E(1_F \cdot \|N_t - N_s\|_{\mathbb{G}}^2) = E(\|\int 1_{F\times]s,t]} X\, dM\|_{\mathbb{G}}^2)$$
$$= \int_{F\times]s,t]} \operatorname{Trace}(X \circ \tilde{Q}_M \circ X^*)\, d\alpha_M .$$

This expresses (22.7.1). The right-hand side of (22.7.4) being clearly a predictable process, (22.7.2) follows readily from (22.7.1).

For every $X \in \mathscr{E}$, it is readily checked that

(22.7.6) $$E\{1_F\ (N_t - N_s)^{\otimes 2}\} = \int_{F\times]s,t]} X \circ \tilde{Q}_M \circ X^*\, d\alpha_M .$$

If X is an element of $\Lambda^2(\mathbb{H}; \mathbb{G}; \mathscr{P}, M)$, it can be approximated by a sequence (X_n) in $\mathscr{E}(\mathscr{L}(\mathbb{H}; \mathbb{G}))$, and an immediate continuity argument shows that (22.7.6) holds for such an X and the corresponding integral process $N := (\int X\, dM)$. Since (22.7.6) can be written

$$\mu_N = X \circ \tilde{Q}_M \circ X^* \alpha_M ,$$

we obtain (22.7.3). The equality (22.7.2) now follows from (22.7.1) and (22.7.3). From the definition of $\langle\langle M\rangle\rangle$ and according to (22.7.3), the Doleans measure of $(\int_{]0,t]} X_s \otimes X_s d\langle\langle M\rangle\rangle_s)_{t\in T}$ is μ_N but, since the latter process is predictable, this is precisely $\langle\langle N\rangle\rangle$.

22.8 Theorem (Tensor quadratic variation). *Let $M \in \mathscr{M}^2(\mathbb{H})$ (resp. $\mathscr{M}^2_\infty(\mathbb{H})$). There exists an $\mathbb{H} \hat{\otimes}_1 \mathbb{H}$ valued R.R.C. process, which is uniquely defined up to P-equality, denoted by $[\![M,]\!]$ and called the tensor quadratic variation of M,* with the following properties.

(1°) *For every increasing sequence (Π_n) of increasing subsequences of $\mathbb{R}^+ : \Pi_n := \{0 < t_0 < t_1 \ldots < t_n < \ldots\}$ such that $\lim_{k\uparrow\infty} t_k = +\infty$ and $\lim_{n\to\infty} \delta(\Pi_n) = 0$, where $\delta(\Pi_n)$ is the "mesh" of Π_n defined by $\delta(\Pi_n) := \sup_{t_i\in\Pi_n} (t_{i+1} - t_i)$, one has for every $t \in \mathbb{R}^+$*

(22.8.1) $$[\![M, M]\!]_t = \lim_n \text{prob.} \sum_{t_i\in\Pi_n} (M_{t_{i+1}\wedge t} - M_{t_i\wedge t})^{\otimes 2} ,$$

where the convergence in probability is the convergence of $\mathbb{H} \hat{\otimes}_2 \mathbb{H}$-valued random variables.

(2°) *$M^{\otimes 2} - [\![M, M]\!]_t$ is an $\mathbb{H} \hat{\otimes}_1 \mathbb{H}$-martingale.*

(4°) For every t,

$$[\![M, M]\!]_t = \langle\langle M^c, M^c\rangle\rangle_t + \sum_{s\leq t} \Delta M_s^{\otimes 2} = [\![M^c, M^c]\!]_t + \sum_{s\leq t} \Delta M_s^{\otimes 2} \quad \text{a.s.}$$

where M^c is the continuous part of M, as defined in 19.2, and the series on the right-hand side are absolutely convergent in $\mathbb{H} \hat{\otimes}_1 \mathbb{H}$.

Proof. We write a formula analogous to the one in the proof of 18.6:

$$M_t^{\otimes 2} - M_0^{\otimes 2} = \sum_{t_i \in \Pi_n} (M_{t_{i+1} \wedge t} - M_{t_i \wedge t})^{\otimes 2} + M_{t_i \wedge t} \otimes (M_{t_{i+1} \wedge t} - M_{t_i \wedge t}) + + (M_{t_{i+1} \wedge t} - M_{t_i \wedge t}) \otimes M_{t_i \wedge t}$$

If we introduce the same process X^n as in 18.6, we may write

$$\sum_{t_i \in \Pi_n} M_{t_i \wedge t} \otimes (M_{t_{i+1} \wedge t} - M_{t_i \wedge t}) = \int_{]0,t]} X^n \otimes dM,$$

where X^n is considered as a process with values in $\mathscr{L}(\mathbb{H}; \mathbb{H} \hat{\otimes}_2 \mathbb{H})$.

The proof may now be performed along the lines of the proof of 18.6. Details are left to the reader.

We only make the following comments. The processes $\int_{]0,t]} X^n \otimes dM$ are $\mathbb{H} \hat{\otimes}_2 \mathbb{H}$-valued process, which, restricted to suitable stochastic intervals $]0, \tau_k]$, converge in $\mathscr{M}^2_\infty(\mathbb{H})$. The convergence in $L^2_{\mathbb{H} \otimes_2 \mathbb{H}}(\Omega, \mathscr{F}, P)$ of $\int_{]0,\tau_k \wedge t]} X^n \otimes dM$ easily implies the convergence in probability.

The convergence of the real series $\langle M^c, M^c \rangle + \sum_{s \leq t} \|\Delta M_s\|^2$, which follows from 20.5, (4°), shows that the convergence of the series in the 4th statement of the theorem actually holds in $\mathbb{H} \hat{\otimes}_1 \mathbb{H}$ and not only in $\mathbb{H} \hat{\otimes}_2 \mathbb{H}$. This shows that $[\![M, M]\!]$ is $\mathbb{H} \hat{\otimes}_1 \mathbb{H}$-valued.

23. Localisation of processes and semimartingales

23.1 Definitions. Let $(X_t)_{t \in \mathbb{R}^+}$ be a process with state-space $\mathscr{V}$ and τ a stopping time. The process $(X_{t \wedge \tau})_{t \in \mathbb{R}^+}$ will be called the *process X stopped at time τ* and denoted by X^τ.

When X has paths with left limits one may define the process

$$X^{\tau^-} := \begin{cases} X & \text{on } [0, \tau[\\ X_{\tau-} & \text{on } [\tau, \infty[\end{cases}$$

This process will be called the process *stopped strictly before time τ*.

Let $\mathscr{H}$ be a given class of processes. We shall say that the process X belongs locally to the class $\mathscr{H}$ when an increasing sequence $(\tau_n)_{n \geq 0}$ of stopping times exists with the following properties.

(i) $\lim_n \tau_n = +\infty$ a.s.

(ii) For every n the process X^{τ_n} belongs to $\mathscr{H}$.

(iii) For every n and every stopping time σ the process $X^{\tau_n \wedge \sigma}$ belongs to $\mathscr{H}$.

Such a sequence $(\tau_n)_{n \geq 0}$ of stopping times will be called a *$\mathscr{H}$-localizing sequence for X*, and the class of processes for which a $\mathscr{H}$-localizing sequence exists will be denoted by $\mathscr{H}_{\text{loc}}$.

23.2 Some spaces of R. R. C. processes

Let $\mathbb{B}$ be a separable Banach space and $p > 0$. $\mathcal{W}^p_\infty(\mathbb{B}; \mathcal{F})$ is the class of regular (with respect to $\mathcal{F}$) right-continuous $\mathbb{B}$-valued processes V with finite variation $|V|_\infty$ (B-V processes as defined in § 15.1) such that $E(|V_\infty|^p) < \infty$.

We also define the following.

$\mathcal{W}^p(\mathbb{B}; \mathcal{F})$: class of R. R. C $\mathbb{B}$-valued processes with finite variation such that $E(|V_t|^p_t) < \infty$ for all $t \in \mathbb{R}^+$

$\mathcal{W}^0(\mathbb{B}; \mathcal{F})$: class of R. R. C. $\mathbb{B}$-valued processes with finite variation

$$\mathcal{W}^p(\mathcal{F}) := \mathcal{W}^p(\mathbb{R}; \mathcal{F}) \quad p \geq 0$$
$$\mathcal{W}^p_\infty(\mathcal{F}) := \mathcal{W}^p_\infty(\mathbb{R}; \mathcal{F}) \quad p > 0$$
$$\mathcal{W}^p_+(\mathcal{F}) := \{V : V \in \mathcal{W}^p(\mathcal{F}),\ V \text{ increasing}\}$$
$$\mathcal{W}^p_{\infty +}(\mathcal{F}) := \{V : V \in \mathcal{W}^p_\infty(\mathcal{F}),\ V \text{ increasing}\}$$

$\mathcal{Q}(\mathbb{B}; \mathcal{F}) := \{X\colon X$ is a $\mathbb{B}$-valued quasimartingale, with respect to $\mathcal{F}$, right-continuous of class [L. D]$\}$

$$\mathcal{Q}(\mathcal{F}) := \mathcal{Q}(\mathbb{R}; \mathcal{F})$$
$$\mathcal{Q}_+(\mathcal{F}) := \{X\colon X \in \mathcal{Q}(\mathcal{F}),\ X \text{ submartingale}\}$$

When $\mathcal{F}$ is clear we omit $\mathcal{F}$ from the above notation.

If $V \in \mathcal{W}^p(\mathbb{B})$, we write $|V|$ for the increasing process such that $|V|_t(., \omega)$ is the variation of $V(., \omega)$ on the interval $[0, t]$. It is easy to see, using the definition of the variation and binary divisions of $\mathbb{R}^+$, that $|V| \in \mathcal{W}^p_+$. As a consequence, any V in $\mathcal{W}^p$ (resp. $\mathcal{W}^p_\infty$) can be written as the difference of two processes V_1 and V_2 in $\mathcal{W}^p_+$ (resp. $\mathcal{W}^p_{\infty,+}$).

According to definitions in 21.1, we may speak of the classes $\mathcal{W}^p_{\text{loc}}(\mathbb{B}; \mathcal{F})$ $\mathcal{M}^p_{\text{loc}}(\mathbb{B}; \mathcal{F})$, $\mathcal{Q}_{\text{loc}}(\mathbb{B}; \mathcal{F})$... etc.

Because of the stopping theorem, Theorem 13.5, as soon as a sequence (τ_n) exists with properties (i) and (ii) for a process X, relative to $\mathcal{H} = \mathcal{M}^p(\mathbb{B}; \mathcal{F})$ or $\mathcal{H} = \mathcal{Q}(\mathbb{B}; \mathcal{F})$, property (iii) is automatically fulfilled.

Let (τ_n) be a $\mathcal{H}$-localizing sequence of stopping times for X. We see immediately that $(\tau_n \wedge n)_{n \in \mathbb{N}}$ is also a $\mathcal{H}$-localizing sequence and this clearly implies

$$\mathcal{W}^p_{+\text{loc}} = \mathcal{W}^p_{\infty,+,\text{loc}} \quad \text{and} \quad \mathcal{M}^p_{\text{loc}}(\mathbb{B}) = \mathcal{M}^p_{\infty,\text{loc}}(\mathbb{B}).$$

Clearly also $\mathcal{W}^0(\mathbb{B}) = \mathcal{W}^0_{\text{loc}}(\mathbb{B})$.

The elements of $\mathcal{M}^1_{\text{loc}}(\mathbb{B})$ (resp. $\mathcal{M}^2_{\text{loc}}(\mathbb{B})$) are called *local martingales* (resp. *locally square integrable martingales*).

If we want to consider processes taking their values in the dual $\mathbb{B}'$ of a separable Banach space $\mathbb{B}$ (the example of point processes illustrates the interest in considering such a situation: see § 15), the space $\mathbb{B}'$ itself, being not necessarily separable, we define $\mathcal{W}^p(\mathbb{B}'; \mathcal{F})$ as the set of R. R. C. $\mathbb{B}'$-valued weak processes V with finite variation (for the norm of $\mathbb{B}'$) and such that $E(|V|^p_t) < \infty$ for all $t \in \mathbb{R}^+$.

Clearly no confusion arises from this notation because, if $\mathbb{B}'$ is separable, weak processes with values in $\mathbb{B}'$ are ordinary processes, the weak measurability of a mapping from $(\Omega, \mathscr{F}_t)$ in $\mathbb{B}'$ implying its measurability when $\mathbb{B}'$ is separable.

In the same way we may define the space $\mathscr{Q}(\mathbb{B}', \mathscr{F})$ of R.R.C. $\mathbb{B}'$-valued weak quasimartingales of class [L. D] (see 13.9), the class $\mathscr{M}^p(\mathbb{B}', \mathscr{F})$ of R.R.C. weak martingales M such that $E(\|M_t\|^p) < \infty$ for every $t \in \mathbb{R}^+$ and the corresponding classes $\mathscr{Q}_{\mathrm{loc}}(\mathbb{B}', \mathscr{F})$ and $\mathscr{M}^p_{\mathrm{loc}}(\mathbb{B}', \mathscr{F})$.

23.3 The processes $\mathbf{M}^c$, $\mathbf{M}^d$, $\langle \mathbf{M} \rangle$, $\langle\langle \mathbf{M} \rangle\rangle$, $[\mathbf{M}]$, $[\![\mathbf{M}]\!]$, of $\mathbf{M} \in \mathscr{M}^2_{\mathrm{loc}}(\mathbb{H})$

We assume that $\mathbb{H}$ is a separable Hilbert space and $M \in \mathscr{M}^2_{\mathrm{loc}}(\mathbb{H})$. Let (τ_n) be a localizing sequence of stopping times for M. According to the definition, M^{τ_n} belongs to $\mathscr{M}^2(\mathbb{H})$ for every n. It is therefore possible to define the processes $M^{\tau_n, c}$, $M^{\tau_n, d}$, $\langle M^{\tau_n}, M^{\tau_n} \rangle$, $\langle\langle M^{\tau_n}, M^{\tau_n} \rangle\rangle$, $[M^{\tau_n}]$, $[\![M^{\tau_n}]\!]$, $\check{M}^{\tau_n}$ as in sections 21 and 22.

Let us consider, for example, $M^{\tau_n, c}$. As an immediate consequence of the definition and the uniqueness (up to P-equality) of $M^{\tau_n, c}$ for every n, the processes $M^{\tau_n, c}$ and $M^{\tau_m, c}$ coincide on $[0, \tau_n]$ for every $m > n$.

Therefore, there is a uniquely (up to P-indistinguishability) defined process M^c with the property that M^c coincides with $M^{\tau_n, c}$ on $[0, \tau_n]$. This process clearly does not depend on the localizing sequence considered.

The same argument holds immediately for the definition of processes $\langle M, M \rangle$, $\langle\langle M, M \rangle\rangle$, $[M, M]$, $[\![M, M]\!]$ and $\check{M}$.

As in the case of L^2-Martingales, we often write $\langle M \rangle$, (resp. $\langle\langle M \rangle\rangle$, resp. $[M]$, resp. $[\![M]\!]$) instead of $\langle M, M \rangle$ (resp. $\langle\langle M, M \rangle\rangle$, resp. $[M, M]$, resp. $[\![M, M]\!]$).

If M and N are two elements of $\mathscr{M}^2_{\mathrm{loc}}(\mathbb{H})$ and (τ_n) and (σ_n) are localizing sequences for M and N respectively, $(\tau_n \wedge \sigma_n)$ is a localizing sequence for M and N and the same reasoning as above shows the existence of unique processes $\langle M, N \rangle$, $\langle\langle M, N \rangle\rangle$ $[M, N]$, $[\![M, N]\!]$, with the property that for every n, $\langle M, M \rangle = \langle M^{\tau_n \wedge \sigma_n}, N^{\tau_n \wedge \sigma_n} \rangle$ on $[0, \tau_n \wedge \sigma_n]$ and the analogous equalities for the three other processes.

23.4 Theorem. *Let $\mathbb{H}$ be a separable Hilbert space and $M \in \mathscr{M}^2_{\mathrm{loc}}(\mathbb{H})$. Then*

(1°) $\|M\|^2 - \langle M \rangle \in \mathscr{M}^1_{\mathrm{loc}}$ *and* $M \otimes M - \langle\langle M \rangle\rangle \in \mathscr{M}^1_{\mathrm{loc}}(\mathbb{H} \hat{\otimes}_1 \mathbb{H})$;

(2°) $\|M\|^2 - [M] \in \mathscr{M}^1_{\mathrm{loc}}$ *and* $M \otimes M - [\![M]\!] \in \mathscr{M}^1_{\mathrm{loc}}(\mathbb{H} \hat{\otimes}_1 \mathbb{H})$.

For every stopping time σ the following inequalities holds (with terms which are possibly equal to $+\infty$).

(3°) $E(\|M_\sigma - M_0\|^2) = E(\langle M \rangle_\sigma) = E([M]_\sigma)$

(4°) $E\{\sup_{t \leqslant \sigma} \|M_t - M_0\|^2\} \leqslant 4\, E(\langle M \rangle_\sigma) = 4\, E([M]_\sigma)$

(5°) $E\{\sup_{t < \sigma} \|M_t - M_0\|^2\} \leqslant 4\, [E(\langle M \rangle_{\sigma^-}) + E([\check{M}])_{\sigma^-}]$

Proof. One considers a localizing sequence (τ_n) for X.

Since $(\|M\|^2 - \langle M \rangle^{\tau_n} = \|M^{\tau_n}\|^2 - \langle M^{\tau_n} \rangle$ and $(M \otimes M - \langle\langle M \rangle\rangle)^{\tau_n} = M^{\tau_n} \otimes M^{\tau_n} -$

$-\langle\langle M^{\tau_n}\rangle\rangle$ for every n, one immediately obtains the relations 1°). The relations 2°) are obtained in the same way.

The equations and inequalities (3°), (4°), (5°) are clear for every stopping time $\tau_n \wedge \sigma$. They follow for σ by an immediate passage to the limit when n tends to infinity. □

23.5 Theorem. *The following three properties are equivalent.*

(a) $X \in \mathcal{Q}_{\text{loc}}$.

(b) *There exists a unique decomposition* $X = M + V$ *of* X *where* $M \in \mathcal{M}^1_{\text{loc}}, V \in \mathcal{W}^1_{\text{loc}}$, V *is predictable and* $V_0 = 0$

(c) *There exists* $N \in \mathcal{M}^2_{\text{loc}}$ *and* $A \in \mathcal{W}^1_{\text{loc}}$ *such that* $X = N + A$.[1])

Proof.

(1°) The implication c) $\Rightarrow$ a) is trivial.

(2°) We prove that a) $\Rightarrow$ b).

The uniqueness of the decomposition is clear because, if (τ_n) is a localizing sequence for X and $X - V$, one has $(X^{\tau_n} - V^{\tau_n}) = (X - V)^{\tau_n} \in \mathcal{M}^1$, and V^{τ_n} is the predictable dual projection of X^{τ_n}, which is uniquely defined. Since $V\,1_{[0,\tau_n]}$ is uniquely defined for all n, V is itself unique. To prove the existence of the decomposition we first observe that, for every $m > n$, the restriction to $[0, \tau_n]$ of λ_X is the Doléans measure of X^{τ_n}. Therefore, the Meyer processes V^m and V^n of X^{τ_m} and X^{τ_n} coincide on $[0, \tau_n]$. One can therefore define V by $V_{[0,\tau_n]} = V^n$ for all n. This process is predictable and has the properties $V^{\tau_n} = V^n \in \mathcal{W}^1_\infty$ and $(X - V)^{\tau_n} = X^{\tau_n} - V^n \in \mathcal{M}^1_\infty$ for all n.

On account of the first statement of the theorem, in order to show that b) $\Rightarrow$ c), it is enough to prove the second statement for a process $M \in \mathcal{M}^1_{\text{loc}}$.

Let (τ_n) be a localizing sequence of stopping times for M. The stopping times

$$\sigma_n := \inf\{t : |M_t| > n\} \wedge \tau_n$$

also constitute a localizing sequence.

We define M^n as being the martingale $M^{\sigma_n} - M^{\sigma_{n-1}}$ and set

$$A^n := M^n_{\sigma_n} 1_{[\sigma_n, \infty[} \in \mathcal{W}^1_\infty .$$

Let $\tilde{A}^n$ be the dual predictable projection of A^n. One has

(23.5.1) $$M^n = M^n 1_{[\sigma_{n-1}, \sigma_n[} + \tilde{A}^n + (A^n - \tilde{A}^n) \in \mathcal{M}^1_\infty$$

This shows that $M^n 1_{[\sigma_{n-1}, \sigma_n[} + \tilde{A}^n$ is a martingale. We will show that this martingale actually belongs to $\mathcal{M}^2_\infty$. Since we already know that $|M^n_{\sigma_n}| \in L^1(\Omega, \mathcal{F}_{\sigma_n}, P)$ and therefore

(23.5.2) $$E|\tilde{A}^n_\infty| = E|A^n_\infty| = E|M^n_{\sigma_n}| < \infty$$

we deduce that $A^n - \tilde{A}^n \in \mathcal{W}^1_{\text{loc}} \cap \mathcal{M}^1_{\text{loc}}$. Since $M^n 1_{[\sigma_{n-1}, \sigma_n[} \leqslant n$ the proof of the theorem now reduces to proving $E(|\tilde{A}^n_\infty|^2) < \infty$.

---

[1]) In general, in this second decomposition N ist not unique.

In fact, if we have this property, then $M^n 1_{[\sigma_{n-1},\sigma_n[} + \tilde{A}^n \in \mathcal{M}^2_\infty$ and if we set

$$N := \sum_{n\geq 1} M^n 1_{[\sigma_{n-1},\sigma_n[} + \tilde{A}^n$$

$$A := \sum_{n\geq 1} A^n - \tilde{A}^n = \sum_{n\geq 1} (A^n - \tilde{A}^n)\, 1_{[\sigma_n, \infty[}$$

N^{σ_n} (resp. A^{σ_n}) is, for every σ_n, the sum of a finite number of processes in $\mathcal{M}^2_\infty$ (resp. $\mathcal{W}^1_\infty$), so N (resp. A) belongs to $\mathcal{M}^2_{\text{loc}}$ (resp. $\mathcal{W}^1_{\text{loc}}$).

Let us therefore prove that $E(|\tilde{A}^n_\infty|^2) < \infty$.

Since the R.R.C. martingale $M^n 1_{[\sigma_{n-1},\sigma_n[} + \tilde{A}^n$ is uniformly integrable, it coincides with the right-continuous version of $(E(\tilde{A}^n_\infty | \mathcal{F}_t))_{t\geq 0}$. According to § 14.5, the left-continuous version of this martingale coincides with the predictable projection of the constant process $\tilde{A}^n_\infty$. Therefore, $(M^n 1_{[\sigma_{n-1}\sigma_n[}(t^-))_{t\geq 0}$ is the predictable projection of $(\tilde{A}^n_\infty - \tilde{A}_{t-})_{t\geq 0}$.

For the process with bounded variation $\tilde{A}^n$, let us write

$$(23.5.3) \qquad |\tilde{A}^n_\infty|^2 \leqslant 2 \int_0^\infty (\tilde{A}^n_\infty - \tilde{A}^n_{t-})\, d\tilde{A}^n_t - |\Delta A_{\sigma_n}|^2 .$$

The definition of the dual predictable projection gives

$$(23.5.4) \qquad E(|\tilde{A}^n_\infty|^2) \leqslant 2\, E\left(\int_0^\infty M^n 1_{[\sigma_{n-1},\sigma_n[}(t^-)\, d\tilde{A}^n_t \right) - E(|\Delta \tilde{A}_{\sigma_n}|^2)$$

but $|M^n 1_{[\sigma_{n-1},\sigma_n[}(t^-)| \leqslant n$ and therefore

$$(23.5.5)\ E(|A^n_\infty|^2) \leqslant 2n\,(E|\tilde{A}^n_\infty|) < \infty$$

This completes the proof of the theorem. □

23.6 Theorem (Vector case). *Let $\mathbb{B}'$ be the dual Banach space of a separable Banach space $\mathbb{B}$, and $X \in \mathcal{Q}_{\text{loc}}(\mathbb{B}')$. Then*

(i) *There exists a unique (up to P-equality) representation $X = M + V$ of X, where $M \in \mathcal{M}^1_{\text{loc}}(\mathbb{B}')$ and $V \in \mathcal{W}^1_{\text{loc}}(\mathbb{B}')$ with $V_0 = 0$ and V weakly predictable. If $\mathbb{B}'$ is separable, V is actually predictable.*

(jj) *There also exists $N \in \mathcal{M}^2_{\text{loc}}(\mathbb{B}')$ and $A \in \mathcal{W}^1_{\text{loc}}(\mathbb{B}')$ such that $X = N + A$. If $\mathbb{B}'$ is separable and reflexive, the properties $X \in \mathcal{Q}_{\text{loc}}(\mathbb{B}')$,* (i) *and* (jj) *are equivalent.*

Proof. The first part of the proof is exactly like the proof of 23.5. b, using the vector version of the Meyer Decomposition Theorem. For the proof of (2°), we define $\sigma_n := \inf\{t : \|M_t\| > n\} \wedge \tau_n$ and A^n, M^n and $\tilde{A}^n$ as in the proof of 23.5. *c*. The only change consits in substituting the following equation for equation (23.5. *c*).

$$(23.6.3) \qquad \|\tilde{A}^n_\infty\|^2_{\mathbb{B}'} = \sup_{x\in\mathbb{B}\,\|x\|\leqslant 1} |\langle x, \tilde{A}^n_\infty\rangle|^2$$

$$\leqslant 2 \sup_{x\in\mathbb{B},\|x\|\leqslant 1} \left[\int_0^\infty \langle x, \tilde{A}^n_\infty - \tilde{A}^n_{t^-}\rangle\, d\langle x, \tilde{A}^n_t\rangle - |\langle x, \Delta\tilde{A}_{\sigma_n}\rangle|^2\right]$$

$$\leqslant 2 \sup_{x\in\mathbb{B},\|x\|\leqslant 1} \left|\int_0^\infty \langle x, M^n 1_{[\sigma_{n-1},\sigma_n[}(t^-)\rangle\, d\langle x, \tilde{A}^n_t\rangle\right|$$

because $(\langle x, M^n 1_{[\sigma_{n-1},\sigma_n[}(t^-)\rangle)_{t\geqslant 0}$ is clearly the predictable projection of $\langle x, \tilde{A}^n_\infty - \tilde{A}^n_{t^-}\rangle$. However, $\|M^n 1_{[\sigma_{n-1},\sigma_n[}(t^-)\| \leqslant n$ implies the inequality

$$E\|\tilde{A}^n_\infty\|^2_{\mathbb{B}} \leqslant 2\,E\Big(\sup_{\substack{x\in B\\ \|x\|\leqslant 1}} n\|x\|^2\,\|\tilde{A}^n_\infty\|\Big)$$

$$\leqslant 2n\,E\|\tilde{A}^n_\infty\|$$

and therefore the same conclusion as in Theorem 23.5. □

23.7 Definition *(Semimartingales) and remarks.* (1°) An RR.C. real process X (resp. a $\mathbb{B}'$-valued R.R.C. process X where $\mathbb{B}'$ is the Banach dual of a separable Banach space $\mathbb{B}$) is called a *semimartingale* with respect to the given filtration $\mathscr{F}$ when X can be written $X = M + V$, where $M \in \mathscr{M}^2_{\mathrm{loc}}(\mathscr{F})$ (resp. $\mathscr{M}^2_{\mathrm{loc}}(\mathbb{B}';\mathscr{F})$ and $V \in \mathscr{W}^0(\mathscr{F})$ (resp. $\mathscr{W}^0(\mathbb{B}';\mathscr{F})$.

We will denote by $S(\mathscr{F})$ (resp. $\mathscr{S}(\mathbb{B}';\mathscr{F})$ the class of real (resp. $\mathbb{B}'$-valued) semimartingales. We clearly have

$$\mathscr{Q}_{\mathrm{loc}} \subset S.$$

Elements of $\mathscr{Q}_{\mathrm{loc}}$ are usually called *special semimartingales* (see [Mey. 3] or [Jac. 2])

(2°) These notions of semimartingales and special semimartingales (at least in the case of real processes) have been introduced by P.A. Meyer. The reason for introducing semimartingales is that they will turn out to be stable for some important operations, in particular (this will be proved in section 25) for the following operation. If Z is a semimartingale, $((Z_t))_{t\geqslant 0}$ is a semimartingale for every twice continuously differentiable function φ.

(3°) It should be emphasized that, as in 23.5. c, the decomposition $X = M + V$, $M \in \mathscr{M}^2_{\mathrm{loc}}$, $V \in \mathscr{W}^0$ is not unique. It is, however, easy to see that, when X is Hilbert valued, the process M^c (cf. 23.3) does not depend on the decomposition. We leave the proof of this as an exercise E. 13. This process will be called the *continuous local martingale part of* X and be denoted by X^c.

23.8 The stochastic integral of elementary processes

By an *elementary predictable* $\mathbb{H}$-*valued process* (for a given filtration $\mathscr{F}$ and $\mathbb{H}$ being a separable Hilbert space) we mean a process X of the form

(23.8.1) $$X = \sum_{i=1}^{n} 1_{]s_i, t_i] \times F_i} \alpha_i, \quad \text{where} \quad F_i \in \mathscr{F}_{s_i}, \; s_i \leqslant t_i, \; \alpha_i \in \mathbb{H}$$

and n is any positve integer. Z being a real R.R.C. process, we denote by $\int X dZ$ (or sometimes $X o Z$) the $\mathbb{H}$-valued process defined for $t \geqslant 0$ by

(23.8.2) $$\left(\int X dZ\right)_t := \sum_{i=1}^{n} 1_{F_i}(Z_{t_i \wedge t} - Z_{s_i \wedge t}) \cdot \alpha_i$$

This process is clearly independent of the particular representation (23.8.1) of X. One also writes $\int_{]0,t]} X dZ$ or $\int_{]0,t]} X_s dZ_s$ for the Random variable $(\int X dZ)_t$.

When $Z \in \mathscr{W}^0$, the random variable $\int_{]0,t]} X dZ$ is nothing but the Stieltjes integral with respect to $t \curvearrowright Z_t(\omega)$ for every ω.

23.9 Theorem (Existence of "control processes"). *Let Z be a real semimartingale. There exists a process $A \in \mathscr{W}_+^0$ with the following property. For every elementary predictable real process X and every stopping time τ, one has*

(23.9.1) $$E\{\|\sup_{t<\tau} \int_{]0,t]} X dZ\|_{\mathbb{H}}^2\} \leqslant E(A_{\tau-} \int_{]0,\tau[} \|X_s\|_{\mathbb{H}}^2 dA_s).$$

We will call such a process A a *control process* for Z. In the terminology of [MeP. 1] this can be stated as follows. *Every real semimartingale is a Π^*-process.*

Proof. Let $Z = M + V$ with $M \in \mathscr{M}_{\text{loc}}^2 V \in \mathscr{W}^0$. One has

$$\int X dZ = \int X dM + \int X dV$$

and therefore

(23.9.2) $$E\{\|\sup_{t<\tau} \int_{]0,t]} X dZ\|^2\} \leqslant 2E\{\|\sup_{t<\tau} \int_{]0,t]} X dM\|^2\} + 2E\{\|\sup_{t<\tau} \int_{]0,t]} X dV\|^2\}.$$

From the definition in 16.8 it follows immediately that, for every stopping time σ, $(\int X dZ)^\sigma = \int X dZ^\sigma$ and that $\int X dM$ is an element of $\mathscr{M}_{\text{loc}}^2$ (use a localizing sequence (σ_n) for M).

Let σ be a stopping time such that $M^\sigma \in \mathscr{M}_\infty^2$. Assuming we take a representation (23.8.1) of X in which the rectangles $]s_i, t_i] \times F_i$ are disjoint, we may write

$$[\![(\int X dM)_t^\sigma]\!]^2 - \int_{]0,t\wedge\sigma]} \|X\|^2 d\langle M\rangle = \sum_{i=1}^{n} \|\alpha_i\|^2 1_{F_i}[(M_{t_i\wedge t}^\sigma - M_{s_i\wedge t}^\sigma)^2 - (\langle M^\sigma\rangle_{t_i\wedge t} - \langle M^\sigma\rangle_{s_i\wedge t})]$$

Since $(M^{\sigma}_{t_i \wedge t} - M^{\sigma}_{s_i \wedge t})^2 - (\langle M^{\sigma} \rangle_{t_i \wedge t} - \langle M^{\sigma} \rangle_{s_i \wedge t})$ is a martingale we obtain

$$\langle \int X dM \rangle \leqslant \int \|X\|^2 d\langle M \rangle$$

Similarly, one finds immediately from the definition (see 19.3) that

$$[\overset{\vee}{\overline{\int X dM}}] \leqslant \int \|X\|^2 d[\overset{\vee}{M}] \tag{23.9.3}$$

Theorem 19.4 then shows that for every stopping time,

$$E(\|\sup_{t<\tau} \int_{]0,t]} X dM\|^2) \leqslant 4E(\int_{]0,\tau[} \|X\|^2 d(\langle M \rangle + [\overset{\vee}{M}])). \tag{23.9.4}$$

Now let $|V|$ be the variation process of V. The Schwarz inequality for the Stieltjes integral gives

$$\|\sup_{t<\tau} \int_{]0,t]} X dV\|^2 \leqslant [\int_{]0,\tau[} \|X\| d|V|]^2 \leqslant |V|_{\tau-} \int_{]0,\tau[} \|X_s\|^2 d|V|_s \tag{23.9.5}$$

If we set

$$A := 8(\langle M \rangle + [\overset{\vee}{M}]) + 2(2 \vee |V|)$$

the inequalities (23.9.2), (23.9.4) and (23.9.5) give the inequality of the theorem. □

23.10 Notation. We shall denote by $\mathscr{A}(Z)$ the set of all control processes of the semimartingale Z. It was proved in [MeP. 3] that every real Π^*-process is a semimartingale. This is expressed by the following theorem.

23.11 Theorem. *Every R. R. C. real process Z which admits a control process is a semimartingale. More precisely, if there exists a positive increasing process A, such that (23.9.1) holds for every real, elementary, predictable X and every stopping time τ, Z is a semimartingale.*

Proof. Let $A \in \mathscr{A}(Z) \neq \emptyset$ and $\sigma_n := \inf\{t: A_t > n\}$. We define

$$\tilde{Z} := Z - \sum_n \Delta Z_{\sigma_n} 1_{\{\sigma_n > \sigma_{n-1}\}} 1_{[\sigma_n, \infty[}. \tag{23.11.1}$$

The process $\tilde{Z}$ has thus no jump at time σ_n and the process $\int X d\tilde{Z}^{\sigma_n}$ has no jump at time σ_n.

Therefore,

$$E\{\sup_{t \leqslant \sigma_n} |\int X d\tilde{Z}^{\sigma_n}|^2\} = E\{\sup_{t<\sigma_n} |\int X d\tilde{Z}^{\sigma_n}|^2\} \leqslant nE(\int_{]0,\sigma_n[} |X|^2 dA). \tag{23.11.2}$$

For every sequence $0 = t_0 < t_1 < \ldots < t_n$ one has

$$\sum_{i=1}^{n} E|E(\tilde{Z}^{\sigma_n}_{t_{i+1}} | \mathscr{F}_{t_i}) - \tilde{Z}^{\sigma_n}_{t_i}| = E(\int_{]0,\sigma_n[} X d\tilde{Z}^{\sigma_n})$$

with

$$X := \sum_{i=1}^{n} (1_{]t_i,t_{i+1}]\times F_i} - 1_{]t_i,t_{i+1}]\times F_i^c}),$$

where

$$F_i := (E(\tilde{Z}^{\sigma_n}_{t_{i+1}} \mid \mathscr{F}_{t_i}) - \tilde{Z}^{\sigma_n}_{t_i} > 0\}$$

This remark and inequality (23.11.2) lead to

$$|\lambda_{\tilde{Z}^{\sigma_n}}|(\mathbb{R}^+ \times \Omega) \leqslant \sup_{|X|\leqslant 1} E\Big|\int_{]0,\sigma_n]} X\,d\tilde{Z}^{\sigma_n}\Big|$$

(23.11.3) $$|\lambda_{\tilde{Z}^{\sigma_n}}|(\mathbb{R}^+ \times \Omega) \leqslant \sup_{|X|\leqslant 1} E\Big(\Big|\int_{]0,\sigma_n]} X\,d\tilde{Z}^{\sigma_n}\Big|^2\Big)\Big]^{\frac{1}{2}} \leqslant [nE(A_{\sigma_n -})]^{\frac{1}{2}}$$

This shows that $\tilde{Z}^{\sigma_n} \in \mathscr{Q}$. As $\sum_n \Delta Z_{\sigma_n} 1_{\{\sigma_n > \sigma_{n-1}\}} 1_{[\sigma_n,\infty[}$ is clearly in $\mathscr{W}^0$, (23.11.3) and (23.11.1) yield the theorem. □

23.12 The stochastic integral of elementary processes (vector case)*

Let $\mathbb{B}'$ be the Banach dual space of a separable Banach space $\mathbb{B}$ and $\mathbb{G}$ a separable Hilbert space. We consider a subspace $\mathbb{L}$ of $\mathscr{L}(\mathbb{B}'; \mathbb{G})$ (the space of bounded linear operators from $\mathbb{B}'$ into $\mathbb{G}$).

By an elementary predictable $\mathbb{L}$-valued process (for a given filtration $\mathscr{F}$) we mean a process X of the form

(23.12.1) $$X = \sum_{i=1}^{n} 1_{]s_i,t_i]\times F_i} a_i,$$

where $F_i \in \mathscr{F}_{s_i}$, $s_i \leqslant t_i$, $a_i \in \mathbb{L}$ and n is any positive integer. Z being a $\mathbb{B}'$-valued R.R.C. process, we denote by $\int X\,dZ$ the $\mathbb{H}$-valued process defined for $t > 0$ by

(23.12.2) $$\Big(\int X\,dZ\Big)_t := \sum_{i=1}^{n} 1_{F_i} a_i (Z_{t_i\wedge t} - Z_{s_i\wedge t}).$$

One also writes $\int_{]0,t]} X\,dZ$ or $\int_{]0,t]} X_s\,dZ_s$ for the $\mathbb{H}$-valued random variable $(\int X\,dZ)_t$.

Since we did not define the quadratic variation for a $\mathbb{B}'$-valued martingale (the definition of this concept was fundamentaly based on the Hilbert structure), we have to restrict ourselves to Hilbert-valued martingales to state the extensions of Theorems 23.9 and 23.11 to vector-valued (finite- or infinite-dimensional) processes.

23.13 Definition. Let $\mathbb{H}$ and $\mathbb{G}$ be two separable Hilbert spaces and $\mathbb{L}$ a closed linear subspace of $\mathscr{L}(\mathbb{H}; \mathbb{G})$. Let Z be a $\mathbb{H}$- valued semimartingale. An increasing, positive, adapted process will be called an $\mathbb{L}$-*control* process for Z if, for every elementary $\mathbb{L}$-valued predictable process X and every stopping time τ, one has

(23.13.1) $$E\Big\{\Big\|\sup_{t<\tau} \int_{]0,t]} X\,dZ\Big\|^2_{\mathbb{G}}\Big\} \leqslant E\Big(A_{\tau-} \int_{]0,\tau[} \|X_s\|^2\,dA_s\Big).$$

A will be called a *control process* if this holds for every $\mathbb{G}$ with $\mathbb{L} = \mathscr{L}(\mathbb{H};\mathbb{G})$. We call $\mathscr{A}(Z)$ the set of control processes for Z.

Let us remark that Theorem 23.11 states the following. Every real process which admits an $\mathbb{R}$-control process is a semimartingale and therefore admits control processes. In the infinite-dimensional case (as noted in [MeP. 1]) the existence of an $\mathbb{R}$-control process is no longer sufficient (see exercise E. 10).

The extension to vector-valued processes is given in the following theorem.

23.14 Theorem. (Characterization of semimartingales)*. *Let* $\mathbb{H}$ *be a separable Hilbert space.*

(1°) *An* R.R.C. $\mathbb{H}$*-valued process is a semimartingale if and only of it admits a control process.*

(2°) *If* Z *is an* R.R.C. $\mathbb{H}$*-valued process admitting an* $\mathbb{H}$*-control process* ($\mathbb{H}$ *being considered as* $\mathscr{L}(\mathbb{H};\mathbb{R})$), *then it is a semimartingale.*

Proof. If Z is an $\mathbb{H}$-valued semimartingale, there is nothing to change in the proof of Theorem 23.9 and a control process is still given by

$$A := 8(\langle M\rangle + [\check{M}]) + 2(2 \vee |V|),$$

where $Z = M + V$ is a decomposition of Z with $M \in \mathscr{M}^2_{\mathrm{loc}}(\mathbb{H})$ and $V \in \mathscr{W}^0(\mathbb{H})$.

To prove the converse we have only to prove part 2° of Theorem 23.14.

Let A be an $\mathbb{H}$-control process for Z. This means that for every $X := \sum_{i=1}^{n} h_i 1_{]s_i,t_i]\times F_i}$, $F_i \in \mathscr{F}_{s_i}$, $s_i < t_i$, $h_i \in \mathbb{H}$, one has for every stopping time τ,

$$E\{\sup_{t<\tau} |\textstyle\int\langle X, dZ\rangle|^2\} \leqslant E\{A_{\tau-} \int_{]0,\tau[} \|X_s\|^2\, dA_s\}. \tag{23.14.1}$$

It is immediately seen that this inequality is still true if we consider processes X of the form

$$X := \sum_{i=1}^{n} U_i\, 1_{]s_i,t_i]},$$

where U_i is a bounded, $\mathbb{H}$-valued, random variable which is $\mathscr{F}_{s_i}$-measurable.

Let us define then $\tilde{Z}$ as in (23.11.1) and remark that

$$\sum_{i=1}^{n} E(\|E(\tilde{Z}^{\sigma_n}_{t_{i+1}}|\mathscr{F}_{t_i}) - \tilde{Z}^{\sigma_n}_{t_i}\|_{\mathbb{H}}) = E\Big(\int_{]0,\sigma_n[} \langle X, d\tilde{Z}^{\sigma_n}\rangle\Big) \tag{23.14.2}$$

if

$$X := \sum_{i=1}^{n} 1_{]t_i,t_{i+1}]} \frac{E(\tilde{Z}^{\sigma_n}_{t_{i+1}}|\mathscr{F}_{t_i}) - \tilde{Z}^{\sigma_n}_{t_i}}{\|E(\tilde{Z}^{\sigma_n}_{t_{i+1}}|\mathscr{F}_{t_i}) - \tilde{Z}^{\sigma_n}_{t_i}\|_{\mathbb{H}}}.$$

The fact that A is an $\mathbb{H}$-control process leads to the inequality

$$\sum_{i=1}^{n} E(\|E(\tilde{Z}^{\sigma_n}_{t_{i+1}}|\mathscr{F}_{t_i}) - \tilde{Z}^{\sigma_n}_{t_i}\|_{\mathbb{H}}) \leqslant [n\, E(A_{\sigma_n^-})]^{\frac{1}{2}} \tag{23.14.3}$$

and therefore to

$$|\lambda_{Z^{\sigma_n}}|(\mathbb{R}^+ \times \Omega) \leqslant n\,[E(A_{\sigma_n^-})]^{\frac{1}{2}},$$

which expresses $\tilde{Z}^{\sigma_n} \in \mathcal{Q}(\mathbb{H})$.

Since $\sum_n \Delta Z_{\sigma_n} 1_{\{\sigma_n > \sigma_{n+1}\}} 1_{[\sigma_n, \infty[} \in \mathcal{W}^0(\mathbb{H})$

we have proved

$$Z = \tilde{Z} + A, \quad \text{when} \quad \tilde{Z} \in \mathcal{Q}_{\text{loc}}(\mathbb{H}),\ A \in \mathcal{W}^0(\mathbb{H}). \quad \square$$

23.15 Remark. (1°) The example given below in exercise E. 10 and taken from [MeP. 3] shows that in order to be a semimartingale, it is not sufficient for an R.R.C. $\mathbb{H}$-valued process Z to admit an $\mathbb{R}$-control process.

(2°) In the language of M. Metivier and J. Pellaumail [MeP. 1], Theorem 23.14 says that every semimartingale is a Π^*-process.

Exercises and supplements

1. Square integrable martingales

E. 1. Let $M \in \mathcal{M}^2$, $X \in \Lambda^2(\mathcal{P}, \mathcal{M})$ and $N = \int X dM$. For every stopping time T one has $\Delta N_T = X_T \Delta M_T$.

(*H*int: Show that the equality holds for every simple predictable process X. Use Lemma 16.3).

E. 2. Let $M = M^c + \sum_n A_n - \tilde{A}_n$ be the decomposition of M in $\mathcal{M}^2_\infty$ and $(T_n)_{n>0}$ as in Theorem 17.7.

Let $X \in L^2(\mathbb{R}^+ \times \Omega, \mathcal{P}, \lambda_M)$ and $N = \int X dM$.

Show that $N^c = \int X dM^c$, that N has no jump outside $\bigcup_n [T_n]$, $\Delta N_{T_n} = X_{T_n} \Delta M_{T_n}$, and that the orthogonal decomposition of N associated with (T_n) by Theorem 17.7 is

$$N = \int X dM^c + \sum_n \int X d(A_n - \tilde{A}_n).$$

(*Hint:* Use Proposition (18.13.1) and $\langle H, \int X dM^d \rangle = \int X d\langle H, M^d \rangle$ to show that $\int X dM^d$ is in the orthogonal subspace of $\mathcal{M}^{2,c}_\infty$, that is, in $\mathcal{M}^{2,d}_\infty$. Then use exercise E. 1.).

E. 3. Let $M \in \mathcal{M}^2$ and $X \in L^2(\mathcal{P}, M)$. Show that $\int X dM$ is the unique element in $\mathcal{M}^2_\infty$ with the property

$$\forall N \in \mathcal{M}^2_\infty, \quad E((\int_0^\infty X dM) N_\infty) = E(\int_0^\infty X_s d\langle M, N\rangle_s) = E(\int_0^\infty X_s d[M, N]_s).$$

(*Hint*: Use Proposition 18.13). This characteristic property of the martingale $\int X dM$ is taken as the definition of the stochastic integral by P.A. Meyer in [Mey. 3].

E. 4. Let $(\Omega, (\mathcal{F}_t)_{t \in \mathbb{R}^+}, P)$ be the stochastic basis as defined in the example of section 4.8 and T be the totally inaccessible stopping time $T(\omega) := \omega$. We set $A := 1_{[T, \infty[}$

(a) What is the dual predictable projection $\tilde{A}$ of A?

(b) Let M be the martingale $A - \tilde{A}$. What is $[M]$?

(c) Evaluate $E(\sup_{t<T} |M_t|^2)$, $E(\langle M\rangle_{T^-})$, $E([M]_{T^-})$.

Answers:

(a) $$\tilde{A}_t(\omega) = \int_0^{t \wedge \omega} \frac{1}{1-u} du$$

(b) $$[M] = A$$

(c) $$\sup_{t<T} |M_t|^2 = 1, \quad \langle M\rangle_{T^-} = 1, \quad [M]_{T^-} = 0$$

This produces a counter-example to formula (19.2.2).

E. 5. We consider for Ω the two points set

$$\Omega := \{1, 2\}$$
$$\mathcal{F}_t := \{\emptyset, \Omega\}, \quad \text{if} \quad t < 1,$$
$$\mathcal{F}_t := \mathfrak{P}(\Omega), \quad \text{if} \quad t \geq 1.$$

We define $u(\omega) = \omega$. Show that u is an accessible non-predictable stopping time. If $P\{1\} = p_1$ and $P\{2\} = p_2 = 1 - p_1$, we define a martingale M by setting

$$M_t := \begin{cases} 0 & \text{if} \quad t < 1 \\ \dfrac{1}{p_1} 1_{\{1\}} - \dfrac{1}{p_2} 1_{\{2\}} & \text{if} \quad t \geq 1. \end{cases}$$

One can easily evaluate

$$E(\sup_{s<u} |M_s|^2) = \frac{1}{p_2} \quad \text{and} \quad E(\langle M\rangle_{u^-}) = \frac{1}{p_1}.$$

This provides us with a counter-exemple to an inequality of the type

$$E(\sup_{s<u} |M_s|^2) \leq C E(\langle M\rangle_{u^-})$$

for some constant C.

E. 6. The following example shows the impossibility of a formula of the type

$$E(\sup_{s<u} |M_s|^2) \leqslant CE([M]_{u-}).$$

Take

$$\Omega := \{1, 2, \ldots, n\}.$$

For $t \in [k, k+1[$, the σ-algebra $\mathscr{F}_t$ is generated by the atoms $\{j\}_{j \leqslant \inf(k,n)}$.

Let q be a positive number with $0 < q < 1$,

$$P\{1\} = q,\ P\{2\} = (1-q)q, \ldots, P(n-1) = (1-q)^{n-2} q,$$
$$P(n) = (1-q)^{n-1}.$$

The martingale M is defined as follows. $M_t(\omega)$ is constant for $t \in [k, k+1[$ and

$$M_0 = 0,\ M_{k+1} = M_k - \frac{1}{q} 1_{\{k+1\}} + \frac{1}{(1-q)} 1_{\{k+2, \ldots, n\}}.$$

Show that the stopping time u, defined by $u(\omega) = \omega$, is accessible but not predictable. Compute $E[M]_{u-}$ and $E(\sup_{0 \leqslant t < u} |M_t|^2)$ and show that

$$\lim_{u \to 0} \frac{E[M]_{u-}}{E(\sup_{0 \leqslant t < u} |M_t|^2)} = \frac{1}{n-1}.$$

The following exercises show that *even in the finite dimensional case* the theory of Hilbert-valued square integrable martingales differs somewhat from the theory of real-valued ones.

E. 7. *Very Strong orthogonality.* (1°) Two Hilbert-valued martingales M and $N \in \mathscr{M}^2_\infty(\mathbb{H})$ are said to be *very strongly orthogonal* if $\lambda_{M \otimes M} = 0$ in which case, one also has $\lambda_{N \otimes M} = 0$.

(a) If M and N are very strongly orthogonal, they are strongly orthogonal (we use the abreviation V. S. O.).

(b) M and N are V. S. O. iff $\langle\langle M, N \rangle\rangle = 0$.

(2°) Let W_1 and W_2 be two real Brownian motions which are independant with respective covariances σ_1^2 and σ_2^2. One defines

$$M = \begin{pmatrix} W_1 \\ W_2 \end{pmatrix} \in \mathscr{M}^2(\mathbb{R}^2), \quad N = \begin{pmatrix} W_2 \\ W_1 \end{pmatrix} \in \mathscr{M}^2(\mathbb{R}^2).$$

One has $\langle M, N \rangle = 0$, while $\langle\langle M, N \rangle\rangle = \begin{pmatrix} 0 & \sigma_1^2 t \\ \sigma_2^2 t & 0 \end{pmatrix}$.

This proves that V. S. O. is not implied by strong orthogonality.

3°) Prove that in the decomposition 20–2 the martingales are V. S. O.

E. 8. *A representation theorem* (see J. Y. Ouvrard [Ouv. 1]). Let $M \in \mathscr{M}^2_\infty(\mathbb{H})$, $\mathscr{H}_1 := \{\int X dM : X \in \Lambda^2(\mathbb{H}; \mathbb{G}, \mathscr{P}, M)\} \subset \mathscr{M}^2_\infty(\mathbb{G})$. Call $\mathscr{H}_2$ the orthogonal subspace of $\mathscr{H}_1$ in $\mathscr{M}^2_\infty(\mathbb{G})$.

Show (compare with 18.15) that every element of $\mathscr{H}_2$ is V.S.O. to every element in $\mathscr{H}_1$ and that every $L \in \mathscr{M}^2_\infty(\mathbb{G})$ has a unique representation

$$L = \int X\,dM + N,\, X \in \Lambda^2(\mathbb{H}; \mathbb{G}; \mathscr{P}, M),\quad N \in \mathscr{H}_2\,.$$

E. 9. *On processes which are integrable in the sense of* §22 *but not in the sense of* §20 (in the finite-dimensional case).

Let f_1 and f_2, h_1, h_2 be four functions on $\mathbb{R}^+$ such that

$$\int_0^\infty f_1^2(s)\,ds < \infty \quad \int_0^\infty f_2^2(s)\,ds < \infty \quad \int_0^\infty h_i^2(s) f_i^2(s) < \infty \quad \text{for} \quad i = 1, 2 \quad \text{and}$$

$$\int_0^\infty h_i^2(s) f_j^2(s)\,ds = \infty \quad \text{if} \quad i \neq j\,.$$

W_1 and W_2 being two independant Brownian motions with covariance 1, we define

$$M := \begin{pmatrix} \int f_1\,dW_2 \\ \int f_2\,dW_2 \end{pmatrix} \in \mathscr{M}^2_\infty(\mathbb{R}^2)\,.$$

Show that the deterministic process $X = \begin{pmatrix} h_1 \\ h_2 \end{pmatrix}$ is not integrable in the sense that $\int \|X\|^2\,d\mu_M = +\infty$, while it belongs to $\Lambda^2(\mathbb{R}^2, R, \mathscr{P}, M)$, the integral being the real Gaussian process

$$\left(\int_0^t h_1(s) f_1(s)\,dW_1(s) + \int_0^t h_2(s) f_2(s)\,dW_2(s)\right)_{t \geqslant 0}\,.$$

E. 10. We consider $\mathbb{H} = L^2([0, 1], \mathscr{B}_{[0,t]}, \lambda)$, λ being the Lebesgue measure, and define the deterministic process $X_t := 1_{]0,t]}$.

Show that this process admits the process t as an $\mathbb{R}$-control process but is not a semimartingale.

2. Semimartingales

E. 11. If Z is a Hilbert-valued semimartingale with bounded jumps, then $Z \in \mathscr{Q}_{\text{loc}}$. (*Hint:* Define $\tau_n := \inf\{t : \|Z_t\| > n\} \wedge \inf\{t : \|N_t\| > n\} \wedge \inf\{t : \|A_t\| > n\}$, where $Z = N + A$, $N \in \mathscr{M}^2_{\text{loc}}(\mathbb{H})$ and $A \in \mathscr{W}^0(\mathbb{H})$. Show that $E\|\Delta A_{\tau_n}\|^2 < \infty$ and therefore $A \in \mathscr{W}^1_{\text{loc}}$).

E. 12. Let X be an $\mathbb{R}^d$-valued process with increments independent of the past for a given filtration. We assume X stochastically continuous, regular, right-continuous (see chapter 2, E. 4.2, for this latter assumption). Then (1°) X is a semimartingale, if and only if $t \curvearrowright E(X_t)$ has finite variation, (2°) if X has bounded jumps, X is a quasi-martingale which can be uniquely written as $X = M + A$ where M is a square integrable martingale and A is a predictable process with integrable variation.

Hints: Take $C > 0$. In each finite interval X has a finite number of jumps with norm greater than C. The process $V_t := \sum_{s \leqslant t} 1_{]C,\infty[} \Delta X_s$ then has finite variation and $X - V$ is a process with independent increments and bounded jumps. The problem is then reduced to proving part (2°). Then use the finiteness of all moments of $\sup_{s \leqslant t} \|X_s - X_0\|$ (see chapter 1, E. 21) to prove that $(X_t - E(X_t))_{t \geqslant 0} \in \mathcal{M}^2(\mathbb{R}^d)$, with $t \curvearrowright E(X_t)$ continuous.

E. 13. *Continuous local martingale part of a semimartingale.* Let X be a semimartingale with values in the Hilbert space $\mathbb{H}$, and let $X = M + V$, $X = N + W$ be two decompositions of X, with M and N in $\mathcal{M}^2_{\text{loc}}(\mathbb{H}, \mathcal{F})$, V and W in $\mathcal{W}^0(\mathbb{H}, \mathcal{F})$. Prove that $M^c = N^c$.

Hint: Show that $N \in \mathcal{M}^2_{\text{loc}}(\mathbb{H}, \mathcal{F}) \cap \mathcal{W}^0(\mathbb{H}; \mathcal{F})$ implies $N = 0$.

E. 14. *Topology on semimartingales* (cf. [MeP. 1], chap. 3). Let us consider the set $\mathcal{S}$ of $\mathbb{H}$-valued semimartingales defined on a given stochastic basis, for $t \in [0, T]$.

For every $Z \in \mathcal{S}$ we set

$$F(Z) := \inf\{E(1 \wedge A_T) : A \in \mathcal{A}(Z)\}.$$

Define for every $Z \in \mathcal{S}$ and $\varepsilon > 0$

$$w(Z, \varepsilon) := \{X : X \in \mathcal{S},\ F(X - Z) < \varepsilon\}.$$

(1°) $w(Z, \varepsilon)$, $\varepsilon > 0$, defines a basis of neighbourhoods of Z for a metrizable topology on $\mathcal{S}$ which makes $\mathcal{S}$ a metrizable topological vector space.

Hint: Remark that $F(Z + Z') \leqslant 2F(Z) + 2F(Z')$.

(2°) The topological vector space $\mathcal{S}$ is complete.

Hints: Consider a sequence (Z^n) and assume that for every n there exists $A^n \in \mathcal{A}(Z^{n+1} - Z^n)$ such that $E(1 \wedge A^n_T) \leqslant 8^{-n}$. Show, using a Borel-Cantelli argument that the sequence (Z^n) converges P. a. s. uniformly on $[0, T]$ to a process Z admitting the process $A := \sum_n 2^n A^n$ as a control process and that

$$\sum_{k \geqslant 0} 2^{k+1} A^{n+k} \in \mathcal{A}(Z - Z^n).$$

Historical and bibliographical notes

The importance of square integrable martingales for stochastic integration was recognized by Doob [Doo. 1]. Ph. Courrege [Cou] and Meyer [Mey. 4] gave the first systematic treatment, which is close to the account given here in section 18. H. Kunita and S. Watanabe gave the theory its present status [KuW]. H. Kunita [Kun] made the first extension of stochastic integrals to Hilbert-valued square integrable martingales. The explicit introduction of the Meyer process $\langle\langle M \rangle\rangle$ was made in [MeP. 5] and the isometric stochastic integral in the vector case (finite- and infinite-dimensional) was made by Metivier-Pistone [MPi. 2]. Galtchouk introduced the same integral in the finite-dimensional case to obtain integral representation theorems [Gal. 1] (see also Ouvrard [OuV. 1] and E. 8). The account given here follows [Met. 4].

The fundamental structure theorem, Theorem 17.7, and everything in section 17 is reproduced from [Mey. 3].

The stopped inequalities in section 19 are due to M. Metivier and J. Pellaumail (MeP. 3]. The account is the same as in [MeP. 1] (see also [Mey. 6].

The definition of semimartingales given here was introduced in the Doléans-Dade-Meyer paper [DoM. 1]. The characterization with the help of control processes was given by M. Metivier and J. Pellaumail [MeP. 1, 3]. There are important questions which are left aside in this book of which we now give a short bibliographical account.

Semimartingales and stochastic measures

In a series of papers M. Metivier and J. Pellaumail studied L^0-valued measure on predictable sets associated with a semimartingale [Pel. 1, 2], [MeP. 4], [Met. 5] (see also Kußmaul [Kus]. It was noticed in [Met. 5] that, as an easy consequence of the Doob-Meyer theorem, every L^1-valued stochastic measure was associated with a semimartingale. The general result, which stated that every L^0-stochastic measure is uniquely associated with a semimartingale and conversely (for finite-dimensional processes), was proved by Dellacherie-Mokobodski-Meyer and rediscovered by K. Bichteler. This gives an interesting characterization of semimartingales and explains why semimartingales are in some sense the "most general processes" for which a stochastic integral has been defined. For an exposition of these theorems see [MeP. 1], chap. 5. See also [Del. 4, 5].

This characterization of semimartingales led to the theory of "formal semimartingales" by L. Schwartz [Schw. 1], which are essentially defined as stochastic measures and where an extended notion of localization is used.

Topologies on spaces of martingales and semimartingales

We essentially considered $\mathcal{M}^2$ and only mentioned $\mathcal{M}^p$. Very interesting spaces of martingales are the spaces $\mathcal{H}^p := \{M : M \text{ martingale } E(\sup|M_t|^p) < \infty\}\ p \geqslant 1$

endowed with the norm $\|M\|_{\mathscr{H}^p} := [E(\sup_t |M_t|^p)]^{1/p}$. Because of the Doob inequality, $\mathscr{M}^p = \mathscr{H}^p$ for $p > 1$. The space $\mathscr{H}^1$ is more interesting than $\mathscr{M}^1$ and its topology may be defined by different norms, the equivalence of which follows from deep and famous inequalities of Burkholder, Garsia, Gundy, Fefferman. A simple synthetic exposition of this may be found in [MeP. 1], chap. 4, but the reader will find an exposition with many bibliographical references in chap. VII of [DeM]. We mention in references only a few classical contributions to this question, e. g., [Bur. 2, 3], [Gar].

Topologies on spaces of semimartingales were studied by M. Emery [Eme. 1, 2, 3], in connection with "stability properties" of stochastic differential equations. See also Ph. Protter [Pro. 3]. For an account of this see the book by Dellacherie and Meyer, again, chap. VII.

The invariance of the semimartingale property

An important question is whether a process which is a semimartingale remains a semimartingale if the filtration or the probability law or the time is changed. We shall later prove the invariance by an absolutely continuous change of probability (chapter 6).

We saw in chapter 2, exercise E. 1.1, that, under an augmentation of the filtration, a martingale does not remain a martingale (example taken in [Ito. 1]), but does remain a semimartingale. The effect of a change in filtration has been studied by Jeulin [Jeu] and Yor [Yor. 3]. Chap. IX of [Jac. 2] is devoted to this problem. To which extend the filtration can be reduced is studied by Stricker in [Str. 3].

The invariance by "change of time", based on results by Kazamaki [Kaz. 2], is described briefly in [Mey. 3], chap. IV, and extensively in [Jac. 1], chap. X.

Representation theorems and stable spaces of martingales

The question is to know when a martingale N can be expressed as a stochastic integral with respect to a martingale M or, in other words, belongs to the stable subspace generated by M (in a given space of martingales). We refer to [Jac. 1], chap. IV, for an exposition of results and references.

Part II:

Stochastic Calculus

Chapter 5

Stochastic integral with respect to semimartingales and the transformation formula

This chapter is devoted to the construction of the stochastic integral of predictable processes with respect to a general semimartingale and to the main formula of stochastic integration, which is the foundation of the stochastic calculus, namely, the transformation formula.

In building the stochastic integral with respect to the semimartingale Z the control processes of Z, as introduced in §23, play a fundamental role. A predictable process X is integrable with respect to Z iff there exists a control process A for which the real process $(\int_{]0,t]} \|X_s\|^2 dA_s)_{s \in \mathbb{R}^+}$ is finite a. s. (Integrals here being Stieltjes integrals along the paths of X and A). The stochastic integral possesses continuity properties for the seminorms associated with the control processes.

The transformation formula, as presented here, originates in the Ito formula, which Ito proved for stochastic integrals with respect to Brownian motion and which was further extended to more general situations by A. Kunita, S. Watanabe, P. A. Meyer, C. Doleans-Dade and others. It essentially says that, if Z is a semimartingale and φ a twice differentiable function on the state-space of Z, $\varphi(Z)$ is also a semimartingale. The formula also gives a representation of $\varphi(Z)$ as a stochastic integral. This formula is to be compared with the classical "differentiation formula"

$$\varphi(Z_t) = \varphi(Z_0) + \int_0^t \varphi'(Z_s)\, dZ_s\,,$$

which is valid when Z is a *continuous* process with paths of finite variation.

The special feature of the "transformation formula" compared with the "differentiation formula" is that it involves second-order derivatives. In the special case of a continuous martingale Z, for example, it will read

$$\varphi(Z_t) = \varphi(Z_0) + \int_0^t \varphi'(Z_s)\, dZ_s + \frac{1}{2}\int_0^t \varphi''(Z_s)\, d\langle Z\rangle_s$$

This appearance of second-order derivatives is sometimes seen as intolerable by people who try to avoid it by inventing other stochastic integration theories (Stratonovitch-type integral: see E. 8). It should be remarked at this point that if one considers a Stieltjes integration with respect to a process Z which is increasing and has isolated jump one has

$$\varphi(Z_t) = \varphi(Z_0) + \int_0^t \varphi'(Z_{s-})dZ_s + \sum_{s \leqslant t} [\varphi(Z_s) - \varphi(Z_{s-}) - \varphi'(Z_{s-})\Delta Z_s],$$

the term under the sign $\sum_{s \leqslant t}$ being of the order of $\frac{1}{2}\varphi''(Z_{s-})(\Delta Z_s)^2$ if the jumps are small.

If one remembers that continuous processes introduced in modeling physical phenomena are often idealized versions (by a limiting procedure) of processes modified by frequent incoherent perturbations of microscopic discontinuous character, the presence of a term $\int_0^t \varphi''(Z_s)d[Z]_s$, where $[Z]$ is an increasing process associated with Z (its quadratic variation), is not only natural but essential for the physical interpretation, the process $[Z]$ having the approximate signification of the accumulated "energy" of perturbating shocks $\sum_{s \leqslant t} |\Delta Z_s|^2$ until time t.

For the comfort of the reader the stochastic integral is presented first in the case of real processes. This is the goal of section 24.

In § 25 we define the quadratic variation $[Z]$ of a real semimartingale and establish the transformation formula in this case.

Sections 26 and 27 deal with the same problems for multidimensional semimartingales.

24. Stochastic integral in the real case

In this section we consider only the stochastic integral of real-valued processes with respect to real-valued semimartingales. This is done for the sake of having simple notation. The multidimensional case will be treated in section 26 in exactly the same way.

24.1 Definition of stochastic integrals. Let Z be a real semimartingale, $A \in \mathscr{A}(Z)$ (see definition in § 23). For every real measurable process X we define

$$\lambda_t^A(X) := A_t \int_{]0,t]} |X_s|^2 \, dA_s$$

(this expression being possibly infinite).

We denote by $\lambda^A(X)$ the real process $(\lambda_t^A(X))_{t \in \mathbb{R}^+}$. If X is adapted, $\lambda^A(X)$ clearly has the same property.

The notion of stochastic integral will be derived from the following sequence of lemmas.

We call $\mathscr{E}$ the vector space of real predictable elementary processes.

Lemma 1. *Let τ be a stopping time with $E(|A_{\tau-}|^2) < \infty$ and Λ_τ^A be the vector space* *of real predictable processes X having the property $E(\lambda_{\tau^-}^A(X)) < \infty$. On this vector space we consider the seminorm*

$$p_\tau^A(X) := [E(A_{\tau^-} \int_{]0,\tau[} |X_s|^2 \, dA_s)]^{\frac{1}{2}}.$$

Let Π_τ be the vector space of real functions on $[0, \tau[$, which are restrictions to $[0, \tau[$ of R.R.C. processes, endowed with the seminorm $Y \curvearrowright [E(\sup_{s<\tau} |Y_s|^2]^{\frac{1}{2}}$. One has $\mathscr{E} \subset \Lambda_\tau^A$ *and there exists one and only one linear continuous extension to Λ_τ^A of the mapping defined on $\mathscr{E}$ by $X \curvearrowright 1_{[0,\tau[} \int X\,dZ$ and with values in Π_τ.*

Proof. If X is a bounded predictable process on $[0, \tau[$, the real number $E(A_{\tau^-} \int_{]0,\tau[} |X_s|^2 \, dA_s)$ is precisely the integral of $|X|^2$ with respect to the measure ν defined on $\mathscr{P}$ by $\nu(G) := E(A_{\tau^-} \int_{]0,\tau[} 1_G(s,.)\,dA_s)$. The function p_τ^A is therefore an L^2-seminorm. Since the hypothesis $E(|A_{\tau^-}|^2) < \infty$ implies ν is bounded, clearly $\mathscr{E} \subset \Lambda_\tau^A$.

The continuity of the mapping $X \curvearrowright 1_{[0,\tau[} \int X\,dZ$ is a straightforward consequence of the defining property of a control process A. Since $\mathscr{E}$ is clearly a dense subset of Λ_τ^A, we have only to prove the completeness of Π_τ.

Let (Y^n) be a Cauchy sequence in Π_τ and let us choose a subsequence $(Y^{n_k})_{k \in \mathbb{N}}$ such that

$$P\left(\sup_{s<\tau} |Y_s^{n_{k+1}} - Y_s^{n_k}|^2 > \frac{1}{2^k}\right) \leqslant \frac{1}{2^k}$$

For every k and $\ell \in \mathbb{N}$ we may write

$$P\left\{\sup_{t<\tau} |Y_t^{n_{k+\ell}} - Y_t^{n_k}|^2 > \frac{1}{2^{k-1}}\right\} \leqslant \frac{1}{2^{k-1}}$$

and the Borel-Cantelli lemma implies

$$P\left(\limsup_{k\to\infty} \left\{\sup_{t<\tau} |Y_t^{n_{k+\ell}} - Y^{n_k}|^2 > \frac{1}{2^{k-1}}\right\}\right) = 0.$$

This says that, for P-almost all ω, the mappings $t \curvearrowright Y^{n_k}(t,\omega)$ converge uniformly on $[0, \tau(\omega)[$ to a function $t \curvearrowright Y(t,\omega)$.

The process $Y \cdot 1_{[0,\tau[}$ is regular, right-continuous and belongs to Π_τ. □

Lemma 2. *Let X be a real predictable process, such that the process $\lambda^A(X)$ is finite. Then there exists an increasing sequence (τ_n) of stopping times with the properties*

(i) $\lim_n \tau_n = +\infty$,

(ii) *for every n one has $E(|A_{\tau_n^-}|^2) < \infty$ and $E(\lambda_{\tau_n^-}^A(X)) < \infty$.*

Moreover, there exists a unique R. R. C. process Y such that, for every n, the mapping $X \curvearrowright 1_{[0,\tau_n[} Y$ is the mapping defined in Lemma 1.
The process Y is therefore defined independently of the sequence (τ_n).

Proof. Let τ and σ be two stopping times with the properties of Lemma 1 and let Y^τ and Y^σ the two corresponding processes, the existence of which follows from the lemma. It is obvious that Y^τ and Y^σ coincide on $[0,\tau \wedge \sigma[$. If (τ_n) is a sequence of stopping times with properties (i) and (ii), it is therefore possible to define Y by setting

$$Y_{[1,\tau_n[} := Y^{\tau_n} 1_{[0,\tau_n[} \quad \text{up to } P\text{-equality of processes.}$$

If (τ_n) and (σ_n) are two sequences of stopping times with properties (i) and (ii) and Y^1 and Y^2 are the corresponding processes, the above reasoning gives: $Y^1 1_{[0,\tau_n \wedge \sigma_n[} = Y^2 1_{[0,\tau_n \wedge \sigma_n[}$ for every n, and therefore $Y^1 = Y^2$ (up to P-equality).

We are left to prove the existence of a sequence (τ_n) with properties (i) and (ii). But, using the finiteness of $\lambda^A(X)$, this is done by setting

$$\tau_n := \inf\{t : A_t \wedge \lambda_t^A(X) \geqslant n\}. \quad \square$$

Lemma 3. *For every $A \in \mathscr{A}(Z)$ such that the process $\lambda^A(X)$ is finite (up to an evanescent process), let us denote by Y^A the unique R. R. C. process, the existence of which follows from Lemma 2. There exists an R. R. C. process Y (uniquely defined up to P-equality) such that $Y = Y^A$ (P-equality) for every A such that $A \in \mathscr{A}(Z)$ and $\lambda^A(X)$ is finite.*

Proof. Let A^1 and A^2 be two elements of $\mathscr{A}(Z)$, with $\lambda^{A^1}(X)$ and $\lambda^{A^2}(X)$ finite (up to evanescent processes). The process $A^1 + A^2$ clearly belongs to $\mathscr{A}(Z)$ and $\lambda^{A^1+A^2}(X)$ is finite. Since the seminorm associated with $A^1 + A^2$ is bigger than the seminorms associated with A^1 and A^2 respectively, Lemma 1 shows that the processes $1_{[0,\sigma_n[} Y^{A_1}$ and $1_{[0,\sigma_n[} Y^{A_2}$ coincide with $1_{[0,\sigma_n[} Y^{A_1+A_2}$ on the stochastic interval $[0, \sigma_n[$, where σ_n is defined as follows.

$$\sigma_n := \inf\{t : (A^1 + A^2) \vee \lambda_1^{A_1+A_2}(X) \geqslant n\}$$

This proves the lemma. $\square$

Definition. Let X be a real predictable process such that there exists a process $A \in \mathscr{A}(Z)$ with the property that $\lambda^A(X)$ is finite (up to an evanescent process). The R. R. C. process Y, which is uniquely defined (up to P-equality), as a consequence of Lemmas 1 to 3 above, is called the *stochastic integral* of X with respect to Z. This process will be denoted by $\int X dZ$ or

$$\left(\int_{]0,t]} X_s dZ_s\right)_{t\in\mathbb{R}^+}.^{1)}$$

We call the class of processes X, for which the stochastic integral exists in this sense, the class of *Z-integrable processes.*

Let u and t be two positive numbers (resp. σ, τ two finite stopping times) with $u \leqslant t$ (resp. $\sigma \leqslant \tau$).

We shall denote by $\int_{]u,t]} X_s dZ_s$ (resp. by $\int_{]\sigma,\tau]} X_s dZ_s$) the random variable $\int_{]0,t]} X_s dZ_s - \int_{]0,u]} X_s dZ_s$ (resp. $(\int X dZ)_\tau - (\int X dZ)_\sigma$).

24.2 Theorem ("Dominated convergence"). *Let $(X^n)_{n\in\mathbb{N}}$ be a sequence of predictable processes, which, except on an evanescent set of couples (t,ω), converge to a predictable process X. We assume the existence of a Z-integrable process Φ such that $|X^n| \leqslant \Phi$ for every n. Then the processes X^n and the process X are Z-integrable and there exists a subsequence $(X^{n_k})_{k\in\mathbb{N}}$ with the following property. For P-almost all $\omega\in\Omega$ and every $t\in\mathbb{R}^+$ the functions $s \rightsquigarrow (\int_{]0,s]} X^{n_k} dZ)(\omega)$ converge uniformly on $[0,t]$ to the function $s \rightsquigarrow (\int_{]0,s]} X dZ)(\omega)$, when k tends to infinity.*

Proof. We consider $A \in \mathscr{A}(Z)$ such that $\lambda^A(\Phi)$ is finite. The inequality $|X^n| \leqslant \Phi$ P.a.e. implies $\lambda^A(X^n) \leqslant \lambda^A(\Phi)$ P.a.e. and also $\lambda^A(X) \leqslant \lambda^A(\Phi)$ P.a.e. This yields the Z-integrability of X^n and X. Let us define, for $r > 0$, the following stopping time σ_r.

$$\sigma_r := \inf\{t : A_t \vee \lambda_t^A(\Phi) \geqslant r\}.$$

Then the processes $1_{[0,\sigma_r[}\int X^n dZ$ converge to $1_{[0,\sigma_r[}\int X dZ$ in Π_{σ_r} (Lemma 1) and there exists a subsequence $(X^{n_k})_{k\in\mathbb{N}}$ with the property

(24.2.1) $$\lim_{k\to\infty} \sup_{t<\sigma_r} \left|\int_{]0,t]} (X - X^{n_k})\, dZ\right| = 0 \quad \text{a.s.}$$

Using a "diagonal-procedure" we immediately see that there exists a subsequence for which the equation (24.2.1) holds for every r.

The equality $\lim_{r\to\infty} \sigma_r = +\infty$ a.s. leads to the statement of the theorem. □

24.3 Proposition. (1°) *Let Z be a semimartingale and X a Z-integrable process. Then, for every finite stopping time σ*

$$\Delta_\sigma\left(\int X dZ\right) = X_\sigma \Delta_\sigma(Z).$$

(2°) *Let Z be a continuous semimartingale. Then the process $\int X dZ$ is continuous for every Z-integrable process X.*

[1]) In the "Strasbourg notation" the process $\int X dZ$ is also denoted by $X \circ Z$. We shall not use this notation here since we want to avoid any confusion with $x \circ z$ where x is an operator acting on a vector z. (See also remark in 26.4).

(3°) *Let* $]s,u] \times F$ *be a predictable rectangle and* X *a* Z*-integrable predictable process. Then*

$$\int 1_{]s,u] \times F} X dZ = 1_F \left(\int 1_{]s,u] \times \Omega} X dZ\right)$$

Proof. These three statements are straightforward consequences of the definition of the stochastic integral of an elementary process, when X itself is a predictable elementary process. Using Theorem 24.2, we immediately obtain these statements with the help of a sequence of elementary processes converging to X. □

24.4 Theorem. *Let* Z *be a semimartingale,* X *a* Z*-integrable process and* $A \in \mathscr{A}(Z)$ *such that* $\lambda_t^A(X) < \infty$ *for all* t P*.a.s.*

(1°) *Then the process* $S := \int X dZ$ *is a semimartingale and for every stopping time* σ

$$E\Big(\sup_{0 \leq t < \sigma} |S_t|^2\Big) \leq E\Big(A_{\sigma-} \int_{]0,\sigma[} X_s^2 dA_s\Big). \tag{24.4.1}$$

The process $\Big(\int_{]0,t]} (1 + |X_s|^2 dA_s)_{t \in \mathbb{R}^+}$ *is a control process for* S.

(2°) *Let* Y *be a predictable process, such that the increasing process* $\Big(\int_{]0,t]} |Y_s|^2 (1 + |X_s|^2) dA_s\Big)_{t \in \mathbb{R}^+}$ *is finite. Then* Y *is* S*-integrable and*

$$\int Y dS = \int Y X dZ \tag{24.4.2}$$

(3°) *Assume* $Z \in \mathscr{W}^0$. *Then the process* $S := \int X dZ$ *belongs to* $\mathscr{W}^0$ *and its paths are* P*.a.s. the same as those defined by the Stieltjes integral*

$$t \curvearrowright \int_{]0,t]} X(s,\omega) dZ(s,\omega)$$

for each fixed ω.

(4°) *Assume* $Z \in \mathscr{M}^1_{\mathrm{loc}}$. *If* X *is predictable and locally bounded one has also* $\int X dZ \in \mathscr{M}^1_{\mathrm{loc}}$.

(5°) *Assume* $Z \in \mathscr{M}^2_{\mathrm{loc}}$. *Then every predictable process* X, *such that the process* $\Big(\int_{]0,t]} |X_s|^2 d\langle Z \rangle_s\Big)_{t \in \mathbb{R}^+}$ *is locally integrable, is* Z*-integrable and the stochastic integral* $\int X dZ$ *belongs to* $\mathscr{M}^2_{\mathrm{loc}}$. *If* $Z \in \mathscr{M}^2$, *this stochastic integral coincides with the one defined in section 18.*

Proof. (1°) When X is an elementary predictable process, inequality (24.4.1) expresses nothing but the fact that A is a control process for Z. If X is a predictable process and σ is a stopping time such that $E(\lambda_{\sigma-}^A(X)) \vee E(|A_{\sigma-}|^2) < \infty$, the extension by continuity in Lemma 1 shows that (24.4.1) also holds for X.

If (τ_n) is the following sequence of stopping times

$$\tau_n := \inf\{t \colon \lambda_t^A(X) \vee |A_t|^2 > n\}$$

and σ is any stopping time, then $\lambda_{(\sigma \wedge \tau_n)^-}^A(X) \vee |A_{(\sigma \wedge \tau_n)^-}|^2 \leq n$ and inequality (24.4.1) holds for the stopping times $\sigma \wedge \tau_n$. Passing to the limit when n tends to $+\infty$, we obtain (24.4.1) for σ, whether the terms of the inequality are finite or not.

Now let X be a Z-integrable process and Y be an elementary predictable process given by

$$Y := \sum_{i=1}^{n} \alpha_i 1_{]s_i, t_i] \times F_i}$$

By definition,

$$\int_{]0,t]} Y dS = \sum_{i=1}^{n} \alpha_i 1_{F_i} \int_{]s_i, t_i]} X dZ$$

Using Proposition 24.3–3° and the linearity of the stochastic integral we get

(24.4.3) $$\int_{]0,t]} Y dS = \int_{]0,t]} Y X dZ .$$

Hence, using the continuity property of the stochastic integral,

(24.4.4) $$E|\sup_{0 \leqslant t < \sigma} \int_{]0,t]} Y dS|^2 \leqslant E(A_{\sigma-} \int_{]0,\sigma]} |Y_s X_s|^2 dA_s)$$

If we set

$$\tilde{A}_t := \int_{]0,t]} (1 + |X_s|^2) dA_s ,$$

we obtain

$$E|\sup_{0 \leqslant t < \sigma} \int_{]0,t]} Y dS|^2 \leqslant E(\tilde{A}_{\sigma-} \int_{]0,\sigma[} |Y_s|^2 d\tilde{A}_s)$$

which says that $\tilde{A}$ is a control process for S.

(2°) From the above, Y is S-integrable and (24.4.3) says that (24.4.2) is true for elementary predictable processes Y. If $\lambda^{\tilde{A}}(Y)$ is finite, the continuity property of the stochastic integral immediately shows the validity of (24.4.2) for such a process Y.

(3°) Let Z be an increasing process. For every elementary predictable process X, $\int X dZ$ is clearly of finite variation and coincides with the Stieltjes integral on every path. Now, if X is a positive Z-integrable process we may find a sequence X^n of elementary processes such that the paths of $\int X^n dZ$ converge uniformly to the paths of $\int X dZ$ on every bounded interval (see Theorem 24.2). This proves that the paths of $\int X dZ$ are increasing on any bounded intervall $[0, t]$ of $\mathbb{R}^+$. Considering now every Z in $\mathscr{W}^0$ as the difference of two increasing processes and X as the difference of two positive processes, we obtain the third statement of the theorem.

(4°) Let Z be a local martingale and X a locally bounded predictable process. Let σ be a finite stopping time such that X^σ is bounded $Z^\sigma \in \mathscr{M}^1_\infty$ and $E(|A_{\sigma-}|^2) < \infty$.

Since there exists, by hypothesis, a sequence (σ_n) of stopping times with these properties such that, moreover, $\lim_n \sigma_n = +\infty$, we reduce the proof to proving the martingale property for the stopped process $(\int X dZ)^\sigma$. If X is an elementary predictable process, the obvious equality $(\int X dZ)^\sigma = \int X dZ^\sigma$ and the martingale property of Z^σ show immediately that $(\int X dZ)^\sigma$ belongs to $\mathcal{M}^1_\infty$.

Now, if X is Z-integrable, we consider (using Lemma 1) a sequence of elementary predictable processes X^n such that

$$\lim_{n\to\infty} E(\sup_{t<\sigma} |\int_{]0,t]} (X - X^n) dZ|^2) = 0$$

We have, using Proposition 24.3,

$$E(\sup_{t\leqslant\sigma} |\int_{]0,t]} (X - X^n) dZ|) \leqslant E(\sup_{t<\sigma} |\int (X - X^n) dZ|) + E(|(X - X^n)_\sigma \Delta_\sigma Z|)$$

Since X is bounded on $[0, \sigma]$, the sequence X^n can be chosen uniformly bounded and, according to Theorem 24.2., noe may also assume that the paths of $\int X^n dZ$ converge uniformly to the paths of $\int X dZ$ on any bounded interval, in particular on $[0, \sigma]$. This yields

$$\lim_{n\to\infty} E(\sup_{t\leqslant\sigma} |\int_{]0,t]} (X - X^n) dZ|) = 0$$

The martingale property for $(\int X dZ)^\sigma$ is then easily derived from the martingale property of $(\int X^n dZ)^\sigma$ for every n.

(5°) We consider $Z \in \mathcal{M}^2_{\text{loc}}$ and $\langle Z \rangle$ the Meyer process of Z as defined in section 23. Considering a localizing sequence (σ_n) of stopping times for Z and a sequence (σ'_n) such that $E(\int_{]0,\sigma'_n]} |X_s|^2 d\langle Z \rangle_s) < \infty$ for every n, we define another localizing sequence (τ_n) with the same properties by setting

$$\tau_n := \inf\{t: [Z]_t \vee \langle Z \rangle_t > n\} \wedge \sigma_n \wedge \sigma'_n .$$

From this definition follows trivially

$$E(\langle Z \rangle_{\tau_n}) = E([Z]_{\tau_n}) = E(|Z|^2_{\tau_n}) < \infty ,$$

$$\langle Z \rangle_{\tau_n^-} \leqslant n \qquad \text{and} \qquad [Z]_{\tau_n^-} \leqslant n .$$

Since

$$E(\int_{]0,\tau_n]} |X|^2 d\langle Z \rangle) = E(\int_{]0,\tau_n]} |X|^2 d[Z])$$

for every predictable process X ($\langle Z \rangle$ and $[Z]$ have the same Doleans measure!), we obtain

$$E(\int_{]0,\tau_n]} |X|^2 d([Z] + \langle Z \rangle)) < \infty$$

and the positive process $(\int_{]0,t]} |X|^2 d(\langle Z \rangle + [Z]))_{t\in\mathbb{R}^+}$ is finite.

We know from section 23 that $1 + 4(\langle Z \rangle + [Z])$ is a control process for Z. This implies the Z-integrability of X.

Now let $\int XdZ^{\tau_n}$ be the stochastic integral with respect to $Z^{\tau_n}\in\mathcal{M}^2_\infty$, as defined in section 18. This defines, for every predictable process X that is bounded on $[0,\tau_n]$, a process $\Phi(X)\in\mathcal{M}^2_\infty$. The mapping $X\curvearrowright\Phi(X)$ has the following continuity property (see section 18).

$$\begin{aligned}E(\sup_{t<\tau_n}|(\Phi(X))_t|^2) &\leqslant E(\sup_{t\leqslant\tau_n}|\int_{]0,t]} XdZ^{\tau_n}|^2)\\ &\leqslant 4E(\int_{]0,\tau_n]}|X|^2 d\langle Z\rangle) = 4E(\int_{]0,\tau_n]}|X|^2 d[Z])\end{aligned}$$

The mapping $X\curvearrowright(\int XdZ)1_{[0,\tau_n[}$, where $\int XdZ$ is the stochastic integral, as defined in this section, and the mapping $X\curvearrowright\Phi(X)1_{[0,\tau_n[}$ are equal for every elementary predictable X. Moreover, for every locally bounded predictable X

$$\begin{aligned}E(\sup_{t<\tau_n}|(\psi(X))_t|^2) &\leqslant 8n(\int_{]0,\tau_n[}|X|^2 d(\langle Z\rangle+[Z]))\\ &\leqslant 8n(\int_{]0,\tau_n[}|X|^2 d(\langle Z\rangle+[Z])).\end{aligned}$$

The above two mappings are continuous for the same topologies and therefore coincide for every predictable X which is bounded on $[0,\tau_n[$. This shows that $\int XdZ$ is a locally square integrable martingale. If Z is a square integrable martingale, $\int XdZ$ coincides with $\Phi(X)$, that is, with the integral process as defined in section 18. □

25. Quadratic variation and the transformation theorem

25.1 Theorem. *Let Z be a real semimartingale and $[Z]$ the following process.*

(25.1.1) $$[Z]_t := Z_t^2 - Z_0^2 - 2\int_{]0,t]} Z_{s-}dZ_s$$

(Since Z is assumed to be regular, the process (Z_{s-}) is locally bounded, predictable and therefore Z-integrable).

(1°) *Then $[Z]$ is increasing adapted right-continuous and $[Z]_0=0$.*

(2°) *Let us consider for every $n\in\mathbb{N}$ an increasing sequence $\Pi^n:=\{0=\tau_1^n<\tau_2^n< <\dots<\tau_k^n<\dots\}$ of stopping times with the properties*

(i) $$\forall n,\ \lim_{k\to\infty}\tau_k^n = +\infty \quad \text{a.s.}$$

(ii) $$\lim_{n\to\infty}\sup_k(\tau_{k+1}^n-\tau_k^n)=0 \quad \text{a.s.}$$

We set

$$v_n(t,Z) := \sum_k (Z_{\tau_{k+1}^n\wedge t} - Z_{\tau_k^n\wedge t})^2.$$

Then the sequence $(v_n(t,Z))_{n\geqslant 0}$ of random variables converges in probability for every *fixed t to the random variable $[Z]_t$. There exists a subsequence $(v_{n_k}(.,Z))_{k\geq 0}$ of processes, the paths of which converge uniformly on any bounded interval to the paths of $[Z]$.*

Proof. Let us write

(25.1.2) $$v_n(t,Z) = \sum_k (Z^2_{\tau^n_{k+1}\wedge t} - Z^2_{\tau^n_k\wedge t}) - \sum_k 2Z_{\tau^n_k\wedge t}(Z_{\tau^n_{k+1}\wedge t} - Z_{\tau^n_k\wedge t})$$
$$= Z_t^2 - Z_0^2 - 2\sum_k (Z_{\tau^n_{k+1}\wedge t} - Z_{\tau^n_k\wedge t})$$

and set

(25.1.3) $$\Phi_n := \sum_k Z_{\tau^n_k} 1_{]\tau^n_k,\tau^n_{k+1}]}.$$

Assumptions (i) and (ii) imply the convergence of $(\Phi_n(t,\omega))_{n\geq 0}$ to $Z_{t^-}(\omega)$ for every t and ω, with

(25.1.4) $$|\Phi_n(t,\omega)| \leq \sup_{s<t} |Z_s(\omega)| < \infty \text{ for every } t \text{ and } \omega.$$

Since we may write (25.1.2) in the form

$$v_n(t,Z) = Z_t^2 - Z_0^2 - 2\int_{]0,t]} \Phi_n dZ,$$

all the convergence properties stated in the theorem follow in a straightforward way from Theorem 24.2 and the monotonicity of the processes $v_n(.,Z)$ implies the monotonicity of $[Z]$. □

25.2 Definition *(Quadratic variation of a semimartingale).* The process defined by (25.1.1) is called the *quadratic variation* of the semimartingale Z.
When $Z\in\mathcal{M}^2_{\text{loc}}$ this definition clearly coincides with the definition given in sections 18 and 19. Theorem 24.4 yields the following immediate consequence of Theorem 25.1.

25.3 Corollary 1. *Let Z be a real semimartingale. Then $|Z|^2$ is a semimartingale and, for every predictable process X that has paths bounded on any bounded interval, one has*

$$\int_{]0,t]} X_s d(Z_s^2) = 2\int_{]0,t]} X_s Z_{s^-} dZ_s + \int_{]0,t]} X_s d[Z]_s.$$

25.4 Corollary 2. *Let Π^n be defined as in Theorem 25.1. Let us define for every* R.R.C. *process X*

$$w_n(t,Z,X) := \sum_k X_{\tau^n_k}(Z_{\tau^n_{k+1}\wedge t} - Z_{\tau^n_k\wedge t})^2.$$

Then the sequence $(w_n(t,Z,X))_{n\in\mathbb{N}}$ of random variables converges in probability to $\int_{]0,t]} X_{s^-} d[Z]_s$. One can even choose a subsequence $(w_{n_k}(.,Z,X))_{k\in\mathbb{N}}$ such that, with *probability one, the paths of the processes $w_{n_k}(.,Z,X)$ converge uniformly on every bounded interval $[0,t]$ to the paths of $\int X_{s-}\,d[Z]_s$, when $k\to\infty$.*

Proof. Let us write

$$w_n(t,Z,X)=\sum_k X_{\tau_k^n}(Z^2_{\tau^n_{k+1}\wedge t}-Z^2_{\tau^n_k\wedge t})-2\sum_k X_{\tau^n_k}Z_{\tau^n_k}(Z_{\tau^n_{k+1}\wedge t}-Z_{\tau^n_k\wedge t})$$

and define

$$\Phi_n:=\sum_k X_{\tau^n_k}1_{]\tau^n_k,\tau^n_{k+1}]}$$

and

$$\psi_n:=\sum_k X_{\tau^n_k}Z_{\tau^n_k}1_{]\tau^n_k,\tau^n_{k+1}]}.$$

We may remark that

$$w_n(t,Z,X)=\int_{]0,t]}\Phi_n d(Z^2)-2\int_{]0,t]}\psi_n dZ.$$

Applying Theorem 24.2, we obtain the convergence in probability of the sequence $(w_n(t,Z,X))_{n\geqslant 0}$ to $\int_{]0,t]}X_{s-}d(Z_s^2)-2\int_{]0,t]}X_{s-}Z_{s-}dZ_s$. The latter expression is nothing but $\int_{]0,t]}X_s d[Z]_s$, according to Corollary 1.

We also see from Theorem 24.2 that we can extract a subsequence $(w_{n_k})_{k\in\mathbb{N}}$ having the property of the last statement of Corollary 2. □

25.5 Corollary 3. *For every semimartingale Z and every $t\in\mathbb{R}^+$ the sum $\sum_{s\leqslant t}|\Delta Z_s|^2$ is P.a.s. finite.*

The process $([Z]_t-\sum_{s\leqslant t}|\Delta Z_s|^2)_{t\in\mathbb{R}^+}$ is a.s. continuous.

Proof. Let us consider the following sequence Π^n. For every n we set

(25.5.1) $$\tau_0^n:=0,\dots,\tau^n_{k+1}:=\inf\left\{t:\ t>\tau^n_{k+1},\ |Z_t-Z_{\tau^n_k}|>\frac{1}{n}\right\}\wedge\left(\tau^n_k+\frac{1}{n}\right).$$

This sequence Π^n clearly satisfies conditions (i) and (ii) of Theorem 25.1. We may write

(25.5.2) $$\begin{aligned}v_n(t,Z)=\sum_k\big[&(\Delta Z_{\tau^n_{k+1}\wedge t})^2+(Z_{(\tau^n_{k+1}\wedge t)^-}-Z_{\tau^n_k\wedge t})^2\\&+2\Delta Z_{\tau^n_{k+1}\wedge t}(Z_{(\tau^n_{k+1}\wedge t)^-}-Z_{\tau^n_k\wedge t})\big].\end{aligned}$$

Let us define

$$D_n(t):=\left\{(s,\omega):s\leqslant t,\ |\Delta Z_s(\omega)|\geqslant\frac{3}{n}\right\}.$$

As a consequence of definition (25.5.1)

(25.5.3) $$D_n(t) \subset \bigcup_k [\![\tau^n_{k+1} \wedge t]\!].$$

Since the inequality $(a+b)^2 \geqslant \frac{a^2}{3}$ holds for $|a| \geqslant 3|b|$ and since $|\Delta Z_s(\omega)| \geqslant \frac{3}{n}$ for $(s,\omega) \in D_n(t)$ and $|Z_{(\tau^n_{k+1} \wedge t)^-} - Z_{\tau^n_k \wedge t}| \leqslant \frac{1}{n}$ by definition, it follows from (25.5.1) that

(25.5.4) $$3\,v_n(t,Z)(\omega) \geqslant \sum_{(s,\omega) \in D_n(t)} |\Delta Z_s(\omega)|^2.$$

Now, for every subsequence $(n_r)_{r \in \mathbb{N}}$ of the set of integers, increasing to infinity,

(25.5.5) $$\sum_{s \leqslant t} |\Delta Z_s(\omega)|^2 = \sup_r \sum_{(s,\omega) \in D_{n_r}(t)} |\Delta Z_s(\omega)|^2.$$

If we consider a subsequence (n_r) with the property that the sequence $(v_{n_r}(.,Z))_{r \in \mathbb{N}}$ of processes converges uniformly P.a.s. on every interval $[0,t]$, we obtain the finiteness of $\sum_{s \leqslant t} |\Delta Z_s(\omega)|^2$ from (25.5.5) and inequality (25.5.4). This proves the first statement of Corollary 3.

Since $\sup_{s \leqslant t} |\Delta Z_s(\omega)| < \infty$ for each ω and $\sup_k |Z_{(\tau^n_{k+1} \wedge t)^-} - Z_{\tau^n_k \wedge t}| \leqslant \frac{1}{n}$, we see that the sums

$$u_n(t,Z)(\omega) := \sum_k 2\,\Delta Z_{\tau^n_{k+1} \wedge t}(\omega)\,(Z_{(\tau^n_{k+1} \wedge t)^-}(\omega) - Z_{\tau^n_k \wedge t}(\omega))$$

converge a.s. to zero uniformly on any finite interval $[0,u]$. Using (25.5.2), we see that for a suitable subsequence $(n_r)_{r \in \mathbb{N}}$ of integers the processes $(w_{n_r}(.,Z))_{r \in \mathbb{N}}$, where

$$w_n(t,Z) := \sum_k (Z_{(\tau^n_{k+1} \wedge t)^-} - Z_{\tau^n_k \wedge t})^2$$

converge a.s. uniformly on any bounded interval $[0,u]$ when n tends to infinity. But according to (25.5.3), the limit can be nothing but the process $([Z]_t - \sum_{s \leqslant t} |\Delta Z_s|^2)_{t \in \mathbb{R}^+}$.

The jumps of $w_n(.,Z)$, being smaller than $\frac{1}{n^2}$ by definition, the limit process $([Z]_t - \sum_{s \leqslant t} |\Delta Z_s|^2)_{t \in \mathbb{R}^+}$ is continuous. □

25.6 Notation. We shall write

$$[Z]^d_t := \sum_{s \leqslant t} |\Delta Z_s|^2,$$

$$[Z]^c_t := [Z]_t - [Z]^d_t.$$

It will be proved as an exercise (see E.5 at the end of the chapter), that if $Z = Z^c + Z^d$, where Z^c is the uniquely determined continuous martingale part of Z (see 23.7), one has $[Z]^c = \langle Z^c \rangle = [Z^c]$ and $[Z]^d = [Z^d]$.

25.7 Theorem (Transformation Theorem). *Let Z be a real semimartingale and φ a twice continuously differentiable function on $\mathbb{R}$. Then the process $(\varphi(Z_t))_{t\in\mathbb{R}^+}$ is a semimartingale which is P-equal to the process $T^\varphi(Z)$ defined by*

$$T^\varphi(Z)_t := \varphi(Z_0) + \int_{]0,t]} \varphi'(Z_{s-})dZ_s + \tfrac{1}{2} \int_{]0,t]} \varphi''(Z_{s-})d[Z]_s$$
$$+ \sum_{s\leq t} (\varphi(Z_s) - \varphi(Z_{s-}) - \Delta Z_s \varphi'(Z_{s-}) - \tfrac{1}{2}|\Delta Z_s|^2 \varphi''(Z_{s-}),$$

the family $(\varphi(Z_s(\omega)) - \varphi(Z_{s-}(\omega)) - \Delta Z_s(\omega)\varphi'(Z_{s-}(\omega)) - \frac{1}{2}|\Delta Z_s(\omega)|^2\varphi''(Z_{s-}(\omega)))_{s\leq t}$ of real numbers occurring in the latter expression being summable for all t, P.a.s.

Proof. Let us first assume that the second derivative φ'' is bounded. Then we have, for every x and $h \in \mathbb{R}$,

(25.7.1) $$|\varphi(x+h) - \varphi(x) - h\varphi'(x) - \tfrac{1}{2}h^2\varphi''(x)| \leq h^2 v(|h|),$$

where v is a positive function on $\mathbb{R}^+$ with the property

$$\lim_{h\to 0} v(|h|) = 0$$

Using a J. Pellaumail's idea we define, for every n,

$$\tau_0^n := 0, \ldots, \tau_{k+1}^n := \inf\left\{t\colon t > \tau_k^n;\ |Z_t - Z_{\tau_k^n}| > \frac{1}{n}\right\} \wedge \left(\tau_k^n + \frac{1}{n}\right).$$

For every n the sequence $(\tau_k^n)_{k\in\mathbb{N}}$ of stopping times satisfies

(25.7.2) $$\lim_{k\to\infty} \tau_k^n = +\infty$$

(25.7.3) $$\sup_k |\tau_{k+1}^n - \tau_k^n| \leq \frac{1}{n}.$$

We write

(25.7.4) $$\varphi(Z_t) - \varphi(Z_0) = \sum_k \varphi(Z_{\tau_{k+1}^n\wedge t}) - \varphi(Z_{\tau_k^n\wedge t}).$$

Let us then define the following processes $u_{n,k}$, $v_{n,k}$, $w_{n,k}$ and $y_{n,k}(t)$

(25.7.5) $$u_{n,k}(t) := (Z_{\tau_{k+1}^n\wedge t} - Z_{\tau_k^n\wedge t})\varphi'(Z_{\tau_k^n\wedge t}) + \tfrac{1}{2}(Z_{\tau_{k+1}^n\wedge t} - Z_{\tau_k^n\wedge t})^2\varphi''(Z_{\tau_k^n})$$

(25.7.6) $$v_{n,k}(t) := \varphi(Z_{\tau_{k+1}^n\wedge t}) - \varphi(Z_{(\tau_{k+1}^n\wedge t)-}) - \Delta Z_{\tau_{k+1}^n}\varphi'(Z_{\tau_k^n}) \cdot 1_{\{\tau_k^n\leq t\}}$$
$$- \tfrac{1}{2}|\Delta Z_{\tau_{k+1}^n}|^2 \varphi''(Z_{\tau_k^n}) \cdot 1_{\{\tau_k^n\leq t\}}$$

(25.7.7) $$w_{n,k}(t) := \varphi(Z_{(\tau^n_{k+1}\wedge t)-}) - \varphi(Z_{\tau^n_k \wedge t}) - (Z_{(\tau^n_{k+1}\wedge t)-} - Z_{\tau^n_k\wedge t})\varphi'(Z_{\tau^n_k}) - \tfrac{1}{2}(Z_{(\tau^n_{k+1}\wedge t)-} - Z_{\tau^n_k\wedge t})^2\varphi''(Z_{\tau^n_k}).$$

(25.7.8) $$y_{n,k}(t) := \Delta Z_{\tau^n_{k+1}}(Z_{(\tau^n_{k+1}\wedge t)-} - Z_{\tau^n_k\wedge t})\varphi''(Z_{\tau^n_k}).$$

With these definitions one may write (25.7.4) in the form

(25.7.9) $$\varphi(Z_t) - \varphi(Z_0) = \sum_k u_{n,k}(t) + \sum v_{n,k}(t) + \sum_k w_{n,k}(t) - \sum_k y_{n,k}(t).$$

Observing that $|Z_{(\tau^n_{k+1}\wedge t)-} - Z_{\tau^n_k\wedge t}| \leqslant \frac{1}{n}$ and using (25.7.1), we get

$$\begin{aligned}|\sum_k w_{n,k}(t)| &\leqslant v\left(\frac{1}{n}\right)\sum_k (Z_{(\tau^n_{k+1}\wedge t)-} - Z_{\tau^n_k\wedge t})^2\\ &\leqslant 2v\left(\frac{1}{n}\right)\left(\sum_k |\Delta Z_{\tau^n_{k+1}\wedge t}|^2 + \sum_k (Z_{\tau^n_{k+1}\wedge t} - Z_{\tau^n_k\wedge t})^2\right).\end{aligned}$$

We know from Theorem 25.1 and Corollary 25.5 that the sum on the right-hand side of the latter inequality is bounded for P-almost all ω. Therefore

(25.7.10) $$\lim_{n\to\infty}\sup_{s\leqslant t}(|\sum_k w_{n,k}(s)|) = 0 \quad \text{a.s.}$$

For the process $y_{n,k}$ we immediately derive

$$|\sum_k y_{n,k}(t)| \leqslant \frac{1}{n}\sup_{x\in\mathbb{R}}|\varphi''(x)|\sum_k |\Delta Z_{\tau^n_{k+1}}|^2 1_{\{\tau^k_n \leqslant t\}}$$

and therefore

(25.7.11) $$\lim_{n\to\infty}\sup_{s\leqslant t}|\sum_k y_{n,k}(t)| = 0 \quad \text{a.s.}$$

Let us now define for all n the predictable process Φ_n.

$$\Phi_n := \sum_k \varphi'(Z_{\tau^n_k})1_{]\tau^n_k,\tau^n_{k+1}]}$$

The regularity property of the process Z implies, for every (s, ω),

(25.7.12) $$\lim_{n\to\infty}\Phi_n(s,\omega) = \varphi'(Z_{s-}(\omega)).$$

Using Theorem 24.2 and Corollary 25.4 we obtain the convergence in probability of $(\sum_k u_{n,k}(t))_{n\in\mathbb{N}}$ to the random variable $\int_{]0,t]}\varphi'(Z_{s-})dZ_s + \frac{1}{2}\int_{]0,t]}\varphi''(Z_{s-})d[Z]_s$. It is even possible to find a subsequence $(\sum_k u_{n_r,k})_{r\in\mathbb{N}}$ such that, for every $u\in\mathbb{R}^+$,

(25.7.13) $$\lim_{r\to\infty}\sup_{t\leqslant u}\left|\int_{]0,t]}\varphi'(Z_{s-})dZ_s + \tfrac{1}{2}\int_{]0,t]}\varphi''(Z_{s-})d[Z]_s - \sum_k u_{n_r,k}(t)\right| = 0 \quad \text{a.s.}$$

Now let us set

$$a_n(s) := \sum_k 1_{\{\tau^n_{k+1} = s\}}\, v_{n,k}(s)$$

and observe that

$$\sum_k v_{n,k}(t) = \sum_{s \leqslant t} a_n(s). \tag{25.7.14}$$

According to formula (25.7.1),

$$a_n(s) \leqslant |\Delta Z_s|^2\, v(|\Delta Z_s|).$$

Using again the finiteness of $\sup_{s \leqslant t} |\Delta Z_s(\omega)|$ for P-almost all ω and Corollary 25.5 we see that the family $(a_n(s))_{n \in \mathbb{N}}$ is dominated by a summable family $(K(t,\omega)\, |\Delta Z_s(\omega)|^2)_{s \leqslant t}$, for some finite $K(t,\omega)$, while

$$\lim_{n \to \infty} a_n(s) = \varphi(Z_s) - \varphi(Z_{s^-}) - \Delta Z_s \varphi'(Z_{s^-}) - \tfrac{1}{2} |\Delta Z_s|^2 \varphi''(Z_{s^-}).$$

In view of (25.7.14) we have therefore P-almost surely

$$\begin{aligned} \lim_{n \to \infty} \sup_{u \leqslant t} \Big| \sum_k v_n(u,\omega) - \sum_{s \leqslant u} \varphi(Z_s(\omega)) - \varphi(Z_{s^-}(\omega)) - \Delta Z_s(\omega) \varphi'(Z_{s^-}(\omega)) \\ - \tfrac{1}{2} |\Delta Z_s(\omega)|^2 \varphi''(Z_{s^-}(\omega)) \Big| = 0, \end{aligned} \tag{25.7.15}$$

the family $(\varphi(Z_s(\omega)) - \varphi(Z_{s^-}(\omega)) - \Delta Z_s(\omega)\varphi'(Z_{s^-}(\omega)) - \tfrac{1}{2}|\Delta Z_s(\omega)|^2 \varphi''(Z_{s^-}(\omega)))_{s \leqslant t}$ being summable.

Equalities (25.7.10), (25.7.11), (25.7.13) and (25.7.15) show the existence of an increasing sequence $(n_r)_{r \in \mathbb{N}}$ of integers such that, with probability one, the paths of the processes $\sum_k (u_{n_r,k} + v_{n_r,k} + w_{n_r,k} + y_{n_r,k})$ converge to the paths of the process $T^\varphi(Z)$ uniformly on each bounded interval $[0,t]$.

This proves the theorem when we assume φ'' bounded. Let us consider now the case where φ'' is not bounded and set $\tau_n := \inf\{t : |Z_t| \geqslant n\}$.

We may, for each n, consider a function φ_n which is equal to φ on the ball $\{x : |x| \leqslant n\}$ and such that its second derivative φ''_n is bounded. On $[0,\tau_n[$ the process $\varphi(Z)$ and $\varphi_n(Z)$ are the same. This also holds for the processes $T^\varphi(Z)$ and $T^{\varphi_n}(Z)$. Applying the first part of the proof, we know that the processes $\varphi_n(Z)$ and $T^{\varphi_n}(Z)$ are P-equal. This means that $\varphi(Z)$ and $T^\varphi(Z)$ are P-equal on the stochastic interval $[0,\tau_n[$. Since $\lim_n \tau_n = +\infty$ a.s. the P-equality of $\varphi(Z)$ and $T^\varphi(Z)$ is proved.

$T^\varphi(Z)$ being the sum of two stochastic integrals and of a process with finite variation, it is a semimartingale and all the assertions of the theorem are established.

26. Stochastic integral with respect to multidimensional semimartingales and tensor quadratic variation

With the help of control processes the theory of sections 24 and 25 has a straightforward extension to Hilbert-valued semimartingales and to processes X which are operator-valued. The only new feature is that, besides the quadratic variation, one has to introduce the tensor quadratic variation in order to write the transformation formula. This has already been done for square integrable martingales in section 22.

We state the results without reproducing the proofs when the proofs are mere retranscriptions of those given for real-valued semimartingales and real-valued integrands.

26.1 Construction of the stochastic integral. We consider separable Hilbert spaces $\mathbb{H}$ and $\mathbb{G}$. Let Z be an $\mathbb{H}$-valued semimartingale. For every control process $A \in \mathscr{A}(Z)$ (see section 23) and every measurable process X with values in $\mathscr{L}(\mathbb{H}; \mathbb{G})$ we set

$$\lambda_t^A(X) := A_t \int_{]0,t]} \|X_s\|^2 dA_s .$$

X being a predictable process such that $\lambda_t^A(X)$ is a.s. finite for every t, we consider an increasing sequence (τ_n) of stopping times having the properties (i) and (ii) of Lemma 2 in section 24.1, namely,

(i) $\lim_n \tau_n = +\infty$,

(ii) for every n $E(|A_{\tau_n^-}|^2) < \infty$ and $E(\lambda_{\tau_n^-}^A(X)) < \infty$. [We are provided with such a sequence by setting for example

$$\tau_n := \inf\{t : \lambda_t^A(X) \geqslant n\} \wedge \inf\{t : |A_t|^2 \geqslant n\}] .$$

Considering, as in section 23, the seminorms

$$p_{\tau_n}^A(X) = [E(A_{\tau_n^-} \int_{]0,\tau_n[} \| X_s\|^2 dA_s)]^{\frac{1}{2}},$$

X can be approximated for each of these seminorms by a sequence of elementary $\mathscr{L}(\mathbb{H}; \mathbb{G})$-valued predictable processes. For each τ_n the stochastic integral of elementary $\mathscr{L}(\mathbb{H}; \mathbb{G})$-valued predictable processes, as defined in section 21.8, has a unique continuous extension as a mapping of the space of $\mathscr{L}(\mathbb{H}; \mathbb{G})$-valued predictable processes with finite seminorm $p_{\tau_n}^A$ into the space of process Y defined on $[0, \tau_n[$, which are restrictions to $[0, \tau_n[$ of R.R.C. $\mathbb{G}$-valued processes with the norm $[E(\sup_{t<\tau_n} \|Y_t\|_{\mathbb{G}}^2)]^{\frac{1}{2}}$.

Lemmas 2 and 3 of section 24.1 show that taking the value of this extension for each n defines unambiguously (up to P-equality) a $\mathbb{G}$-valued R.R.C. process on each $[0, \tau_n[$ and therefore on $\mathbb{R}^+ \times \Omega$. This process is by definition the *stochastic integral process* (or for short, *the stochastic integral*) of the process X with respect to Z. This process will be denoted by $\int XdZ$ or $\int X_s dZ_s$, while the value of this process at time t is a $\mathbb{G}$-valued random variable denoted by $\int_{]0,t]} XdZ$ or $\int_{]0,t]} X_s dZ_s$.

In some cases we shall consider processes X with values in a Banach space $\mathbb{K}$, the elements of which act linearly and continuously on $\mathbb{H}$. If, for $k \in \mathbb{K}$, the linear operation is denoted by $h \curvearrowright k \circ h \in \mathbb{G}$ or $h \curvearrowright k * h$ etc., the corresponding stochastic integral of X with respect to Z ($\mathbb{K}$ being identified with a subspace of $\mathscr{L}(\mathbb{H}; \mathbb{G})$) will be denoted by $\int X \circ dZ$ (resp. $\int X * dZ$ etc.).

This is the case, in particular, when $\mathbb{K} = \mathbb{H}$ is identified with the dual of $\mathbb{H}$ itself, the linear operation being the scalar product. When the scalar product is written $x \cdot y$ (resp. $\langle x, y \rangle$) the stochastic integral of an $\mathbb{H}$-valued process X with respect to Z is denoted by $\int X \cdot dZ$ (resp. $\int \langle X, dZ \rangle$).

The class of predictable processes X (with valued in $\mathscr{L}(\mathbb{H}; \mathbb{G})$ for which an $A \in \mathscr{A}(Z)$ exists with the property (ii) above will be called the *class of Z-integrable processes*.

26.2 Remark. It is immediate that when Z is a square integrable martingale the above integral and the one defined in sections 20.4 and 22.6 coincide for a process X which is integrable in both senses. However, we have already noticed (see 22.5) that we integrated in 22.6 processes X having possibly unbounded operators as their values.

26.3 Theorem *(Dominated convergence). Let $(X^n)_{n \in \mathbb{N}}$ be a sequence of predictable $\mathscr{L}(\mathbb{H}; \mathbb{G})$-valued processes, which, except on an evanescent set of pairs (t, ω), converge to a predictable process X. We assume the existence of a Z-integrable real process Φ such that $\|X^n\| \leqslant \Phi$ for every n. Then the process X^n and X are Z-integrable and there exists a subsequence $(X^{n_k})_{k \in \mathbb{N}}$ with the following property. For P-almost all $\omega \in \Omega$ and every $t \in \mathbb{R}^+$ the functions $s \curvearrowright (\int_{]0,s]} X^{n_k} dZ)(\omega)$ converge uniformly on $[0, t]$ to the function $s \curvearrowright (\int_{]0,s]} XdZ)(\omega)$ when $k \to \infty$.*

Proof. One has only to replace $|\cdot|$ by $\|\cdot\|$ in the proof of Theorem 24.2. □

26.4 Other properties of the stochastic integral. We do not rephrase the statements 24.3 and 24.4 (1°) for Hilbert-valued semimartingales Z and $\mathscr{L}(\mathbb{H}; \mathbb{G})$-valued processes X. They are valid if we replace everywhere in the statements and in the proofs $|X_s|$ by $\|X_s\|$.

The same is true for statements 24.4 (3°), 24.4 (4°) and 24.4 (5°). As to the analogue of statement 24.4 (2°), it reads:

Let Y be a predictable $\mathscr{L}(\mathbb{G}; \mathbb{K})$-valued process, where $\mathbb{K}$ is a separable Hilbert space, and assume that the increasing process $(\int_{]0,t]} \|Y_s\|^2 (1 + \|X_s\|^2) dA_s)_{t \in \mathbb{R}^+}$ is finite.

Then Y is S-integrable and $\int Y dS = \int Y \circ X dZ$, where $Y \circ X(t,\omega)$ [1]*) is, for each (t,ω), the composition of operators $Y(t,\omega)$ and $X(t,\omega)$.*

26.5 Theorem and definition (Quadratic variation). *Let Z be an $\mathbb{H}$-valued semimartingale, $[Z]$ and $v_n(.,Z)$ the following processes.*

$$[Z]_t := \|Z_t\|^2 - \|Z_0\|^2 - 2 \int_{]0,t]} Z_{s^-} \cdot dZ_s .$$

(Denoting by $x \cdot y$ the scalar product in $\mathbb{H}$).

$$v_n(t,Z) := \sum_k \|Z_{\tau^n_{k+1} \wedge t} - Z_{\tau^n_k \wedge t}\|^2 ,$$

where, for each n, $\Pi^n := \{0 = \tau^n_1 < \tau^n_2 < \ldots < \tau^n_k < \ldots\}$ is an increasing sequence of stopping times with the properties

(i) $\forall n \lim_{k \to \infty} \tau^n_k = \infty$ *a.s.*

(ii) $\lim_{n \to \infty} \sup_k (\tau^n_{k+1} - \tau^n_k) = 0$ *a.s.*

Then the processes $[Z]$ and $v_n(.,Z)$ have the same properties as in Theorem 25.1.

The process $[Z]$ is called the *quadratic variation of the semimartingale Z.*

Proof. Exactly the same as for Theorem 25.1. □

26.6 Corollary 1. *Let Z be an $\mathbb{H}$-valued semimartingale. Then $\|Z\|^2$ is a real semimartingale and for every $\mathbb{G}$-valued predictable process, the paths of which are bounded on any bounded interval, one has*

$$\int_{]0,t]} X_s d(\|Z\|^2_s) = 2 \int_{]0,t]} X_{s^-} \otimes Z_{s^-} dZ_s + \int_{]0,t]} X_s d[Z]_s .$$

where, for $h \in \mathbb{H}$, $g \in \mathbb{G}$, $g \otimes h$ denotes the linear mapping $k \curvearrowright (k \cdot h) g$ from $\mathbb{H}$ into $\mathbb{G}$.

26.7 Corollary 2. *Let Π^n be defined as in 21.4 and X be a $\mathbb{G}$-valued R.R.C. process. If we define*

$$w_n(t,Z,X) := \sum_k X_{\tau^n_k} \|Z_{\tau^n_{k+1} \wedge t} - Z_{\tau^n_k \wedge t}\|^2 ,$$

the sequence $(w_n(t,Z,X))_{n \in \mathbb{N}}$ of $\mathbb{G}$-valued random variables converges in probability to $\int_{]0,t]} X_{s^-} d[Z]_s$. One can even choose a subsequence $(w_{n_k}(.,Z,X))_{k \in \mathbb{N}}$ such that, with probability one, the paths of the processes $(w_{n_k}(.,Z,X))_{k \in \mathbb{N}}$ converge uniformly on any bounded interval $[0,t]$ to the paths of $\int X_{s^-} d[Z]_s$, when $k \to \infty$.

---

[1]) Because of the confusion which could arise from this classical notation for the composition of operators and the Strasbourg notation $X \circ Z$ for stochastic integrals, we never use the latter.

26.8 Corollary 3. *For every* $\mathbb{H}$*-valued semimartingale* Z *and every* $t \in \mathbb{R}^+$ *the sum* $\sum_{s \leqslant t} \|\Delta Z_s\|^2$ *is P.a.s. finite.*

The process $([Z]_t - \sum_{s \leqslant t} \|\Delta Z_s\|^2)_{t \in \mathbb{R}^+}$ *is continuous.*

26.9 Mutual variation of two semimartingales – Integration by part formula. Let Z^1 and Z^2 be two real semimartingales. Using the equality

$$Z^1 Z^2 = \tfrac{1}{4}((Z^1 + Z^2)^2 - (Z^1 - Z^2)^2),$$

we derive immediately from Theorem 25.1, applied to semimartingales $Z^1 + Z^2$ and $Z^1 - Z^2$, the following formula.

$$(26.9.1) \qquad Z_t^1 Z_t^2 = Z_0^1 Z_0^2 + \int_{]0,t]} Z_{s-}^1 dZ_s^2 + \int_{]0,t]} Z_{s-}^2 dZ_s^1 + \tfrac{1}{4}((Z^1 + Z^2) - (Z^1 - Z^2))$$

If we consider a family (Π^n) of sequences of stopping times as in Theorem 25.1 and set

$$v_n(t, Z^1, Z^2) := \sum_k (Z^1_{\tau^n_{k+1} \wedge t} - Z^1_{\tau^n_k \wedge t})(Z^2_{\tau^n_{k+1} \wedge t} - Z^2_{\tau^n_k \wedge t}),$$

we immediately obtain from Theorem 25.1 that the sequence $(v_n(t, Z^1, Z^2))_{n \in \mathbb{N}}$ of real random variables converges in probability to $\frac{1}{4}([Z^1 + Z^2]_t - [Z^1 - Z^2]_t)$ and that a subsequence can be chosen in such a way that the convergence holds with probability one uniformly on bounded intervals.

We set

$$(26.9.2) \qquad [Z^1, Z^2] := \tfrac{1}{4}([Z^1 + Z^2] - [Z^1 - Z^2])$$

and call $[Z^1, Z^2]$ the *mutual variation of the two real semimartingales* Z^1 *and* Z^2.

If Z^1 and Z^2 are two $\mathbb{H}$-valued semimartingales, the definition (26.9.2) still makes sense and formula (26.9.1) reads

$$(26.9.3) \qquad Z_t^1 Z_t^2 = Z_0^1 Z_0^2 + \int_{]0,t]} Z_{s-}^1 dZ_s^2 + \int_{]0,t]} Z_{s-}^2 dZ_s^1 + [Z^1, Z^2]$$

this formula is known as the *integration by part formula.*

Using the above sequence (v_n) and a technique already used in 25.5, it is easy to see that *when* V *is a real R.R.C. process with paths of finite variation*, the following formula holds,

$$(26.9.4) \qquad [Z, V]_t = \sum_{s \leqslant t} \Delta Z_s \Delta V_s.$$

We leave the proof of this formula as an exercise (see E.3 at the end of the chapter).

In particular, for *a real R.R.C. process with paths of finite variation* V,

$$(26.9.5) \qquad [V]_t = \sum_{s \leqslant t} |\Delta V_s|^2.$$

26.10 Tensor quadratic variation of a finite dimensional semimartingale. Let us assume that $\mathbb{H}$ is a finite-dimensional Hilbert space with an orthonormal basis $(h_1, \dots h_d)$. If Z is an $\mathbb{H}$-valued semimartingale, we denote by Z^i $i = 1 \dots d$ the coordinate processes of Z. For each (t, ω) we consider the matrix

$$(26.10.1) \qquad [\![Z]\!](t,\omega) := ([Z^i, Z^j](t,\omega))$$

The matrix-valued process $[\![Z]\!]$, which is defined by (26.10.1), is called the *tensor quadratic variation of* Z.

The following relation between $[Z]$ and $[\![Z]\!]$ is immediate.

$$(26.10.2) \qquad [Z](t,\omega) = \operatorname{Trace}([\![Z]\!](t,\omega)).$$

We shall now turn to the definition of the process $[\![Z]\!]$ when $\mathbb{H}$ is any separable Hilbert space. This will give at the same time an intrinsic definition of $[\![Z]\!]$, independent of the choice of a particular orthonormal basis.

In what follows, instead of matrices, we shall speak of elements of the tensor product $\mathbb{H} \hat{\otimes}_2 \mathbb{H}$ (see section 21) with the Hilbert-Schmidt norm.

26.11 Theorem*. *Let Z be an $\mathbb{H}$-valued semimartingale. We denote by $\int Z_{s-} \otimes dZ_s$ (resp. $\int dZ_s \otimes Z_{s-}$) the stochastic integral of the process $(Z_{t-})_{t \in \mathbb{R}^+}$ when $\mathbb{H}$ is identified with a subspace of $\mathscr{L}(\mathbb{H}; \mathbb{H} \hat{\otimes}_2 \mathbb{H})$ in the following way. With each $h \in \mathbb{H}$ is associated the operator $g \curvearrowright h \otimes g$ (resp. $g \curvearrowright g \otimes h$).*

Let us then define the following processes $[\![Z]\!]$ and $\hat{v}_n(.,Z)$.

$$(26.11.1) \qquad [\![Z]\!]_t := Z_t^{\otimes^2} - Z_0^{\otimes^2} - \int_{]0,t]} Z_{s-} \otimes dZ_s - \int_{]0,t]} dZ_s \otimes Z_{s-}$$

$$(26.11.2) \qquad \hat{v}_n(t,Z) := \sum_k (Z_{\tau^n_{k+1} \wedge t} - Z_{\tau^n_k \wedge t})^{\otimes^2},$$

where, for each n, $\Pi^n := \{0 = \tau^n_1 < \tau^n_2 < \dots < \tau^n_k < \dots\}$ is an increasing sequence of stopping times with the properties (i) and (ii) of Theorems 25.1 and 21.5.

Then

(1°) *$[\![Z]\!]$ is a positive $\mathbb{H} \hat{\otimes}_2 \mathbb{H}$-valued process. Almost all its paths are actually in $\mathbb{H} \hat{\otimes}_1 \mathbb{H}$ and have finite variation on every bounded interval $[0, t]$ for the norm of $\mathbb{H} \hat{\otimes}_1 \mathbb{H}$.*

(2°) *For every t the sequence $(\hat{v}_n(t,Z))_{n \in \mathbb{N}}$ of $\mathbb{H} \hat{\otimes}_2 \mathbb{H}$-valued random variables converges in probability to the random variable $[\![Z]\!]_t$ and one can even choose a subsequence $(\hat{v}_{n_k}(.,Z))_{k \in \mathbb{N}}$ of processes, the paths of which converge uniformly on any bounded interval to the paths of $[\![Z]\!]$.*

(3°) $[Z] = \operatorname{Trace} [\![Z]\!]$.

Proof. Let us write

(26.11.3) $$\hat{v}_n(t,Z)=\sum_k \left(Z^{\otimes^2}_{\tau^n_{k+1}\wedge t}-Z^{\otimes^2}_{\tau^n_k\wedge t}\right)-\sum_k Z_{\tau^n_k\wedge t}\otimes\left(Z_{\tau^n_{k+1}\wedge t}-Z_{\tau^n_k\wedge t}\right)$$
$$-\sum_k\left(Z_{\tau^n_{k+1}\wedge t}-Z_{\tau^n_k\wedge t}\right)\otimes Z_{\tau^n_k\wedge t}$$
$$=Z_t^{\otimes^2}-Z_0^{\otimes^2}-\sum_k Z_{\tau^n_k\wedge t}\otimes\left(Z_{\tau^n_{k+1}\wedge t}-Z_{\tau^n_k\wedge t}\right)$$
$$-\sum_k\left(Z_{\tau^n_{k+1}\wedge t}-Z_{\tau^n_k\wedge t}\right)\otimes Z_{\tau^n_k\wedge t}$$

and set

(26.11.4) $$\Phi_n:=\sum_k Z_{\tau^n_k} 1_{]\tau^n_k,\tau^n_{k+1}]}.$$

Assumptions (i) and (ii) imply the convergence of $(\Phi_n(t,\omega))_{n\in\mathbb{N}}$ to $Z_{t^-}(\omega)$ for every t and ω.

We have, moreover, for the processes Φ_n considered as $\mathscr{L}(\mathbb{H};\mathbb{H}\hat{\otimes}_2\mathbb{H})$ valued processes,

(26.11.5) $$\|\Phi_n(t,\omega)\|\leqslant\sup_{s<t}\|Z_s(\omega)\|_{\mathbb{H}}<\infty \text{ for every } t \text{ and } \omega.$$

Writing (26.11.3) in the form

$$\hat{v}_n(t,Z)=Z_t^{\otimes^2}-Z_0^{\otimes^2}-\int_{]0,t]}\Phi_n\otimes dZ-\int_{]0,t]}dZ\otimes\Phi_n,$$

we obtain all the convergences stated in the theorem by applying Theorem 26.3. Since the $\hat{v}_n(t,Z)$ are clearly positive valued in $\mathbb{H}\otimes_2\mathbb{H}$ (see section 21.2), the same holds for $[\![Z]\!]_t$.

If we observe that

$$v_n(t,Z)=\operatorname{Trace}(\hat{v}_n(t,Z)),$$

where the v_n are the real-valued processes defined in Theorem 26, we see that, for the above subsequence $(\hat{v}_{n_k})_{k\in\mathbb{N}}$ for all t and ω,

$$\lim_k \operatorname{Trace}(\hat{v}_{n_k}(t,Z)(\omega))=[Z]_t(\omega).$$

This implies that the sequence $(\hat{v}_{n_k}(t,Z)(\omega))_{k\in\mathbb{N}}$ actually converges in $\mathbb{H}\hat{\otimes}_2\mathbb{H}$ to a limit $[\![Z]\!]_t(\omega)\in\mathbb{H}\hat{\otimes}_1\mathbb{H}$ with $\|[\![Z]\!]_t(\omega)\|_1\leqslant[Z]_t(\omega)$. (Considering an orthonormal basis in $\mathbb{H}$ the reader may view this fact as a property of sequences in $\ell^1(\mathbb{N})$ and $\ell^2(\mathbb{N})$. If a sequence $(x^n)_{n\in\mathbb{N}}$ of elements $(\chi^n_k)_{k\in\mathbb{N}}$ of $\ell^2(\mathbb{N})\cap\ell^1(\mathbb{N})$ converges in $\ell^2(\mathbb{N})$ to $(\chi^\infty_k)_{k\in\mathbb{N}}$ and is such that $\lim_n\|\chi^n\|_1$ exists, then $\chi^\infty\in\ell^1(\mathbb{N})$ and $\|\chi^\infty\|_1\leqslant\lim_n\|\chi^n\|_1$).

In the same way one may write $\|[\![Z]\!]_t-[\![Z]\!]_s\|_1\leqslant[Z]_t-[Z]_s$ a.s. This shows that the variation of the paths for the norm of $\mathbb{H}\hat{\otimes}_1\mathbb{H}$ is smaller than the variation of the paths of $[Z]$ and therefore finite.

This completes the proof of the theorem. □

Taking an orthonormal basis $(h_i)_{i\in\mathbb{N}}$ in $\mathbb{H}$ and writing $Z^i := Z\cdot h_i$ we see that $[\![Z]\!]$ can be represented in this basis by the infinite matrix

$$[\![Z]\!]_t^{i,j} := Z_t^i Z_t^j - Z_0^i Z_0^j - \int_{]0,t]} Z_{s-}^i dZ_s^j - \int_{]0,t]} Z_{s-}^j dZ_s^i .$$

The trace of this matrix is

$$\begin{aligned}\sum_i [\![Z]\!]^{i,i} &= \|Z_t\|^2 - \|Z_0\|^2 - 2\int_{]0,t]} Z_{s-}\, dZ_s \\ &= [Z]_t .\end{aligned}$$

This completes the proof of the theorem. □

$[\![Z]\!]$ is called the tensor quadratic variation of Z.

—

26.12 Corollaries of Theorem 26.11. *Let Z be an $\mathbb{H}$-valued semimartingale. Then* (1°) *$Z^{\otimes^2}$ is an $\mathbb{H}\hat{\otimes}_2\mathbb{H}$-valued semimartingale and for every predictable process X with values in $\mathscr{L}(\mathbb{H}\hat{\otimes}_2\mathbb{H};\mathbb{G})$, with paths bounded on any bounded interval, one has*

$$\int_{]0,t]} X_s d(Z_s^{\otimes^2}) = \int_{]0,t]} X_s Z_{s-}\otimes dZ_s + \int_{]0,t]} X_s dZ_s\otimes Z_{s-} + \int_{]0,t]} X_s d[\![Z_s]\!] ,$$

where $h \curvearrowright X_s(\omega)Z_{s-}(\omega)\otimes h$ (resp. $h\curvearrowright X_s(\omega)h\otimes Z_{s-}(\omega)$) denotes the operator $h\curvearrowright X_s(\omega)(Z_{s-}(\omega)\otimes h)$ (resp. $h\curvearrowright X_s(\omega)(h\otimes Z_{s-}(\omega))$ (see the end of section 26.1).

(2°) *If Π^n is defined as in Theorem 26.11, X is an R.R.C. process with values in $\mathscr{L}(\mathbb{H}\hat{\otimes}_2\mathbb{H};\mathbb{G})$ and if we define*

$$w_n(t,Z,X) := \sum_k X_{\tau_k^n}(Z_{\tau_{k+1}^n\wedge t} - Z_{\tau_k^n\wedge t})^{\otimes^2} ,$$

the sequence of $\mathbb{G}$-valued random variables $(w_n(t,Z,X))_{n\in\mathbb{N}}$ converges in probability to $\int_{]0,t]} X_{s-} d[\![Z]\!]_s$. One can even find a subsequence $(w_{n_k}(.,Z,X))_{k\in\mathbb{N}}$ such that with probability one the paths of the processes $w_{n_k}(.,Z,X)$ converge uniformly on every bounded interval $[0,t]$ to the paths of $\int X_{s-} d[\![Z]\!]_s$ when $k\to\infty$.

(3°) *The family $(\Delta Z_s(\omega))_{s\leq t}$, being summable in $\mathbb{H}\hat{\otimes}_2\mathbb{H}$ with probability one, the $\mathbb{H}\hat{\otimes}_2\mathbb{H}$-valued process $[\![Z]\!]^c := ([\![Z]\!]_t - \sum_{s\leq t}(\Delta Z_s)^{\otimes^2})_{t\in\mathbb{R}^+}$ is a.s. continuous.*

Proof. These properties follow from Theorem 26.11 exactly as Corollaries 1, 2 and 3 follow from Theorem 25.1. □

27. The transformation formula in the multidimensional case

27.1 Theorem (Transformation formula in finite dimension). *Let Z be an $\mathbb{R}^d$-valued semimartingale and φ a real function on $\mathbb{R}^d$ which admits continuous derivatives of first- and second-order, denoted respectively by $(D^i\varphi)_{i=1..d}$ and $(D^{ij}\varphi)_{\substack{i=1..d\\ j=1..d}}$.*

Then the process $(\varphi(Z_t))_{t\in\mathbb{R}^+}$ *is a semimartingale which is P-equal to the following process* $(T_t^\varphi(Z))_{t\in\mathbb{R}^+}$.

$$T_t^\varphi(Z) := \varphi(Z_0) + \sum_{i=1}^{d} \int_{]0,t]} D^i\varphi(Z_{s-})dZ_s^i + \tfrac{1}{2}\sum_{i,j=1}^{d} \int_{]0,t]} D^{ij}\varphi(Z_{s-})d[Z^i, Z^j]_s$$

$$+ \sum_{s\leqslant t} [\varphi(Z_s) - \varphi(Z_{s-}) - \sum_{i=1}^{d} D^i\varphi(Z_{s-})\Delta Z_s^i$$

$$- \tfrac{1}{2}\sum_{i,j=1}^{d} D^{ij}\varphi(Z_{s-})\Delta Z_s^i \Delta Z_s^j]$$

One also has

$$T_t^\varphi(Z) = \varphi(Z_0) + \sum_{i=1}^{d} \int_{]0,t]} D^i\varphi(Z_{s-})dZ_s^i + \tfrac{1}{2}\sum_{i,j=1}^{d} \int_{]0,t]} D^{ij}\varphi(Z_{s-})d[Z^iZ^j]_s^c$$

$$+ \sum_{s\leqslant t} [\varphi(Z_s) - \varphi(Z_{s-}) - \sum_{i=1}^{d} D^i\varphi(Z_{s-})\Delta Z_s^i].$$

In this formulas the family occuring under the symbol $\sum_{s\leqslant t}$ *is a.s. summable.*

Proof. As in the proof of Theorem 25.7 we reduce the situation to the case where the second-order derivatives are bounded functions on $\mathbb{R}^d$.

Let us set, for every $x\in\mathbb{R}^d$ and $h\in\mathbb{R}^d$,

$$A^\varphi(x,h) := \sum_{i=1}^{d} h^i D^i\varphi(x) + \tfrac{1}{2}\sum_{i,j=1}^{d} h^i h^j D^{ij}\varphi(x).$$

As a consequence of the Taylor formula, we obtain

(27.1.1) $$\varphi(x+h) - \varphi(x) - A^\varphi(x,h) \leqslant \|h\|^2 v(\|h\|),$$

where v is a positive function on $\mathbb{R}^+$ with the property

$$\lim_{\|h\|\searrow 0} v(\|h\|) = 0.$$

As in the proof of Theorem 25.7 we define the stopping times

$$\tau_0^n := 0, \ldots, \tau_{k+1}^n := \inf\left\{t\colon t > \tau_k^n \|Z_t - Z_{\tau_k^n}\| > \frac{1}{n}\right\} \wedge \left(\tau_k^n + \frac{1}{n}\right)$$

We set

(27.1.2) $$u_{n,k}(t) := A^\varphi(Z_{\tau_k^n\wedge t}, Z_{\tau_{k+1}^n\wedge t} - Z_{\tau_k^n\wedge t})$$

(27.1.3) $$v_{n,k}(t) := -A^\varphi(Z_{\tau_k^n}, \Delta Z_{\tau_{k+1}^n})1_{\{\tau_k^n\leqslant t\}} + \varphi(Z_{\tau_{k+1}^n\wedge t}) - \varphi(Z_{(\tau_{k+1}^n\wedge t)-})$$

(27.1.4) $$w_{n,k}(t) := \varphi(Z_{(\tau_{k+1}^n\wedge t)-}) - \varphi(Z_{\tau_k^n\wedge t}) - A^\varphi(Z_{\tau_k^n\wedge t}, Z_{(\tau_{k+1}^n\wedge t)-} - Z_{\tau_k^n\wedge t})$$

(27.1.5) $$y_{n,k}(t) := \sum_{i,j=1}^{d} \Delta Z_{\tau_{k+1}^n}^i (Z_{(\tau_{k+1}^n\wedge t)-}^j - Z_{\tau_k^n\wedge t}^j)D^{ij}\varphi(Z_{\tau_k^n})$$

We may write

(27.1.6) $$\varphi(Z_t) - \varphi(Z_0) = \sum_k [\varphi(Z_{\tau^n_{k+1} \wedge t}) - \varphi(Z_{\tau^n_k \wedge t})] = \\ = \sum_k [u_{n,k}(t) + v_{n,k}(t) + w_{n,k}(t) - y_{n,k}(t)].$$

Exactly as in the proof of Theorem 25.7 we obtain, for every $t \in \mathbb{R}^+$,

(27.1.7) $$\lim_{n \to \infty} \sup_{s \leqslant t} (|\sum_k w_{n,k}(s)|) = 0 \quad \text{a.s.}$$

(27.1.8) $$\lim_{n \to \infty} \sup_{u \leqslant t} |\sum_k v_{n,k}(u) - \sum_{s \leqslant u} [\varphi(Z_s) - \varphi(Z_{s^-}) - A^\varphi(Z_{s^-}, \Delta Z_s)]| = 0 \quad \text{a.s.}$$

(27.1.9) $$\lim_{n \to \infty} \sup_{s \leqslant t} (|\sum_k y_{n,k}(s)|) = 0 \quad \text{a.s.,}$$

and, for a suitable subsequence $(n_r)_{r \in \mathbb{N}}$,

(27.1.10) $$\lim_{r \to \infty} \sup_{u \leqslant t} | \sum_{i=1} \int_{]0,t]} D^i \varphi(Z_{s^-}) dZ^i_s + \tfrac{1}{2} \sum_{i,j=1}^{d} \int_{]0,u]} D^{ij} \varphi(Z_{s^-}) d[Z^i, Z^j]_s \\ - \sum_k u_{n_r,k}(u)| = 0 \quad \text{a.s.}$$

The first equality of the theorem follows from the inequalities (27.1.6) to (27.1.10). The second is then implied by Corollary (26.12.3°). □

27.2 Theorem* (Transformation formula in infinite dimension). *Let Z be an $\mathbb{H}$-valued semimartingale and φ a mapping from $\mathbb{H}$ into $\mathbb{G}$ admitting continuous derivatives of first- and second-order denoted by φ' and φ''. We assume that for each $x \in \mathbb{H}$ the derivative $\varphi''(x)$ is an element of $\mathscr{L}(\mathbb{H} \hat{\otimes}_2 \mathbb{H}; \mathbb{G})$ and that the mapping $x \curvearrowright \varphi''(x) \in \mathscr{L}(\mathbb{H} \hat{\otimes}_2 \mathbb{H}; \mathbb{G})$ is uniformly continuous on any bounded subset of $\mathbb{H}$.*

Then the process $(\varphi(Z_t))_{t \in \mathbb{R}^+}$ is a $\mathbb{G}$-valued semimartingale which is P-equal to the following process $(T^\varphi_t(Z))_{t \in \mathbb{R}^+}$, defined by

$$T^\varphi_t(Z) := \varphi(Z_0) + \int_{]0,t]} \varphi'(Z_{s^-})\ dZ_s + \tfrac{1}{2} \int_{]0,t]} \varphi''(Z_{s^-})\ d[\![Z]\!]_s \\ + \sum_{s \leqslant t} [\varphi(Z_s) - \varphi(Z_{s^-}) - \varphi'(Z_{s^-}) \Delta Z_s \\ - \tfrac{1}{2} \varphi''(Z_{s^-})(\Delta Z_s)^{\otimes_2}]$$

We also have

$$T^\varphi_t(Z) = \varphi(Z_0) + \int_{]0,t]} \varphi'(Z_{s^-})\ dZ_s + \tfrac{1}{2} \int_{]0,t]} \varphi''(Z_{s^-})\ d[\![Z[\![^c_s \\ + \sum_{s \leqslant t} [\varphi(Z_s) - \varphi(Z_{s^-}) - \varphi'(Z_{s^-}) \Delta Z_s]$$

In these formulas the family of $\mathbb{G}$-valued random variables occuring under the symbol $\sum_{s \leqslant t}$ is a.s. summable.

Proof. If for every $x \in \mathbb{H}$ and $h \in \mathbb{H}$ we set

$$A^{\varphi}(x,h) := \varphi'(x)h + \tfrac{1}{2}\varphi''(x)h^{\otimes_2}$$

we can reproduce the proof of Theorem 27.1 word for word. We do not repeat it. □

27.3 Corollary (Ito-formula: continuous case). *Let Z be an* $\mathbb{R}^d$*-valued continuous semimartingale and* φ *a real function on* $\mathbb{R}^d$, *which admits continuous derivatives of first- and second-order. Then the process* $(\varphi(Z_t))_{t\in\mathbb{R}^+}$ *is a continuous semimartingale which is P-equal to the process* $(T_t^{\varphi}(Z))_{t\in\mathbb{R}^+}$ *defined by*

$$T_t^{\varphi}(Z) := \varphi(Z_0) + \int_{]0,t]} \sum_{i=1}^{d} D^i\varphi(Z_s)dZ_s^i + \tfrac{1}{2}\int_{]0,t]} \sum_{i,j=1}^{d} D^{ij}\varphi(Z_s)d[Z^i,Z^j]_s.$$

An analogous corollary to Theorem 27.2 can clearly be written.

Exercises and supplements

E.1. *Application of the Ito-Formula to the Langevin equation and Ornstein-Uhlenbeck process.* Langevin introduced the following equation to describe the movement of a particle with coordinate $x(t)$ at time t submitted to small independent shocks.

$$\text{(1)} \qquad \text{“}\ \frac{d^2x(t)}{dt^2} + \alpha\frac{dx}{dt} = \sigma^2\frac{d\beta}{dt}\ \text{”},$$

where α and σ^2 are positive constants and β is a Brownian motion with variance 1.

Because of the non-differentiability of the paths of β, equation (1) is to be interpreted as

$$\text{(2)} \qquad \begin{cases} V(t) = V(0) - \int_{]0,t]} \alpha V(s)ds + \sigma^2\beta(t) \\ x(t) = x(0) + \int_{]0,t]} V(s)ds, \end{cases}$$

$V(t)$ being the speed of the particle at time t.

(1°) Call M the martingale $M_t := \int_0^t \alpha e^{-\alpha s}d\beta(s)$ and U the process with finite variation $U_t := e^{\alpha t}$. Show, using the integration-by-part formula that the following expression gives a solution of (1).

$$\text{(3)} \qquad V(t) = e^{\alpha t}\left[V(0) + \int_0^t \sigma e^{-\alpha s}\mathrm{d}\beta(s)\right]$$

(2°) The process V is called the *Ornstein-Uhlenbeck speed process.* Show that every couple $(V(s), V(t))$ is a Gaussian two-dimensional variable (use a step approximation of $e^{-\alpha s}$) and that its covariance function is given, $s < t$, by

$$C(s,t) := E(V(s)\,V(t)) = \int_{]0,s]} e^{\alpha(t-u)}\sigma^2 e^{\alpha(s-u)}\,du. \tag{4}$$

E.2. *Transformation formula for jump processes.* This transformation formula is very trivial when one considers processes Z, the paths of which have only finitely many jumps and finite variation on any finite interval. Show directly that, in this case, for any function φ which has one continuous derivative, the following formula

$$\varphi(Z_t) = \varphi(Z_s) + \int_0^t \varphi'(Z_{s^-})dZ_s + \sum_{s \leqslant t} \varphi(Z_s) - \varphi(Z_{s^-})$$

holds and compare this with the transformation formula as stated in 25.7.

E.3. *Proof of formula (26.9.4).* Let Z be a real semimartingale and V a real R.R.C. process with paths of finite variation on any finite interval.

Define for every n the increasing sequence Π^n of stopping times as in the proof of Corollary 25.5.

$$\tau_0^n := 0, \ldots \tau_{k+1}^n := \inf\left\{t\colon t > \tau_{k+1}^n, |Z_t - Z_{\tau_k^n}| > \frac{1}{n}\right\} \wedge \left(\tau_k^n + \frac{1}{n}\right).$$

Use the random variables

$$v_n(t, Z, V) := \sum_k (Z_{\tau_{k+1}^n \wedge t} - Z_{\tau_k^n \wedge t})(V_{\tau_{k+1}^n \wedge t} - V_{\tau_k^n \wedge t})$$

to prove that, in this case,

$$[Z, V]_t = \sum_{s \leqslant t} \Delta Z_s \Delta V_s.$$

(*Hint*: Remark that, since V has finite variation on $[0, t]$, one has

$$\lim_n \text{prob} \sum_k (Z_{(\tau_{k+1}^n \wedge t)^-} - Z_{\tau_k^n \wedge t})(V_{\tau_{k+1}^n \wedge t} - V_{\tau_k^n \wedge t}) = 0).$$

If Z and V are Hilbert-valued semimartingales, V having finite variation,

$$[\![Z, V]\!]_t = \sum_{s \leqslant t} \Delta Z_s \otimes \Delta V_s.$$

E.4. *Another form of the integration-by-part formula* (special case). Let Z be a real semimartingale and V a real R.R.C. process with finite variation on finite intervals. Using formula (26.9.4) (proved in E.3), show that

$$V_t Z_t = V_0 Z_0 + \int_{]0,t]} V_{s^-} dZ_s + \int_{]0,t]} Z_s dV_s,$$

where the last integral is to be understood as a Stieltjes integral on every path.

E.5. *Quadratic variation of $Z = X + V$.* Let X be a real semimartingale and V be a real R.R.C. process with paths of finite variation on any finite interval. Show that

$$[X+V] = [X] + \sum_{s \leqslant t} |\Delta V_s|^2 + 2 \sum_{s \leqslant t} |\Delta V_s| \, |\Delta X_s| \, .$$

In particular, if X is continuous $[X+V] = [X] + \sum_{s \leqslant t} |\Delta V_s|^2$.

E.6. *A special form of the transformation formula.* Let φ be a twice continuously differentiable function of two variables, let Z be a real semimartingale and A a real R.R.C. process with finite variation. Prove the following formula.

$$\begin{aligned}\varphi(Z_t, A_t) = \varphi(Z_0, A_0) &+ \int_{]0,t]} D^1\varphi(Z_{s^-}, A_{s^-}) dZ_s + \int_{]0,t]} D^2\varphi(Z_{s^-}, A_{s^-}) dA_s \\ &+ \tfrac{1}{2} \int_{]0,t]} D^{11}\varphi(Z_{s^-}, A_{s^-}) d[Z]_s + \sum_{s \leqslant t} [\varphi(Z_s, A_s) - \varphi(Z_{s^-}, A_{s^-}) \\ &- D^1\varphi(Z_{s^-}, A_{s^-}) \Delta Z_s - D^2\varphi(Z_{s^-}, A_{s^-}) \Delta A_s \\ &- \tfrac{1}{2} D^{11}\varphi(Z_{s^-}, A_{s^-}) |\Delta Z_s|^2]\end{aligned}$$

(*Hint*: Apply the transformation theorem to the bidimensional semimartingale $(Z_t, A_t)_{t \in \mathbb{R}^+}$).

E.7. *Some insight into the meaning of the Ito-Formula.* Let us consider the following vector field $V(x^1, x^2)$ in the plane with coordinates x^1 and x^2.

$$\vec{V}(x^1, x^2) = (-x^2, x^1) \, .$$

Let β be a one-dimensional Brownian motion with variance 1 and let $\varepsilon > 0$.

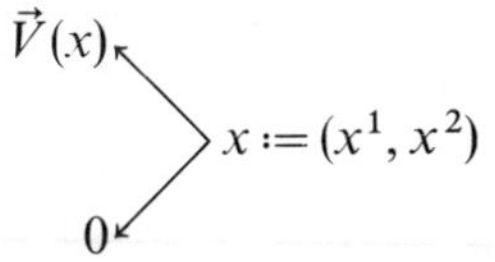

Let $X(t)$ be a "solution" of the "differential equation" $d\vec{X}(t) = \vec{V}(X(t))(dt + \varepsilon d\beta(t))$.
This equation should be interpreted as

(1) $$X(t) = X(0) + \int_0^t \vec{V}(X(s)) ds + \varepsilon \int_0^t V(X(s)) d\beta(s) \, .$$

It will be proved in chapter VIII that such a process X exists and is unique.

(1°) Let X be a (continuous) process such that (1) holds.
Applying the Ito formula to $\|X(t)\|^2$ show that

$$\|X(t)\|^2 = \|X(0)\|^2 e^{\varepsilon^2 t}$$

Compare this with the behaviour of $\|X(t)\|^2$ when $X(t)$ is the solution of the deterministic (usual differential) equation obtained by setting $\varepsilon = 0$ in (1).

(2°) If in (1) β is replaced by a functions $F(t) = \int_0^t \sum_i \alpha_i \delta_{\tau_i}(ds)$, where δ_{τ_i} are Dirac masses at times $\tau_1 < \tau_2 < \ldots \tau_n < \ldots$ representing shocks, and $\alpha_i \in \mathbb{R}$, describe the solution $X(t)$ of the differential equation $\dfrac{d\vec{X}(t)}{dt} = \vec{V}(X(t))\left(1 + \varepsilon \dfrac{dF(t)}{dt}\right)$.
(Observe that the particle jumps from a circle to a bigger one at each shock).

(3°) The suggestion is to see from this example that, despite its continuity, Brownian motion essentially describes a limit model of a cumulative sum of "infinitely small discontinuities". Basically, term involving φ'' in the Ito formula expresses this. (Think also of the formula in E.2 introducing terms other than those involved in the chain-rule formula for differentiation).

E.8. *The differentiation formula and the Stratonovitch integral.* Let us consider the expression of $Z_t^2 - Z_0^2$ as in section 25.1. Instead of writing this difference

$$\sum_k (Z_{\tau_{k+1}^n \wedge t} - Z_{\tau_k^n \wedge t})^2 + 2 \sum_k Z_{\tau_k^n \wedge t}(Z_{\tau_{k+1}^n \wedge t} - Z_{\tau_k^n \wedge t}),$$

as in (25.1.2) let us write it in the form

$$Z_t^2 = Z_0^2 + 2 \sum_k \tfrac{1}{2} (Z_{\tau_{k+1}^n \wedge t} + Z_{\tau_k^n \wedge t})(Z_{\tau_{k+1}^n \wedge t} - Z_{\tau_k^n \wedge t}).$$

The sum occuring in the right-hand side of this equality may be interpreted as some sort of Riemann-Stieltjes sum $\sum_k \xi_k(Z_{\tau_{k+1}^n \wedge 1} - Z_{\tau_k^n \wedge t})$. One naturally wants to know for which processes Y the sums of the type $\sum_k \frac{1}{2} (Y_{\tau_{k+1}^n \wedge t} + Y_{\tau_k^n \wedge t})(Z_{\tau_{k+1}^n \wedge t} -$ $- Z_{\tau_k^n \wedge t})$ converge in some sense to a process. This has indeed be studied and when the limiting process exists it is called the Stratonovitch integral. Let us denote it by $\oint Y dZ_s$. It is clear from the above decomposition that the Stratonovitch integral leads, for Z continuous, to a formula which is analogous to the classical differentiation formula. Thus,

$$Z_t^2 = Z_0^2 + \oint_0^t 2 Z_s dZ_s.$$

And if Z is a continuous process the transformation formula reads

$$\varphi(Z_t) = \varphi(Z_0) + \oint_0^t 2\varphi'(Z_s) dZ_s,$$

which has exactly the same form as the usual chain-rule differentiation formula.

For more details on the Stratonovitch Integral the reader is referred to [Mey. 3], chap. 6.

The simplicity of the differentiation rule with the Stratonovitch integral has the following counterparts.

(1°) The theory has the same "lourdeurs" as the theory of the Riemann Integral compared to the Lebesgue integral. The class of integrable processes is not so simple to define and is more restrictive.

(2°) The integral with respect to a martingale is no longer a martingale. One of the interesting features of the Ito-formula is that it automatically separates the martingale part and the part with finite variation of $\varphi(Z_t)$. Putting into evidence the martingale part of a process is actually an important tool of stochastic calculus (see the following chapter).

In any case, the comparison of the Ito formula and the Stratonovitch differentiation formula makes it clear how we pass from one integral to the other. The following formula may be found in [Mey. 3], chap. 6. If H and X are two *continuous* semi-martingales,

$$\oint_0^t H_u dX_u = \int_0^t H_u dX_u + \tfrac{1}{2} < H^m, X^m > t,$$

where H^m and X^m are the martingale parts of H and X.

E.9. *Extension of the stochastic integral to optional and progressively measurable processes.* (1°) Let us assume that the semimartingale Z admits a control process A which is continuous.

Show that if we set $\mathscr{J} := \{X: X \text{ optional}, \int_0^t |X_s|^2 dA_s \text{ finite all } t\}$ the mapping $X \curvearrowright \int XdZ$ has a unique linear extension to $\mathscr{J}$ such that, for every $X \in \mathscr{J}$ and stopping time τ,

$$(1) \qquad E(\sup_{s<\tau} || \int_{]0,t]} X_s dZ_s ||^2) \leqslant E(A_\tau \int_{[0,\tau[} ||X_s||^2 dA_s),$$

which is called the stochastic integral.

Hint: Show that, for every $X = 1_{[0,\sigma[}$ and then for every $x \in \mathscr{J}$, there exists a predictable Y such that

$$E(A_\tau \int_{[0,\tau[} ||X_s - Y_s||^2 dA_s) = 0.$$

(2°) Let us assume that the semimartingale Z admits a control process A such that $A_t(\omega) = \int_0^t f(s, \omega) ds$, f bounded. Show that if

$$\tilde{\mathscr{J}} := \{X: X \text{ progressively measurable } \int_0^t ||X_s||^2 ds \text{ finite all } t\},$$

the mapping $X \curvearrowright \int XdZ$ has a unique linear extension to $\tilde{\mathscr{J}}$ such that, for every $X \in \tilde{\mathscr{J}}$ and stopping time τ, (1) holds. This extension is called the stochastic integral.

Hint: Same method as in (1°) but with $X \in \tilde{\mathscr{J}}$; take $Y_t^h(\omega) = \frac{1}{h} \int_{t-h}^t X_s(\omega) ds$ and let $h \to 0$ to obtain Y.

(3°) Brownian motion, Poisson process belong to case 2°). We are in case 1°) when we take the following particular semimartingale considered by McShane [McS]. There exists an increasing continuous process A (in McShane it is even deterministic) such that

$$|E(X_t - X_s | \mathscr{F}_s)| \leqslant A_t - A_s \quad \text{for every } s \leqslant t$$
$$|E(X_t^2 - X_s^2 | \mathscr{F}_s) \leqslant A_t - A_s \quad \text{for every } \quad s \leqslant t.$$

E. 10. *Tanaka Formula (Local time of a real semimartingale).* Let X be a real continuous semimartingale. We define $\operatorname{sgn} x = \frac{x}{|x|}$ if $x \neq 0$ and $= 0$ if $x = 0$. Show that the process L defined by $L := |X| - \int_{]0,t]} \operatorname{sgn}(X_s) dX_s$ is increasing with the property $L_t = \int_{]0,t]} 1_{\{X_s = 0\}} dL_s$.

This process is called *local time in* 0 of X.

Hint: Consider a sequence (φ_n) of C^2-functions converging uniformly to $|x|$ with the properties $\varphi_n'(0) = 0$; $\varphi_n''(x) = 0$ for $|x| \geqslant \frac{1}{n}$; $\varphi_n''(x) \geqslant 0$ for $x \leqslant \frac{1}{n}$ and $\int_{-1/n}^{+1/n} \varphi_n''(x) dx = 1$.

Apply the transformation formula and dominated convergence theorem in the form 24.2 for stochastic integrals.

Historical and bibliographical notes

Some landmarks in the evolution of the theory. It is usually said that stochastic integrals go back to N. Wiener who integrated deterministic functions with respect to Brownian motions. Many later theories of stochastic integration are copies of Wiener's in a slightly generalized context. All of them are recognizable by the fact that they start with a process with orthogonal increments or a random setfunction having orthogonal values for disjoint sets, and that the integral is obtained as an easy isometric extension to a suitable L^2-space.

Ito developed the integration of processes with respect to Brownian motion [Ito. 1, 2, 3, 4]. Doob recognized the importance of integrating with respect to square integrable martingales and, as we said at the end of chapter IV, integration of predictable processes with respect to square integrable martingales was performed by Meyer, Courreges, Kunita, Watanabe and Skorokhod [Sko]. The extension to local martingales and semimartingales was made by Meyer [Mey. 8] and Doleans-Dade [Dom. 1]. In this general case, the processes which are integrated are predictable and locally bounded.

When the integral is defined with respect to Brownian motion or with respect to continuous semimartingales, the class of integrated processes is apparently larger.

In the first case, it contains, for example, all progressively measurable processes X such that $\int_0^t |X_s|^2 ds$ is finite (cf. McKean [McK]). We show in exercise E.9 that this class is not "essentially bigger" than the class of processes we integrated in this chapter. We also indicate in the same exercise how the stochastic integral theory of McShane [McS] is included in the Meyer-Doleans-Dade-theory. (See also [Pop. 2], [Pro. 4]).

The presentation of the stochastic integration given in this chapter is not absolutely classical in as much as it relies on the use of control processes, but it is very close in spirit to the account given in [MeP. 1].

Stochastic integrals as vector-valued integrals. This point of view consists in considering the L^0-valued measure on predictable sets and defining the stochastic integral as an integral of an operator-valued function with respect to this vector-valued measure. Stochastic integration becomes a particular case of vector-valued integration. (See Pellaumail's Thesis [Pel. 2] and Metivier [Met. 9], also K. Bichteler [Bic], L. Schwartz [Schw. 2] and the review paper [Del. 5]. We did not take this point of view here because of the necessary prerequisites in vector-valued measures. (See also comments at the end of chap. 4, concerning the characterization of semimartingales).

Transformation formula. The transformation formula was first proved by K. Ito for the stochastic integrals of Brownian motion ("Ito Formula": [Ito. 4]). The present form in the real case as described here is the result of works already mentioned by Kunita, Watanabe, Meyer, Doleans-Dade. An extension to Hilbert-valued martingales was made by H. Kunita but in a rather "weak" form. The full generalization, as presented here, was made possible by the introduction of the tensor quadratic variation [Met. 5]. There have also been attempts to establish an analogous formula for Banach-valued processes (Gravereau and Pellaumail [GrP]).

Chapter 6

First applications of the transformation theorem

This chapter gives some of the most classical examples of applications of the stochastic calculus which was set up in the preceding chapter. Many others will come later.

The examples treated here are intended to show the importance of the transformation formula (or some of its particular forms, like the integration-by-part formula) as a tool for dealing with the laws of processes.

In section 28 we see how the law of a process can in some cases be characterized by a suitable martingale property for some functionals of the process. This idea, which can be applied to a very wide class of processes, will be met again several times in the subsequent chapters and is of great importance in the modern study of processes.

Section 29 studies the exponential of a semimartingale Z, defined as the solution X of the equation $dX = XdZ$.

Section 30 introduces the important problem of the absolutely continuous change of probability and gives a classical application in signal detection theory.

28. Characterizations of Brownian and Poisson processes

This section will provide two fundamental examples of situations where the law of a process X is entirely determined by its semimartingale property and the knowledge of a "characteristic" of the process, that is to say, the knowledge of a specific functional Φ of X such that $\Phi(X)$ is a martingale.

28.1 Theorem (Characterization of finite-dimensional Brownian motion). *A continuous d-dimensional process X, adapted to a given filtration $(\mathscr{F}_t)_{t\in\mathbb{R}^+}$ on a probability space $(\Omega,\mathscr{A},P)$ is a Brownian motion with variance matrix Q if and only if X is a locally square integrable martingale with respect to $(\mathscr{F}_t)_{t\in\mathbb{R}^+}$ and P such that $(X_t^\tau \circ X_t - tQ)_{t\in\mathbb{R}^+}$ (where $X_t^\tau(\omega)$ is the transposed matrix of the column matrix $X_t(\omega)$) is a local martingale. In this case, X is actually an L^2-martingale.*

Proof. One already knows that, if X is a Brownian motion with variance-matrix Q, it is a square integrable martingale with Meyer process $\langle\langle X \rangle\rangle = tQ$ (see 17.3 for the real case and 21.7 for the Hilbert-valued Brownian motion). We have thus only to prove the "sufficient" part of the statement.

Let us assume that X is a continuous locally square integrable martingale with $\langle\langle X \rangle\rangle = tQ$. Since $E(\|X_t\|^2) = E(\langle X \rangle_t) = E(Trace \langle\langle X \rangle\rangle) = t\,Trace(Q)$ for every t (see Theorem 23.4), $E(\|X\|_t^2) < \infty$ and $X \in \mathcal{M}^2(\mathbb{R}^d; \mathcal{F})$. Let s, $t \in \mathbb{R}^+$ with $s < t$. The theorem will be proved if we show that the conditional law of X_t, given $\mathcal{F}_s$, is the Gaussian law with mean zero and variance matrix $(t-s)Q$. This is equivalent to proving that, for every $u \in \mathbb{R}^d$,

$$E(\exp(iu \cdot (X_t - X_s)) \mid \mathcal{F}_s) = \exp\left(-\frac{t-s}{2} u^\tau \circ Q \circ u\right), \qquad (28.1.1.)$$

where u^τ denotes the transposed matrix of the column matrix u, and $u \cdot X$ the scalar product of "column vectors" u and X.

To prove this relation we consider an $F \in \mathcal{F}_s$ and set

$$f(r) := \int_F \exp(iu \cdot X_{s+r}) dP \qquad (28.1.2)$$

Applying the Ito formula to $(e^{iu \cdot X_t})_{t \geq 0}$, we obtain

$$\exp iu \cdot X_t - \exp iu \cdot X_s = \int_{]s,t]} \exp(iu \cdot X_r) iu \cdot dX_r + \\ - \tfrac{1}{2} \int_{]s,t]} u^\tau \circ Q \circ u \exp(iu \cdot X_r) dr.$$

Noticing that the first process on the right-hand side of this equation is a martingale we may write, if we set $t = s + t'$,

$$f(t') = f(0) - \tfrac{1}{2} E\{1_F \cdot \int_s^t u^\tau \circ Q \circ u \exp(iu \cdot X_r) dr\}$$

and thus

$$f(t') = f(0) - \tfrac{1}{2} \int_0^{t'} (u^\tau \circ Q \circ u) f(r) dr. \qquad (28.1.3)$$

The unique solution of this latter equation is

$$f(t') = f(0) \exp\left(-\frac{t'}{2} u^\tau \circ Q \circ u\right).$$

Coming back to the definition of f, this gives

$$E(1_F \exp(iu \cdot X_t)) = \left[E\, 1_F \exp(iu \cdot X_s) \exp\left(-\frac{t-s}{2} u^\tau \circ Q \circ u\right)\right] \qquad (28.1.4)$$

and, since F is any element of $\mathcal{F}_s$,

(28.1.5) $$E(\exp(iu\cdot X_t)\,|\,\mathscr{F}_s)=\exp(iu\cdot X_s)\exp\left(-\frac{t-s}{2}u^{\tau}\circ Q\circ u\right)$$

This immediately yields the relation (28.1.1) and therefore the theorem. □

28.2 Theorem* (Characterization of infinite-dimensional Brownian motion). *Let $\mathbb{H}$ be a separable Hilbert space. A continuous $\mathbb{H}$-valued process X, adapted to a given filtration $(\mathscr{F}_t)_{t\in\mathbb{R}^+}$ on a probability space $(\Omega,\mathscr{A},P)$ is a Brownian motion with nuclear covariance Q if and only if X is a locally square integrable martingale with respect to $(\mathscr{F}_t)_{t\in\mathbb{R}^+}$ and P, with Meyer process $\langle\langle M\rangle\rangle = tQ$. In this case, X is actually an L^2-martingale.*

Proof. Entirely analogous to 28.1. Instead of writing the scalar product $u\cdot X$ in $\mathbb{R}^d$ we write it in $\mathbb{H}$ and we write $\langle u\otimes u, Q\rangle$ instead of $u^{\tau}\circ Q\circ u$ since Q is properly speaking an element of $\mathbb{H}\hat{\otimes}_1\mathbb{H}$. (One could also write $\langle u,\tilde{Q}(u)\rangle_{\mathbb{H}}$, where $\tilde{Q}$ is the nuclear operator associated with Q (see 21.4). □

28.3 Theorem (Characterization of a Poisson process). *Let $(\Pi_t)_{t\in\mathbb{R}^+}$ be a right-continuous real process which can be written $\sum_n 1_{[\tau_n,\infty[}$, where $\{\tau_n\}$ is an increasing sequence of stopping times. Π is a Poisson process adapted to the filtration $(\mathscr{F}_t)_{t\geq 0}$, with parameter λ, if and only if $(\Pi_t-\lambda t)_{t\in\mathbb{R}^+}$ is a local martingale. $(\Pi_t-\lambda t)_{t\in\mathbb{R}^+}$ is then in fact an L^2-martingale.*

Proof. We already know that if Π is a Poisson process, $(\Pi_t-\lambda t)_{t\in\mathbb{R}^+}$ is a martingale. Let us prove the converse statement assuming only that $\Pi_t-\lambda t$ is a local martingale (and therefore – since the jumps of this process are bounded – a locally square integrable martingale). From $E|\Pi_t-\lambda t|^2=\lambda t$ (see 23.4), $\Pi_t-\lambda t\in\mathscr{M}^2$.
Let us write $M_t:=\Pi_t-\lambda t$ and apply the transformation theorem, Theorem 25.7 to the process $\exp(\mathrm{i}u M_t)$ with $u\in\mathbb{R}$. We may write

(28.3.1) $$\exp iuM_{\tau_n}=\exp iuM_{\tau_{n-1}}+\int_{]\tau_{n-1},\tau_n]} iu\exp(iuM_{s^-})\,dM_s + \exp(iuM_{\tau_n})-\exp(iuM_{\tau_n^-})-iu\exp(iuM_{\tau_n^-})\Delta M_{\tau_n}$$

Since $M_{\tau_n^-}=M_{\tau_{n-1}}-\lambda(\tau_n-\tau_{n-1})$ and $\Delta M_{\tau_n}=1$, we have

(28.3.2) $$\exp(iuM_{\tau_{n-1}})[(1+iu)\exp(-iu\lambda(\tau_n-\tau_{n-1}))-1]= =\int_{]\tau_{n-1},\tau_n]} iu\exp(iuM_{s^-})\,dM_s$$

Since the stochastic integral with respect to M of a bounded process is a martingale with mean value zero, we obtain from (28.3.2) for every $F\in\mathscr{F}_{\tau_{n-1}}$

$$\int_F (1+iu)\exp(-iu\lambda(\tau_n-\tau_{n-1}))\,dP=P(\mathrm{F}),$$

which gives

$$\text{(28.3.3)} \qquad E[\exp(-iu\lambda(\tau_n - \tau_{n-1})) \mid \mathscr{F}_{\tau_{n-1}}] = \frac{1}{1+iu}.$$

The random variable $\tau_n - \tau_{n-1}$ is therefore independent from $\mathscr{F}_{\tau_{n-1}}$ and has an exponential distribution with parameter λ. This shows that the process $X = \sum_{n \geqslant 0} 1_{[\tau_n, \infty[}$ is a Poisson process with parameter λ. □

28.4 Example. Let $(M_t)_{t \in \mathbb{R}^+}$ be a d-dimensional, continuous, local martingale with

$$\langle\!\langle M \rangle\!\rangle_t = \int_0^t \Phi(s)\,ds,$$

where Φ is a positive matrix-valued process with the property that Φ^{-1} is defined and can be written $\Phi^{-1}(s,\omega) = \sigma(s,\omega) \circ \sigma^*(s,\omega)$, where σ is a predictable matrix-valued process. If we set

$$w_t := \int_{]0,t]} \sigma_s \, dM_s,$$

the process w is a martingale with $\langle\!\langle w \rangle\!\rangle_t = \int_{]0,t]} \sigma_s \circ \Phi_s \circ \sigma_s{}^* ds$ and therefore $\langle\!\langle w \rangle\!\rangle_t = t \cdot I_d$. This shows that w *is a Brownian motion with the matrix* I_d *as variance matrix.*

28.5 Theorem (A second characterization of finite-dimensional Brownian motion). *A continuous d-dimensional process X, adapted to a given filtration $(\mathscr{F})$ on a probability space $(\Omega, \mathscr{A}, P)$ is a Brownian motion with variance matrix Q and initial point x if and only if for every bounded twice continuously differentiable function φ on $\mathbb{R}^d$, with bounded derivatives, the process M^φ defined by*

$$M_t^\varphi := \varphi(X_t) - \varphi(x) - \tfrac{1}{2} \int_{]0,t]} \sum_{j,k=1}^{d} Q^{jk} D^{jk} \varphi(X_s)\,ds$$

is a martingale with $M_0^\varphi = 0$ P. a. s.

Proof. To prove the necessity we need only remember that $\langle\!\langle X \rangle\!\rangle = [\![X]\!]$, $\langle\!\langle X \rangle\!\rangle_t = [\![X]\!]_t = tQ$ and to use the transformation formula 27.1, which gives

$$\varphi(X_t) - \varphi(x) - \tfrac{1}{2} \int_{]0,t]} \sum_{j,k=1}^{d} D^{jk} \varphi(X_s) Q^{jk} ds = \int_{]0,t]} \sum_{j=1}^{d} D^j \varphi(X_s)\,dX_s^i.$$

Since the right-hand member of this equation is a martingale, we see that M^φ is a martingale.

Conversely, if M^φ is a martingale for every bounded, twice continuously differentiable φ with bounded derivatives we obtain, taking $\varphi(x) = e^{iu \cdot x}$ $u \in \mathbb{R}^d$ that, for every $s < t$ and $F \in \mathscr{F}_s$,

$$E(1_F e^{iu\cdot X_t}) = E(1_F e^{iu\cdot X_s}) - \tfrac{1}{2}\int_s^t E[1_F u^\tau \circ Q \circ u \exp(iu\cdot X_r)]\,dr\,.$$

The argument in the proof of Theorem 28.1 can be reproduced to obtain relation (28.1.1), which says that $X_t - X_s$ is independent of $\mathscr{F}_s$ and has a Gaussian law which is centered and has variance matrix $(t-s)Q$.

Since by hypothesis $0 = E(M_0^\varphi)$ one has $E(\varphi(X_0)) = \varphi(x)$ for every φ. This shows $P\{X_0 = x\} = 1$. □

29. Exponential formulas and linear stochastic differential equations

In this section we show that, (Z_t) being a given semimartingale, there is a unique process X on the same stochastic basis such that $X_t = X_0 + \int_0^t \lambda X_s dZ_s$ $(\lambda \in \mathbb{R})$ and we give an explicit representation of X as a function of Z.

For further easy references we state and prove a lemma as it was established in [Met. 7]. This "Gronwall-type" lemma will be used many times in the sequel.

29.1 Lemma. *Let A be an adapted, right-continuous, increasing, positive process, defined on the stochastic interval $[0,\tau[$ and such that $\sup_{t<\tau} A_t \leqslant \ell \in \mathbb{R}^+$. Let Φ be a real-increasing, adapted process such that, for every stopping time $\sigma \leqslant \tau$,*

$$E(\Phi_{\sigma^-}) \leqslant K + \varrho E\{\int_{[0,\sigma[} \Phi_{s^-} ds_s\}\,,$$

where K and ϱ are two constants.

Then

$$E(\Phi_{\tau^-}) \leqslant 2K \sum_{j=0}^{[2\varrho\ell]} (2\varrho\ell)^j\,,$$

where $[x]$ denotes the integer part of $x \geqslant 0$.

Proof. Let us define recursively the following increasing sequence of stopping times.

$$\sigma_0 := 0 \ldots \sigma_{k+1} := \inf\{t\colon A_t - A_{\sigma_k} \geqslant \frac{1}{2\varrho};\ t > \sigma_k\} \wedge \tau\,.$$

If we set $x_k := E(\Phi_{\sigma_{\bar{k}}})$ and remember that $A_{\tau_-} \leqslant \ell$, we may write

$$\begin{aligned} x_{k+1} &\leqslant K + \varrho E\{\int_{]0,\sigma_k]} \Phi_{s^-} dA_s\} + \varrho E\{\int_{]\sigma_k,\sigma_{k+1}[} \Phi_{s^-} dA_s\} \\ &\leqslant K + \varrho\ell x_k + \tfrac{1}{2} E(\Phi_{\sigma_{\overline{k+1}}}) \\ &\leqslant K + \varrho\ell x_k + \tfrac{1}{2} x_{k+1}\,. \end{aligned}$$

This gives

$$x_{k+1} \leqslant 2K + 2\varrho\ell x_k$$

and therefore the formula of the lemma. □

29.2 Theorem (The exponential of a semimartingale). *Let Z be a real semimartingale defined on a stochastic basis $(\Omega, \mathfrak{A}, (\mathscr{F}), P)$.*

(1°) *For every $\mathscr{F}_0$-measurable random variable ξ_0 there exists a unique (up to P-equality) real R.R.C. process such that*

(29.2.1) $$X_t = \xi_0 + \int_0 X_{s-} dZ_s$$

(2°) *When Z is continuous one has*

(29.2.2) $$X_t = \xi_0 \exp(Z_t - \tfrac{1}{2}[Z]_t)$$

(3°) *In general,*

(29.2.3) $$X_t = \xi_0 \exp(Z_t - \tfrac{1}{2}[Z]_t^c) \prod_{0<s\leqslant t} (1 + \Delta Z_s) \exp(-\Delta Z_s),$$

where the infinite product is a.s. absolutely convergent.

Proof. We organize the proof in the following way. We first prove the uniqueness of X and then show that the process defined by (29.2.3) is a solution of (29.2.1) (i.e., a process X which is R.R.C. and satisfies (29.2.1)). As to statement (2°), this is a trivial consequence of (3°).

Uniqueness of X. If X^1 and X^2 are two processes for which (29.2.1) holds, one has for the process $X = X^1 - X^2$,

(29.2.4) $$X_t = \int_{]0,t]} X_{s-} dZ_s.$$

Let A be a control process of Z. For every stopping time

$$E(\sup_{t<\sigma} |X_t|^2) \leqslant E(A_{\sigma-} \int_{[0,\sigma[} |X_s|^2 dA_s).$$

and, a fortiori,

(29.2.5) $$E(\sup_{t<\sigma} |X_t|^2) \leqslant E(A_{\sigma-} \int_{[0,\sigma[} \sup_{r<s} |X_r|^2 dA_s).$$

If we define $\tau_\ell := \inf\{t: A_t > \ell\}$ for every $\ell \in \mathbb{N}$, we obtain for every stopping time $\sigma < \tau_\ell$

(29.2.6) $$E(\sup_{t<\sigma} |X_t|^2) \leqslant \ell E\{\int_{[0,\sigma[} \sup_{r<s} |X_r|^2 dA_s\}.$$

Setting $\Phi_t := \sup_{r<t} |X_r|^2$ and applying lemma 29.1 we obtain $E(\sup_{t<\tau_\ell} |X_t|^2) = 0$ for every ℓ and therefore $X = 0$.

Proof of (3°). We consider a real R.R.C. process with finite variation A and show it can be determined in such a way that the process

(29.2.7) $$X_t := \xi_0 A_t \exp Z_t$$

is a solution of (29.2.1).

Let φ be a twice continuously differentiable function of two real variables x and y.

It is an easy consequence of the transformation formula applied to φ and the bi dimensional semimartingale (Z_t, A_t) that, if A is a process with finite variation, we have

$$\begin{aligned}\varphi(Z_t, A_t) = \varphi(Z_0, A_0) &+ \int_{]0,t]} D^1\varphi(Z_{s^-}, A_{s^-})dZ_s + \int_{]0,t]} D^2\varphi(Z_{s^-}, A_{s^-})dA_s \\ &+ \tfrac{1}{2} \int_{]0,t]} D^{11}\varphi(Z_{s^-}, A_{s^-})d[Z]_s + \sum_{s\leqslant t} [\varphi(Z_s, A_s) - \varphi(Z_{s^-}, A_{s^-}) \\ &- D^1\varphi(Z_{s^-}, A_{s^-})\Delta Z_s - D^2\varphi(Z_{s^-}, A_{s^-})\Delta A_s \\ &- \tfrac{1}{2} D^{11}\varphi(Z_{s^-}, A_{s^-})|\Delta Z_s|^2]\end{aligned}$$

Taking $\varphi(x, y) = ye^x$ we obtain

(29.2.8) $$\begin{aligned}X_t = X_0 &+ \int_{]0,t]} X_{s^-} dZ_s + \int_{]0,t]} \exp(Z_{s^-})dA_s + \tfrac{1}{2} \int_{]0,t]} X_{s^-} d[Z]_s \\ &+ \sum_{s\leqslant t} [X_{s^-}(\exp(\Delta Z_s) - 1 - \Delta Z_s - \tfrac{1}{2}|\Delta Z_s|^2) \\ &\qquad + \Delta A_s \exp(Z_{s^-})(\exp(\Delta Z_s) - 1)],\end{aligned}$$

the series on the right-hand side of this equation being a. s. summable, which follows from the transformation theorem.

As a consequence of Corollaries 25.5 and 25.6,

(29.2.9) $$\int_{]0,t]} X_{s^-} d[Z]_s - \sum_{s\leqslant t} X_{s^-}|\Delta Z_s|^2 = \int_{]0,t]} X_{s^-} d[Z]_s^c.$$

Since A has finite variation, A^c defined by $A_t^c = A_t - \sum_{s\leqslant t} \Delta A_s$ is continuous and we may write (29.2.8) as follows.

(29.2.10) $$\begin{aligned}X_t = X_0 &+ \int_{]0,t]} X_{s^-} dZ_s + \int_{]0,t]} \exp(Z_{s^-})dA_s^c + \tfrac{1}{2} \int_{]0,t]} X_s d[Z]_s^c \\ &+ \sum_{s\leqslant t} [X_{s^-}(\exp(\Delta Z_s) - 1 - \Delta Z_s) + \Delta A_s \exp(Z_{s^-})\exp(\Delta Z_s)]\end{aligned}$$

If we take A in such a way that

(29.2.11) $$\begin{cases} A_t^c = \exp -\frac{1}{2}[Z]_t^c \\ \Delta A_s = A_{s^-}(-1+(1+\Delta Z_s)\exp(-\Delta Z_s)), \end{cases}$$

we obtain

$$\int_{]0,t]} \exp(Z_{s^-})dA_s^c + \frac{1}{2}\int_{]0,t]} X_s d[Z]_s^c = 0$$

and

$$\sum_{s \leqslant t} [X_{s^-}(\exp(\Delta Z_s) - 1 - \Delta Z_s) + \Delta A_s \exp(Z_{s^-})\exp(\Delta Z_s)] = 0.$$

Relation (29.2.10) then implies that for such a process A, X, defined by (29.2.7), verifies (29.2.1).

The two relations (29.2.11) are clearly satisfied if we set

$$A_t := (\exp - \tfrac{1}{2}[Z_t^c]) \prod_{s \leqslant t} (1 + \Delta Z_s)\exp(-\Delta Z_s).$$

On the one hand, this process is well-defined because, for every $\alpha < 1$, the number of s, $s < t$, such that $|\Delta Z_s(\omega)| > \alpha$ is finite and the infinite product in this formula is a.s. convergent. On the other hand, the only discontinuities of A are at times s, where $\Delta Z_s \neq 0$ and $A_s = A_{s^-}(1 + \Delta Z_s)\exp(-\Delta Z_s)$. This proves the theorem. □

29.3 Definition *(Exponential of a semimartingale)*. The solution X of equation (29.2.1) for $\xi_0 = 1$, which is given by $X_t := \exp(Z_t - \frac{1}{2}[Z]_t^c) \prod_{0<s \leqslant t} (1+\Delta Z_s)\exp(-\Delta Z_s)$ is called the *exponential of* Z and denoted by $\mathscr{E}(Z)$.

29.4 Theorem (Exponential formula for continuous processes). *Let Z and A be two real adapted processes with continuous paths. We assume A increasing and $A_0 = Z_0 = 0$. For every $\lambda \in \mathbb{R}$ we set*

$$\Phi_t^\lambda := \exp\left(\lambda Z_t - \frac{\lambda^2}{2} A_t\right)$$

Then the following hold.

(1°) *Φ^λ is a local martingale for every $\lambda \in \mathbb{R}$ if and only if Z is a local martingale and $[Z] = A$.*

(2°) *If Z is a local martingale with $[Z] = A$ and $E(\int_{]0,t]} e^{2\lambda M_s} dA_s) < \infty$, then Φ^λ is a square integrable martingale.*

(3°) *If Φ^λ is a martingale and if there exists $\lambda_0 \in]0, \infty[$ such that for every $t \in \mathbb{R}^+$ $E[\exp(\lambda_0[Z_t])] < \infty$, then Z is a martingale and $[Z] = A$. If $E(A_t) < \infty$ for all t, Z is an L^2-martingale.*

Proof. If Z is a continuous local martingale and $A = [Z]$ we know from 29.2 that

$\Phi_t^\lambda = 1 + \int_0^t \Phi_s^\lambda dZ_s$. This implies that Φ_t^λ is a local martingale, as a stochastic integral of a locally bounded process with respect to a locally square integrable martingale.

Moreover, in this case

$$E|\Phi_t^\lambda|^2 \leqslant 2 + 2\int_0^t |\Phi_s^\lambda|^2 dA_s \leqslant 2 + 2\int_0^t \exp(2\lambda Z_s) dA_s .$$

This proves the second statement of the theorem.

We now prove the third statement. We recall the following elementary inequalities. There exists a constant K such that, for all $\lambda \in \left[-\frac{\lambda_0}{2}, \frac{\lambda_0}{2}\right]$ and $(x, y) \in \mathbb{R} \times \mathbb{R}^+$,

$$|x - \lambda y| \exp\left(\lambda x - \frac{\lambda^2}{2} y\right) \leqslant K \exp(\lambda_0 |x|) \tag{29.4.1}$$

and

$$|x - \lambda y|^2 \exp\left(\lambda x - \frac{\lambda^2}{2} y\right) \leqslant K \exp(\lambda_0 |x|) . \tag{29.4.2}$$

The martingale property of Φ^λ shows that, for every $s < t$ and $F \in \mathscr{F}_s$,

$$\int_F \exp\left(\lambda Z_s - \frac{\lambda^2}{2} A_s\right) dP = \int_F \exp\left(\lambda Z_t - \frac{\lambda^2}{2} A_t\right) dP . \tag{29.4.3}$$

On account of (29.4.1), (29.4.2) and the hypothesis $E[\exp(\lambda_0 |Z_t|)] < \infty$, we are allowed to differentiate twice under the integral in (29.4.3). This gives

$$\int_F (Z_s - \lambda A_s) \Phi_s^\lambda dP = \int_F (Z_t - \lambda A_t) \Phi_t^\lambda dP \tag{29.4.4}$$

and, assuming for a moment that $E(A_t) < \infty$ for every t,

$$\int_F [(Z_s - \lambda A_s)^2 - A_s] \Phi_s^\lambda dP = \int_F [(Z_t - \lambda A_t)^2 - A_t] \Phi_t^\lambda dP . \tag{29.4.5}$$

For $\lambda = 0$ we obtain

$$\int_F Z_s dP = \int_F Z_t dP ,$$

which expresses the martingale property for Z and, when $E(A_t) < \infty$ for all t, we obtain for $\lambda = 0$

$$\int_F (Z_s^2 - A_s) dP = \int_F (Z_t^2 - A_t) dP ,$$

which expresses the martingale property for $(Z_t^2 - A_t)_{t \in \mathbb{R}^+}$. Therefore, $A = [Z]$.

If we don't assume $E(A_t) < \infty$, we can find a localizing sequence (τ_n) such that

$E(A_{t \wedge \tau_n}) < \infty$. The above reasoning applied to the stopped processes $(A_{t \wedge \tau_n})_{t \in \mathbb{R}^+}$ shows that $(Z^2_{t \wedge \tau_n} - A_{t \wedge \tau_n})_{t \in \mathbb{R}^+}$ is a martingale and again that $A = [Z]$.

In order to complete the proof, we show now that, if Φ^λ is a local martingale for every $\lambda \in \mathbb{R}$, Z is a local martingale and $[Z] = A$. If we define the increasing sequence of stopping times

$$\tau_n := \inf\{t: |Z_t| + |A_t| > n\},$$

we clearly have $\lim_n \tau_n = \infty$ and the stopped processes $(\Phi^\lambda_{t \wedge \tau_n})_{t \in \mathbb{R}^+}$ are bounded martingales. Statement 3°, proved above, then shows that $(Z_{t \wedge \tau_n})_{t \in \mathbb{R}^+}$ is a square integrable martingale and that $[Z] = A$. □

30. Absolutely continuous changes of probability

30.1 Introduction. One frequently has to simultaneously consider different probability laws on the same probability space. It is immediately clear from the definition that, in general, a martingale does not remain a martingale under a change of probability law. Conversely, given a process which is not a martingale under a law P_1, one may ask whether there is a law P_2 for which it becomes a martingale. A nice situation is when the various laws which are to be considered are absolutely continuous with respect to a given one and when the densities can be expressed explicitly.

An example of the necessity of considering several laws is given by statistical problems dealing with the possible laws of an observed stochastic process. An interesting case is the case where the different laws have densities with respect to a fixed one. These densities are called likelihood ratios. We will give an example of this in 30.5 below.

The goal of this section is to show how the stochastic calculus turns out to be helpful in obtaining expressions for the densities.

30.2 Densities and Martingales. Let $(\Omega, (\mathscr{F}_t)_{t \in [0,T]}, \mathfrak{A}, P)$ be a stochastic basis with $T \in \overline{\mathbb{R}}^+$ and $\mathfrak{A} = \mathscr{F}_T$. (When $T = +\infty$ we recall the convention $\mathscr{F}_T = \mathscr{F}_\infty :=$ σ-algebra generated by

$$\bigcup_{t \in \mathbb{R}_+} \mathscr{F}_t).$$

If Q is a probability law on $\mathfrak{A}$, absolutely condinuous with respect to P and with density $f \in L^1(\Omega, \mathfrak{A}, P)$, one may write for every $t \in \mathbb{R}^+$ and $F \in \mathscr{F}_t$

$$(30.2.1) \qquad Q(F) = E_P(f \cdot 1_F) = E_P(E_P(f|\mathscr{F}_t) 1_F),$$

where E_P denotes the expectation with respect to P.

If one calls M the right-continuous version of the martingale $(E_P(f|\mathscr{F}_t))_{t \in [0,T]}$ the process M is uniquely defined up to P-equivalence and formula (30.2.1) expresses the density of Q restricted to $\mathscr{F}_t$ as a variable depending only on the past before t. Thus,

$$(30.2.2) \qquad \forall F \in \mathscr{F}_t, Q(F) = E_P(M_t \cdot 1_F).$$

Absolutely continuous changes of measures on $\mathscr{F}_T$ are therefore in a one-to-one correspondance with positive R.C. martingales M on $[0, T]$ such that $E_P(M_T) = 1$.

When $T = \infty$ (or, more generally, when $\mathscr{F}_T = \mathscr{F}_{T^-}$) we may also say, as a consequence of the martingale convergence theorem, that absolutely continuous changes of measures on $\mathscr{F}_T$ are in a one-to-one correspondance with positive right-continuous, uniformly integrable martingales on $[0, T[$ such that $E_P(M_0) = 1$.

30.3 Invariance of the semimartingale property under an absolutely continuous change of law. We have the following.

Theorem. *Let S be an $\mathbb{H}$-valued semimartingale on a stochastic basis $(\Omega, (\mathscr{F}_t)_{t \in \mathbb{R}^+}, \mathfrak{A}, P)$, $\mathbb{H}$ being a separable Hilbert space. If Q is a probability law on $\mathscr{A}$ which is equivalent to P, S is still a semimartingale on $(\Omega, (\mathscr{F}), \mathfrak{A}, Q)$.*

Proof. We write $S = L + V$, where L is a local martingale and V a process with finite variation.

Let M be the right-continuous P-martingale on $[0, \infty]$ generated by the density f of Q with respect to P, as described in 30.2 above. If we can show the existence of a process B with finite variation such that $(L - B)M$ is a local martingale for P, we may write for every stopping time τ_n of a localizing sequence, for every $s < t \in \mathbb{R}^+$ and every $F \in \mathscr{F}_s$

$$\begin{aligned} E_Q(1_F(L-B)_{s \wedge \tau_n}) &= E_P(1_F(L-B)_{s \wedge \tau_n}\, M_{s \wedge \tau_n}) = E_P(1_F(L-B)_{t \wedge \tau_n} M_{t \wedge \tau_n}) \\ &= E_Q(1_F(L-B)_{t \wedge \tau_n}). \end{aligned}$$

This shows that $(L - B)$ is a local martingale for Q and $S = (L - B) + V + B$ is a semimartingale on $(\Omega, (\mathscr{F}), \mathfrak{A}, Q)$.

The theorem is a straightforward consequence of the following lemma, since the equivalence of P and Q implies $P\{f > 0\} = 1$ and M can therefore be assumed strictly positive. □

Lemma ([MeP. 4]). *Let $\mathbb{H}$ be a separable Hilbert space, L a right-continuous, $\mathbb{H}$-valued local martingale, M a strictly positive uniformly integrable martingale with $E(M_0) = 1$. Then there exists an $\mathbb{H}$-valued, right-continuous process B such that $(L - B)M$ is a local martingale.*

Proof. M being strictly positive, the process $\dfrac{1}{M_{s^-}}$ is locally bounded. Let us indeed consider the increasing sequence (τ_n) of stopping times defined by $\tau_n := \inf$

$\left\{t\colon M_t < \frac{1}{n}\right\} \wedge T$. The right continuity and strict positivity of M imply $\sup_n \tau_n \geqslant T$ a.s. (since $M = 0$ on $[\sup_n \tau_n, T[$). We have, therefore, $1/M_{\tau_n}^- \leqslant n$. The Stieltjes integral $B_t := \int_{]0,t]} \frac{1}{M_s} d[\![L, M]\!]_s$ is well defined for P-almost all $\omega \in \Omega$ and the integration-by-part formula (26.9.3) gives

$$\begin{aligned} BM &= \int B_{s^-} dM_s + \int M_{s^-} dB_s + [\![M, B]\!] \\ &= \int B_{s^-} dM_s + \int M_s dB_s \qquad \text{(use 26.9.4)} \\ &= \int B_{s^-} dM_s + [\![L, M]\!], \end{aligned}$$

which implies

$$(L - B)M = \int L_{s^-} dM_s + \int M_{s^-} dL_s - \int B_{s^-} dM_s .$$

Since M and L are local martingales with respect to P, the same is true for $(L - B)M$. □

30.4 Theorem. (Girsanov). *Let W be a Brownian motion in $\mathbb{R}^d$ with variance matrix C and let Φ be an $\mathbb{R}^d$-valued bounded predictable process. We assume these processes to be defined on a stochastic basis $(\Omega, (\mathscr{F}_t)_{t \in [0,T]}, P)$ with $T < \infty$ and $E[\exp(2 \int_{]0,t]} \Phi_s \cdot dW_s)] < \infty$ for every $t \leqslant T$.*

Let Y be the process defined by

$$Y_t := \int_{]0,t]} C(\Phi_s) ds + W_t \qquad t \leqslant T$$

and

$$\xi_t := \exp\Big(-\int_{]0,t]} \Phi_s \cdot dW_s - \tfrac{1}{2} \int_{]0,t]} \Phi_s \cdot C(\Phi_s) ds\Big) \tag{30.4.1}$$

Then ξ is a positive martingale on the interval $[0, T]$ with $E(\xi_0) = 1$. For the probability $Q := \xi_T \cdot P$ Y is a Brownian process with variance matrix C on $(\Omega, (\mathscr{F}_t)_{t \in [0,T]}, Q)$.

Proof. Since $\int \Phi \cdot dW$ is a real continuous local martingale with $\langle \int \Phi \cdot dW \rangle = \int \Phi_s \cdot C\Phi_s ds \leqslant K \int \|\Phi_s\|^2 ds$ for some constant K, the fact that ξ is a martingale on $[0, T]$ (and even a square integrable martingale) follows from 29.4. 3°. Moreover, $E(\xi_0) = 1$.

For every $\Theta \in \mathbb{R}^d$, one defines

$$Z_t^\Theta := \exp\Big(-\int_{]0,t]} (\Phi + \Theta) \cdot dW - \tfrac{1}{2} \int_{]0,t]} (\Phi_s + \Theta) \cdot C(\Phi_s + \Theta) ds\Big).$$

This can be written

$$Z_t^\Theta = \xi_t \exp\Big\{-\Big(\Theta \cdot \Big(W_t + \int_{]0,t]} C(\Phi_s) ds\Big)\Big) - \tfrac{1}{2} \Theta \cdot C(\Theta) t\Big\}$$

As a consequence of Theorem 29.4 3°, Z^Θ is a martingale.

Therefore, for every $F \in \mathscr{F}_t$ and $t \leqslant T$,

$$\int_F \xi_T \exp\{-\Theta \cdot Y_T - \tfrac{1}{2}\Theta \cdot C(\Theta)T\}\, dP = \int_F \xi_t \exp\{-\Theta \cdot Y_t - \tfrac{1}{2}\Theta \cdot C(\Theta)t\}\, dP$$

or

$$\int_F \exp\{-\Theta \cdot Y_T - \tfrac{1}{2}\Theta \cdot C(\Theta)T\}\, dQ = \int_F \exp\{-\Theta \cdot Y_t - \tfrac{1}{2}\Theta \cdot C(\Theta)t\}\, dQ\,.$$

This expresses the martingale property of $(\exp\{-\Theta \cdot Y_t - \frac{1}{2}\Theta \cdot C(\Theta)t\})_{t \leqslant T}$ for the probability Q. According to Theorem 29.4, $\Theta \cdot Y$ is a Q-local martingale with $\langle \Theta \cdot Y \rangle = \Theta \cdot C(\Theta)$. For every $\Theta \in \mathbb{R}^d$ the process $\Theta \cdot Y$ is then a Brownian motion with variance $\Theta \cdot C(\Theta)$, according to Theorem 28.1, for the probability Q. This proves the theorem. $\square$

30.5 Application to signal-theory: a likelihood-ratio. We consider the following situation, Let r be a signal with the form

$$r(t) = S(t) + \text{noise}\,,$$

where $S(t)$ is the transmitted signal to be detected, while r is the received signal mixed with noise. Let us assume that the noise is a white noise and that the observation through measuring instruments is a process Y, where

$$Y(t) = \int_0^t r(s)\, ds = \int_0^t S(s)\, ds + W(t) \tag{30.5.1}$$

and W is a Brownian motion with variance σ^2. Let $\mathscr{F}^Y$ be the filtration generated by the process Y (the filtration of events observable by a measurement of Y).

The estimate of S_t minimizing the quadratic risk is

$$\hat{S}_t = E(S_t \mid \mathscr{F}_t^Y)\,.$$

We set

$$\hat{W}(t) := W(t) + \int_0^t (S(s) - \hat{S}(s))\, ds \tag{30.5.2}$$

This can be written

$$\hat{W}(t) = Y(t) - \int_0^t \hat{S}(s)\, ds \tag{30.5.3}$$

let us assume that S is bounded. For $F \in \mathscr{F}_t^Y$ we have

$$E(1_F \hat{W}(t+h)) = E(1_F W(t+h)) + \int_0^t E(1_F(S(s) - \hat{S}(s)))\, ds + \int_t^{t+h} E(1_F(S(s) - \\ - \hat{S}(s)))\, ds\,.$$

The latter integral is null from the definition of $\hat{S}$. Since $E(1_F W(t+h)) = E(1_F W(t))$, the relation just written expresses the martingale property of $\hat{W}$ with respect to the filtration $(\mathscr{F}^Y)$. (Notice that, in view of (30.5.3), $\hat{W}$ is already adapted to this filtration). Since the process $\int_0^t (S(s) - \hat{S}(s))ds$ is continuous and of finite variation, formula (30.5.2) shows that

$$[\hat{W}]_t = [W]_t = \sigma^2 t .$$

Therefore $\hat{W}$ is a Brownian motion with variance σ^2.

We now consider the *following statistical problem.* It is not known whether the observed process contains the signal S or is pure noise. In other words there are two possibilities:

(a) $$Y(t) = W(t) \qquad \text{(“Null hypothesis”)}$$

and

(b) $$Y(t) = \int_0^t S(s)ds + W(t) \qquad \text{(“Alternative”)}.$$

We call μ the law of the process Y in the space $(C[0, T], (\mathscr{C}_t)_{t \leqslant T})$ of continuous functions with the “canonical filtration” $\mathscr{C}_t$ (see chapter 1, section 2).

In case (a), μ is equal to the Wiener measure μ_0. In case (b), μ is equal to another probability μ_1.

Considering the formula (30.5.3) and applying the Girsanov theorem, we see that the probability law on $(C[0, T], (\mathscr{C}_t)_{t \leqslant T})$ admitting the density

$$\exp\{- \int_{]0,T]} \hat{S}(s)[dY(s) - \hat{S}(s)ds] - \tfrac{1}{2} \int_{]0,T]} (\hat{S}(s))^2 ds\}$$

with respect to μ_1 is precisely μ_0. We thus obtain the following expression for the “likelihood ratio”, in term of the canonical process ξ.

$$\frac{d\mu_0}{d\mu_1} = \exp(- \int_0^T \hat{S}_s d\xi_s + \tfrac{1}{2} \int_0^t \hat{S}_s^2 ds).$$

This formula leads to a Neyman-Pearson test called the estimation-detection test, because the detection test is preceeded by an estimation $\hat{S}$ of S. [Kai]

30.6 Remark on the hypothesis of Theorem 30.3. The hypotheses that Φ is bounded and $E[\exp(2 \int_{]0,t]} \Phi_s \cdot dW_s)] < \infty$, which implied the integrability of ξ_t were introduced in 30.3 for simplicity. There are much weaker conditions. For these the reader is referred to [LeM] and [MSh].

Exercises and supplements

E. 1. *An exponential inequality.* Let M be a continuous local martingale with $M_0 = 0$. Show that for every $c > 0, K > 0, t > 0$ the following inequality holds.

$$P\{\sup_{0 \leqslant s \leqslant t} |M_s| > c\} \leqslant P\{\langle M_t \rangle > K\} + 2\exp\left(\frac{-c^2}{2K}\right).$$

(*Hints*: Using a stopping procedure show that $Z^\lambda := \exp(\lambda M - \frac{\lambda^2}{2}\langle M \rangle)$ is, for every λ, a positive supermartingale with $E(Z_t^\lambda) \leqslant E(Z_0^\lambda) = 1$. Then apply the Doob inequality).

E. 2. *Majorization in L^p.* Let M be a continuous $\mathbb{H}$-valued local martingale with $M_0 = 0$. Then, for every $p \geqslant 2$, there exists a constant c_P such that, for every $t > 0$,

$$E[\sup_{s \leqslant t} ||M_s||^p] \leqslant c_P(E(\langle M_t \rangle^{p/2}).$$

Hints: First reduce the situation to the case where M is a bounded martingale. Apply the Ito formula to $||M||^p$ to get

$$E(||M_t||^p) \leqslant \frac{p(p-1)}{2} E(\int_{]0,t]} ||M_s||^{p-2} d\langle M \rangle_s)$$

$$\leqslant \frac{p(p-1)}{2} E[(\sup_{s \leqslant t} ||M_s||^{p-2})\langle M \rangle_t].$$

Apply the Hölder inequality to the second member of this inequality to replace it by $\frac{p(p-1)}{2}[E|\langle M_t \rangle|^{p/2}]^{2/p}[E(\sup_{s \leqslant t} ||M_s||^p)]^{\frac{p-2}{2}}$. Apply the Doob inequality in L^p and finally get

$$C_p = [\tfrac{1}{2} p^{(p+1)}(p-1)^{(1-p)}]^{p/2}.$$

E. 3. *Majorization in L^P (continued).* Let M be an $\mathbb{H}$-valued local martingale with $M_0 = 0$. Then for every $p \geqslant 2$ and $t \geqslant 0$ one has for some constant C_p (depending only on p, not on M)

$$E(\sup_{s \leqslant t} ||M_s||^p) \leqslant C_p(E[M_t]^{p/2}).$$

Hint: Apply the Ito formula to $||M||^p$ in the second form of 27.2 and use the fact, derived from the Taylor formula, that

$$\sum_{s \leqslant t} (||M_s||^p - ||M_{s-}||^p - p||M_{s-}||^{p-2} M_{s-} \cdot \Delta M_s) \leqslant \frac{p(p-1)}{2} \sup_{s \leqslant t} ||M_s||^{p-2} \sum_{s \leqslant t} ||\Delta M_s||^2$$

$$= \frac{p(p-1)}{2} \sup_{s \leqslant t} ||M_s||^{p-2} [M_t^d].$$

Then argue as in E.2 and use a proper stopping procedure with stopping times (τ_n) such that $\sup_{s \leqslant t} \|M_{s-}\| \leqslant n$.

Remark. For an analogous inequality with $1 \leqslant p < 2$, see for example [MeP. 1], chapter XI.

E. 4. *The space H^p.* Let H^p, $p \geqslant 2$, be the set of martingales such that $E(\sup_{s \in \mathbb{R}^+} \|M_s\|^p) < \infty$, with the topology defined by the norm $[E(\sup_s \|M_s\|^p)]^{1/p}$. Show that this norm is equivalent to $(E[M]_\infty^{p/2})^{1/p}$.
(This is a trivial consequence of E.3 and of the opposite easy inequality).
This space of martingales is called H^p.

E. 5. *Majorization in L^p (continued).* (1°) If M is a locally square integrable martingale, for every locally bounded predictable process Y and every stopping time τ

$$(\text{E.5.1}) \qquad E(\sup_{t \leqslant \tau} \| \int_{]0,t]} Y dM \|^p) \leqslant C_p \, E(\langle M \rangle_\tau^{\frac{p-2}{2}} \int_{]0,\tau]} \|Y\|^p d\langle M \rangle)$$

with $C_p := [\frac{1}{2} P^{(p+1)}(p-1)^{(1-p)}]^{p/2}$.

(*Hints*: Apply E.2 and transform $\int_{]0,\tau]} \|Y\|^2 d\langle M \rangle$ by a proper Hölder inequality).

(2°) If M is a locally square integrable martingale and if we set $A := 4(\langle M \rangle + [\check{M}])$ (see section 19.3), then for every locally bounded predictable Y and every stopping time τ,

$$(\text{E.5.2}) \qquad E(\sup_{t < \tau} \| \int_{]0,t]} Y dM \|^p) \leqslant C_p \, E(A_\tau^{\frac{p-2}{2}} \int_{]0,\tau[} \|Y\|^p dA).$$

Hints: Use (E.5.1) and the existence of a locally square integrable martingale satisfying properties (i) to (iii) of section 19.4.

(3°) If Z is a semimartingale, there exist two R.R.C. positive increasing processes A and $\tilde{A}$ such that, for every locally bounded predictable Y and every stopping time τ,

$$(\text{E.5.3}) \qquad E(\sup_{t < \tau} \| \int_{]0,t]} Y dZ \|^p) \leqslant E(\tilde{A}_{\tau-} \int_{[0,\tau[} \|Y_s\|^p dA_s).$$

Hints: Use a decomposition $Z = M + V$, where M is a locally square integrable martingale and V a process with finite variation.
The couple $(A, \tilde{A})$ is a "*p-control couple*" for Z.

E. 6. *Solutions of a Linear stochastic differential equation.* (1°) Let M be a locally square integrable $\mathbb{R}^d$-valued martingale and A a $d \times d$ matrix. Show that an R.R.C., $\mathbb{R}^d$-valued process X satisfies

$$X_t = X_0 + \int_0^t AX_s ds + M_t$$

if and only if

$$X_t = e^{tA} X_0 + \int_0^t e^{(t-s)A} dM_s .$$

Hints: Use the integration-by-part formula as in the solution of the Langevin equation (see exercise E.1, chapter 5).

(2°) Conclude that $e^{-tA} X_t$ is a local martingale. If M is a martingale

$$E(X_t) = e^{tA} E(X_0).$$

E. 7. *Cameron-Martin Theorem.* Let ξ be the canonical process on $(C^d; (\mathscr{C}^d)_{t \geq 0})$, where C^d is the space of continuous mappings from $\mathbb{R}^+$ into $\mathbb{R}^d$.

Let $\sigma(x)$ be, for every $x \in \mathbb{R}^d$, an invertible $d \times d$ matrix and $b(x)$ be a bounded vector in $\mathbb{R}^d$. We assume $x \curvearrowright \sigma(x)$ and $x \curvearrowright b(x)$ measurable. Let x_0 be fixed in $\mathbb{R}^d$.

(1°) There exists a law P and a Brownian motion w^1 on $(C^d, (\mathscr{C}^d))$ such that

$$\xi_t = x + \int_0^t \sigma(\xi_s) dw_s^1 \quad \text{and} \quad \langle\!\langle w^1 \rangle\!\rangle_t = t I_d$$

iff there exists P such that ξ_t is a local martingale for P with $\xi_0 = x_0$ P.a.s. and $\langle\!\langle \xi \rangle\!\rangle_t = \int_0^t a(\xi_s) ds$, where $a = \sigma \circ \sigma^*$ (σ^* = transpose of σ).

Hint: Consider

$$\int_0^t \sigma^{-1}(\xi_s) d\xi_s \qquad \text{and use 28.1.}$$

(2°) Let (P, w^1) as in (1°). Consider the measure Q defined on $(C^d, (\mathscr{C}_t^d))$ as follows. For every $T > 0$, the restriction of Q to $\mathscr{C}_T^d$ has the density

$$Z_T := \exp\left(\int_0^T a^{-1} \circ b(\xi_s) d\xi_s - \tfrac{1}{2} \int_0^t b(\xi_s) \cdot a^{-1} \circ b(\xi_s) ds\right)$$

with respect to P

(a) Show that there exists a Brownian motion w^2 for Q such that

$$\xi_t = x_0 + \int_0^t \sigma(\xi_s) dw_s^2 + \int_0^t b(\xi_s) ds \quad \textit{and} \quad \langle\!\langle w_t^2 \rangle\!\rangle = t I_d .$$

(b) If, conversely, Q and w^2 satisfy (a), there exist P and w^1 as in (1°), P being unique.

Hint: Apply the Girsanov formula.

E. 8. *Skorokhod Representation Theorem.* Let W be a $\mathbb{R}^d$-dimensional Brownian movement defined on a stochastic basis $(\Omega, \mathfrak{A}, (F_t)_{t\in[0,T]}, P)$, where $(\mathscr{F}_t)_{t\in[0,T]}$ is the smallest filtration satisfying the "usual hypotheses" and for which $(W_t)_{t\in[0,T]}$ is adapted. Every $\mathbb{R}^q$-valued martingale L on the same stochastic basis has a unique representation as $L_t = L_0 + \int_{[0,t]} X\,dW$.

(Remark we don't assume L to be continuous. This is here implied by the nature of the stochastic basis).

Hints. Use proposition 18.15 or exercise E. 8 chap. 4 to write $L = L_0 + \int X\,dW + N$ and prove that $N = 0$, using the following argument due to Kunita-Watanabe. Let N be a martingale such that $N = 0$ and $\langle N^k, (\int X\,dW)\rangle = 0$ for all k, ℓ, X. Setting $\Phi(r) = E(1_F N_r^k e^{iu\, B_r})$ for $F \in \mathscr{F}_s, s < r$, argue as in the proof of 28.1 to show for every $s < t$

$$E(N_t^k e^{iu\cdot(B_t - B_s)} \mid \mathscr{F}_s) = N_s^k e^{-(t-s)\|u\|^2\backslash 2} \quad .$$

Deduce from this that for every $0 \leqslant t_1 < \ldots < t_n \leqslant T$, $u_1, \ldots u_n \in \mathbb{R}^d$

$$E(N_T^k e^{iu_1\cdot B_{t_1}} e^{iu_2\cdot(B_{t_2}-B_{t_1})} \ldots e^{iu_n\cdot(B_{t_n}-B_{t_{n-1}})}) = E(N_0 e^{-t_1\|u_1\|^2\backslash 2 - (t_n - t_{n-1})\|u_n\|^2\backslash 2}) = 0$$

and therefore $N_T^k = 0$ a. s.

Extend this result to infinite dimensional Brownian movement and martingales (see [Met. 10]).

Historical and bibliographical notes

The classical characterization of Brownian motion given in section 28 is due to P. Levy and the proof follows Kunita-Watanabe.

The exponential formula for Brownian motion is due to K. Ito. It was introduced by Stroock and Varadhan [StV. 1] and Maisonneuve [Mai] in the continuous case and established by C. Doleans-Dade for a general semimartingale. The Girsanov formula was first proved in [Gir. 1] then extended by Van Shuppen and Wong [VsW]. The general result, due to J. Jacod and J. Memin [JaM. 2], can be found in the general account given by J. Jacod in his book [Jac. 2], chap. 7. The proof given here of the invariance of the semimartingale property under an absolutely continuous change of variable is essentially theirs.

Absolute continuity properties based on a Girsanov-type formula for continuous processes have been studied by S. Orey [Ore. 3], Lipcer and Shyryaev [LiS], Ershov [Ers], Strook and Varadhan [StV. 1]. Many authors have contributed to the study of the problem for the laws of discontinuous processes, e.g., Grigelionis [Grig. 4], Kabanov, Lipcer and Shyryaev [KLS], Jacod and Memin [JaM. 2] and many others. The reader interested in details and the most general results will find them with an extensive bibliography in [Jac. 1], chap. 8 and 9. We have already mentioned the practical importance of [LeM] and [MSh] (Remark 30.6).

We should mention the problem of multiplicative decomposition of a semimartingale. If $Z = M + A$ is a continuous semimartingale with martingale part M, the exponential $\mathscr{E}(Z) = e^A e^{M - \frac{1}{2}[M]}$ is written as the product of a process of finite variation and a local martingale. Such a decomposition is called a multiplicative decomposition. For information on this problem strongly related to the previous problems the reader is referred to the paper of J. Memin [Mem] and to [Jac. 1] for a more systematic account (chap. 6).

Chapter 7

Random measures and local characteristics of a semimartingale

At the end of chapter 3 we introduced the notion of point processes. These processes play an important role in the description of jumps of discontinuous processes. For example, let X be an $\mathbb{R}^d$-valued R.R.C. process. For each path ω the set $\{(t, \Delta X_t(\omega)) : \Delta X_t(\omega) \neq 0\}$ defines a discrete distribution of mass $\mu(\omega, ds, dx)$ in the space $\mathbb{R}^+ \times (\mathbb{R}^d - \{0\})$. (See formula (15.12.1)). Many functionals of the process X can be formulated in terms of integrals with respect to the measure $\mu(\omega, ., .)$. If we substract from μ its "dual predictable projection" ν, as defined in section 15, integrals with respect to μ can be decomposed into an integral with respect to ν (which may be non random, as in the case of Poisson point processes) and an integral with respect to $\mu - \nu$, which is usually a martingale or a local martingale. Once more, we have a method of decomposing functionals of processes into a martingale part (therefore with mean zero) and a part of finite variation, which is more or less "explicit".

This chapter is devoted to the study of stochastic integrals with respect to $\mu - \nu$ and to applications giving a new expression of the Ito formula and a characterisation of the laws of some semimartingales. The latter problem leads to the notion of "local characteristics of a semimartingale", which is introduced in the last section of the chapter.

This chapter can be considered as an introduction to Jacod's lecture notes [Jac. 2], which deal with these problems more completely and in more generality but also with many more technicalities.

As always the "usual hypotheses are assumed for the contidered stochastic basis".

31. Stochastic integral with respect to a white random measures

31.1 Random measures – Definition. (a) In section 15.1 we defined $\mathbb{B}'$-valued random measures on $(I, \mathscr{B}_I)$, where I is an interval in $\mathbb{R}^+$. In the present section we consider a particular case, where $\mathbb{B}'$ is a Banach space of measures on an open subset $\mathbb{E}$ of $\mathbb{R}^d$. Such a situation was considered in section 15.11 when we introduced point processes.

The particular $\mathbb{B}'$-valued random measures considered here will briefly be called *random measures.* This terminology agrees with the usual one (see [Jac. 2]). Point processes and associated white random measures (see below) are, for us, the most important examples of random measures.

Let $\mathbb{E}$ be an open subset of $\mathbb{R}^d$ and let us first recall the definitions of the Banach spaces $\mathscr{C}^p(\mathbb{E})$ and its dual $\mathfrak{M}^p(\mathbb{E})$ as introduced in section 15.11. We consider a strictly positive, continuous bounded function p on $\mathbb{E}$. We denote by $\mathfrak{M}^p$ the space of real measures m on $(\mathbb{E}; \mathscr{B}(\mathbb{E}))$ such that $\int p(x)|m|(dx) < \infty$ [1]) with the norm $\|m\|_p := \int p(x)|m|(dx)$. This Banach space is the dual of the space $\mathscr{C}^p$ of continuous functions ϕ on $\mathbb{E}$ such that $\sup_{x \in \mathbb{E}} (|\Phi(x)|/_{p(x)} \in \mathscr{C}_0(E)$ with the norm $\|\Phi\|_p := \sup_{x \in \mathbb{E}} (|\Phi(x)|/_{p(x)})$.
A typical example of p when $\mathbb{E} = \mathbb{R}^d - \{0\}$ is

$$(31.1.1) \qquad p(x) = \frac{\|x\|^\alpha}{1 + \|x\|^\alpha}, \quad \alpha > 0 .$$

(b) A real random measure μ on $\mathbb{R}^+ \times \mathbb{E}$ is a family $\{\mu(\omega; ds, dx) : \omega \in \Omega\}$ of Borel measures on the measure-space $(\mathbb{R}^+ \times \mathbb{E}, \mathscr{B}(\mathbb{R}^+) \otimes \mathscr{B}(\mathbb{E}))$ such that, for every relatively compact set $I \times B \in \mathscr{B}(\mathbb{R}^+) \otimes \mathscr{B}(\mathbb{E})$, the function $\omega \curvearrowright \mu(\omega; I \times B)$ is a random variable. It is said to be positive if $\mu(\omega; ds, dx)$ is a positive Borel measure for every $\omega \in \Omega$.

If Y is a real function on $(\mathbb{R}^+ \times \Omega \times \mathbb{E})$ such that for every $(\omega, t) \in \Omega \times \mathbb{R}^+$ the integral $\iint_{]0,t] \times E} Y(s, \omega, x)\mu(\omega; ds, dx)$ is defined (this means $Y(., \omega, .)$ is $\mathscr{B}(\mathbb{R}^+) \otimes \mathscr{B}(\mathbb{E})$-measurable and $\iint_{]0,t] \times E} |Y(\omega, s, x)| \, |\mu|(\omega; ds, dx) < \infty$), we *denote by* $\int Y d\mu$ the process obtained by setting

$$(31.1.2) \qquad \left(\int Y d\mu\right)(t, \omega) = \iint_{]0,t] \times \mathbb{E}} Y(s, \omega, x)\mu(\omega; ds, dx) .$$

If Y takes its values in a separable Hilbert space, the same definition can be given for every Y such that

$$\iint_{]0,t] \times \mathbb{E}} \|Y(\omega, s, x)\| \, |\mu|(\omega; ds, dx) < \infty .$$

(c) We call *random measure of order p* on $\mathbb{R}^+ \times \mathbb{E}$ a random measure $\mu(\omega; ds, dx)$, which can be expressed as the difference of two positive-valued random measures and such that the process $\left(\iint_{]0,t] \times \mathbb{E}} p(x)|\mu|(\omega, ds, dx)\right)_{t \in \mathbb{R}^+}$ is locally integrable.

From this definition it follows that the mapping $I \curvearrowright \mu(\omega, I, .)$ from $\mathscr{B}_{\mathbb{R}^+}$ into the set of measures on $\mathbb{E}$ takes its values in $\mathfrak{M}^p(\mathbb{E})$ and has locally integrable variation (the variation being taken for the norm in $\mathfrak{M}^p(\mathbb{E})$). It is therefore a weak $\mathfrak{M}^p(\mathbb{E})$-valued random measure on $(\mathbb{R}^+, \mathscr{B}_{\mathbb{R}^+})$ in the sense of (15.1).

---

[1]) $|m|$ denoting as usual the variation measure $m^+ + m^-$ of m.

Conversely, if a weak $\mathfrak{M}^p(\mathbb{E})$-valued random measure is the difference of two positive weak $\mathfrak{M}^p(\mathbb{E})$-valued random measures it is a random measure of order p.

When p has the particular form (31.1.1) we speak of a *random measure of order* α.

We recall that the primitive process F_μ of μ (see 15.1) is the $\mathfrak{M}^p$-valued process defined by $F_\mu(t, \omega) := \mu(\omega;]0, t]), .)$.

The random measure μ will be called *adapted* (resp. optional, resp. predictable) if the process F_μ is the difference of two $\mathfrak{M}^p_+$-valued processes F_μ and F_μ^- such that for any $\Phi \in \mathscr{C}^p$ the real-valued processes $\int_{\mathbb{E}} \Phi(x) F_\mu^+(., t, dx)$ and $\int_{\mathbb{E}} \Phi(x) F_\mu^-(., t, dx)$ are adapted (resp. optional, resp. predictable).

An optional random measure μ will be called *white* (resp. locally white) if F_μ is a weak martingale; e.g. for every $\Phi \in \mathscr{C}^p$ $(\int_{\mathbb{E}} \Phi(x) F(., t, dx))_{t \in \mathbb{R}^+}$ is a real martingale (resp. a local martingale).

Since F_μ has, by definition, a locally integrable variation as a process with values in $\mathfrak{M}^p(\mathbb{E})$, F_μ is a weak quasimartingale whenever μ is adapted. We also know that the *dual predictable projection* of F_μ is defined as a consequence of Theorem 15.8. It is clearly the difference of the dual predictable projections of F_μ^+ and F_μ^- and is therefore the primitive process of a unique random measure ν, which is called the *dual predictable projection of* μ.

The *dual predictable projection of the random measure* μ *is therefore the unique predictable random measure such that* $\mu - \nu$ *is locally white.*

31.2 Example. *Locally white random measure of a semimartingale.* We introduced in 15.12 the point process of jumps of an R.R.C. process X by setting

$$\mu_X(\omega, ds, dx) = \sum_t 1_{\{\Delta X_t \neq 0\}}(t) \varepsilon_{(t, \Delta X_t(\omega))}(ds, dx)$$

If X is an R.R.C. $\mathbb{R}^d$-valued semimartingale, Corollary 25.5 in the real case and 26.12 in the vector case, state that for every $t \geqslant 0$

$$\iint_{]0,t] \times \mathbb{E}} \|x\|^2 \mu_X(\omega, ds, dx) := \sum_{s \leqslant t} \|\Delta X_s\|^2 < \infty \qquad \text{a.s.}$$

(Recall that $\mathbb{E} := \mathbb{R}^d - \{0\}$).

We see that μ *is a point process of order* p if we set $p(x) = \dfrac{\|x\|^2}{1 + \|x\|^2}$ (i.e. μ_X is a *point process of order* 2).

Since the jumps of the right-continuous increasing process $\iint_{]0,t]} p(x) \mu_X(., ds, dx)$ are $\leqslant 1$, this process is locally integrable and one may state that *the point process of jumps of an R.R.C. semimartingale is a random measure of order* p *with* $p(x) = \dfrac{\|x\|^2}{1 + \|x\|^2}$.

If ν_X is the dual predictable projection of μ_X, $\mu_X - \nu_X$ is a uniquely defined locally white random measure associated with X.

We may also remark that μ_X may be used to express the following transformation formula. If φ is a twice continuously differentiable function, the function

$$x \curvearrowright \varphi(X_{s^-}(\omega)+x)-\varphi(X_{s^-}(\omega))-\sum_{k=1}^{d} D^k\varphi\,(X_{s^-}(\omega))x^k$$

is integrable with respect to $\mu(\omega, ds, dx)$ for every ω and, by definition,

$$\sum_{s\leqslant t}\varphi(X_s(\omega))-\varphi(X_{s^-}(\omega))-\varphi'(X_{s^-}(\omega))\cdot\Delta X_s =$$

$$\iint_{]0,t]\times\mathbb{E}}[\varphi(X_{s^-}(\omega)\cdot x)-\varphi(X_{s^-}(\omega))-\varphi'(X_{s^-}(\omega))\cdot x]\,\mu_X(\omega, ds, dx).$$

The transformation formula of section 27.1 in its second form can therefore be written

(31.2.1) $$\varphi(X_t)=\varphi(X_0)+\sum_{k=1}^{d}\int_{]0,t]} D^k\varphi(X_{s^-})dX_s^k$$

$$+\tfrac{1}{2}\sum_{k,\ell=1}^{d}\int_{]0,t]} D^{k,\ell}\varphi(X_{s^-})d[X^k, X^\ell]_s^c$$

$$+\iint_{]0,t]\times\mathbb{E}}[\varphi(X_{s^-}+x)-\varphi(X_{s^-})-\sum_{k=1}^{d} D^k\varphi(X_{s^-})x^k]\,\mu_X(.,ds,dx).$$

31.3 Proposition. *Let μ be a random measure of order p and ν its dual predictable projection.*

(1°) *If μ is optional (resp. predictable), positive, for every $\mathscr{G}\otimes\mathscr{B}_{\mathbb{E}}$-measurable (resp. $\mathscr{P}\otimes\mathscr{B}_{\mathbb{E}}$-measurable) bounded function Y on $\mathbb{R}^+\times\Omega\times\mathbb{E}$, the process $\int Y d\mu$ defined by $(\int Y d\mu)(t,\omega):=\iint_{]0,t]\times\mathbb{E}} Y(s,\omega,x)\mu(\omega,ds,dx)$ is optional (resp. predictable).*

If μ is the difference of two optional (resp. predictable) positive random measures and Y is $\mathscr{G}\otimes\mathscr{B}_{\mathbb{E}}$-measurable (resp. $\mathscr{P}\otimes\mathscr{B}_{\mathbb{E}}$-measurable) such that the integrals $\iint_{]0,t]\times\mathbb{E}} Y(s,\omega,x)\mu(\omega,ds,dx)$ are defined for P-almost all ω and all t, the process $\int Y d\mu$ defined as above is optional (resp. predictable).

(2°) *If Y is a $\mathscr{P}\otimes\mathscr{B}_{\mathbb{E}}$-measurable function on $\mathbb{R}^+\times\Omega\times\mathbb{E}$ such that $\int Y d\mu$ defined as above exists and is integrable, then $E[(\int Y d\mu)(t,.)]=E[(\int Y d\nu)(t,.)]$ for all t, and the process $\int Y d(\mu-\nu)$ is a local martingale.*

Proof. (1°) Consider the case $Y(t,\omega,x)=\varphi(x)1_{[0,T[}(t,\omega)$, where $\varphi\in\mathscr{C}^p(\mathbb{E})$ and T is a stopping time. By definition, $(\int Y d\mu)(t,\omega)=\langle\varphi, F_\mu(t)\rangle 1_{[0,T[}+\langle\varphi, F_\mu(T^-)\rangle 1_{[T,\infty[}$, where the bracket $\langle .,.\rangle$ expresses the duality between $\mathscr{C}^p(\mathbb{E})$ and $\mathfrak{M}^p(\mathbb{E})$. If F_μ is optional, the process $\int Y d\mu$ is clearly optional. Since the σ-algebra $\mathscr{G}$ is generated by the stochastic intervals $[0,T[$ the classical extension principle in measure theory gives the optionality of $\int Y d\mu$ for every $\mathscr{G}\otimes\mathscr{B}_{\mathbb{E}}$-measurable positive Y. The case of a non-positive μ and a non-positive Y is treated by decomposing μ and Y into their positive and negative parts.

The case where μ is predictable and Y is $\mathscr{P} \otimes \mathscr{B}_{\mathbb{E}}$-measurable is handled analogously by considering stochastic intervals $[0, T]$ instead of $[0, T[$.

(2°) If Y is of the form $Y(t, \omega, x) = \varphi(x) 1_{]u, v] \times F}(t, \omega)$, where $\varphi \in \mathscr{C}^p(\mathbb{E})$ and $]u, v] \times F$ is a predictable set, the martingale property of the process $\langle \varphi, F_\mu - F_\nu \rangle$ implies

$$E(1_F \langle \varphi, F_\mu(v \wedge t) - F_\mu(u \wedge t) \rangle) = E(1_F \langle \varphi, F_\nu(v \wedge t) - F_\nu(u \wedge t) \rangle),$$

which can be written

$$E[(\textstyle\int Y d\mu)(t, .)] = E[(\textstyle\int Y d\nu)(t, .)].$$

The extension by measurability principle shows that the equality holds for every function Y which is $\mathscr{P} \otimes \mathscr{B}_{\mathbb{E}}$-measurable and positive and also for every Y for which the integral process exists and is integrable. □

31.4 Further examples. (1°) Let T be a predictable stopping time and h an $\mathbb{E}$-valued $\mathscr{F}_T$-measurable random variable. We consider

$$\mu(\omega, ds, dx) = \varepsilon_{(T(\omega), h(\omega))}(ds, dx).$$

$\forall \varphi \in \mathscr{C}^p(\mathbb{E})$ $\langle \varphi, F_\mu(t) \rangle = \varphi(h) 1_{[T, \infty[}(t)$ and μ is clearly an optional random measure.

Since the dual predictable projection of the process $\varphi(h) 1_{[T, \infty[}$ is $E(\varphi(h) | \mathscr{F}_{T-}) 1_{[T, \infty[}$ (see 15.7), the measure $\nu(\omega,]0, t], dx)$ has the form $1_{[T, \infty[}(t) G_T(\omega, dx)$ with

$$\int \varphi(x) G_T(\omega, dx) = E(\varphi(h) | \mathscr{F}_{T-}) \qquad \text{a.s.}$$

This shows the existence of a conditional law of h with respect to $\mathscr{F}_{T-}$ given by $G_T(\omega, dx)$ and

(31.4.1) $$\nu(\omega, ds, dx) = \varepsilon_{T(\omega)}(ds) G_T(\omega, dx)$$

Assume, in particular, that h is integrable and $E(h | \mathscr{F}_{T-}) = 0$. (This implies that the process $X = h 1_{[T, \infty[}$ is a martingale: see 13.8). Then $X = \int x d\mu$ and $\int x d\nu = 0$.

(2°) Let T be a totally inaccessible stopping time and h an $\mathscr{F}_T$-measurable $\mathbb{E}$-valued random variable. The random measure $\mu(\omega, ds, dx) = \varepsilon_{(T(\omega), h(\omega))}(ds, dx)$ is optional and for every $\varphi \in \mathscr{C}(\mathbb{E})$ we know that the dual predictable projection of $\varphi(h) 1_{[T, \infty[}$ is a continuous process. The process $\int_{\mathbb{E}} \varphi(x) \nu(.,]0, t], dx)$ is continuous in t. This implies that

(31.4.2) $$\nu(., \{s\}, dx) = 0 \qquad \text{P.a.s. for all } s \in \mathbb{R}^+.$$

31.5 Quadratic variation and Meyer process of a purely discontinuous locally square integrable martingale. Let M be a purely discontinuous square integrable martingale which can be written (see Theorems 7.6 and 17.7) as a sum in $\mathscr{M}^2_\infty$:

(31.5.1) $$M = \sum_n \Delta M_{\tau_n} 1_{[\tau_n, \infty[} + \sum_n (\Delta M_{\sigma_n} 1_{[\sigma_n, \infty]} - \tilde{A}_n),$$

where the τ_n are predictable stopping times, the σ_n are totally inaccessible stopping times and $\tilde{A}_n$ is the (continuous) dual predictable projection of $\Delta M_{\sigma_n} 1_{[\sigma_n, \infty[}$.

Let us call μ the point process of jumps of M. Since $\sum_n \|\Delta M_{\tau_n}\|^2 + \sum_n \|\Delta M_{\sigma_n}\|^2 < \infty$ a.s. one may write

$$\sum_{s \leqslant t} \|\Delta M_s\|^2 = \iint_{]0,t] \times \mathbb{E}} \|x\|^2 \mu(\omega, ds, dx).$$

From Theorem 18.6 we obtain the formula

$$[M]_t = \iint_{]0,t] \times \mathbb{E}} \|x\|^2 \mu(\omega, ds, dx). \tag{31.5.2}$$

Since $\langle M \rangle$ is the dual predictable projection of $[M]$ we have, if ν is the dual predictable projection of μ,

$$\langle M \rangle_t = \iint_{]0,t]} \|x\|^2 \nu(\omega, ds, dx). \tag{31.5.3}$$

The decomposition (31.5.1) and the examples in 31.4 immediately yield the following formula for the random measure $q = \mu - \nu$.

$$\sum_{s \leqslant t} q(\omega, \{s\}, dx = \sum_{\tau_n \leqslant t} \varepsilon_{\Delta M_{\tau_n}}(dx) - G_{\tau_n}(\omega, dx) + \sum_{\sigma_n \leqslant t} \varepsilon_{\Delta M_{\sigma_n}}(dx), \tag{31.5.4}$$

where $G_{\tau_n}(\omega, dx)$ is the conditional law of ΔM_{τ_n} given $\mathscr{F}_{\tau_n^-}$.

Using a localizing sequence of stopping times shows that decomposition (31.5.1) and formulas (31.5.2) to (31.5.4) also hold for a locally square integrable martingale.

31.6 Theorem. (Structure of optional white random measures). *Let q be a white random measure of order p. It can be written as*

$$1_{\{s \leqslant t\}} q(\omega, ds, dx) = \sum_n 1_{\{T_n \leqslant t\}} (\mu^{T_n} - \nu^{T_n})(\omega, ds, dx), \tag{31.6.1}$$

where the T_n are either predictable or totally inaccessible with

$$\mu^{T_n}(\omega, ds, dx) := 1_{\{T_n = s\}} q(\omega, ds, dx)$$

$$\nu^{T_n} := \textit{dual predictable projection of } \mu^{T_n}$$

$$(\nu^{T_n} = 0 \quad \textit{if } T_n \textit{ is predictable}).$$

The sum on the right-hand side of (31.6.1) *converges a.s. for the norm*

$$\|q(\omega, ds, dx\|_p = \iint_{]0,t] \times \mathbb{E}} p(x) |q|(\omega, ds, dx).$$

And the sum

$$\sum_n (\mu^{T_n} - \nu^{T_n})(.,]0, t], dx)$$

converges in $L^1(\mathfrak{M}^p)$.

Proof. The $\mathfrak{M}^p$-valued process F_q, being right-continuous with left limits, has discontinuities concentrated on a denumerable family $\{T_n\}$ of stopping times with disjoint graphs. It is possible, as was done in chapter *I* for an R.R.C. process, to introduce a family (τ_n) of predictable stopping times and $\{\sigma_n\}$ of totally inaccessible stopping times, all of them with mutually disjoint graphs, such that

$$\Delta F_q \quad \text{is null outside} \quad \bigcup_n [\tau_n] \cup \bigcup_n [\sigma_n].$$

Let $\varphi \in \mathscr{C}^p(\mathbb{E})$. The process

$$M_t^\varphi := \langle \varphi, F_q(t) \rangle := \iint_{]0,t] \times \mathbb{E}} \varphi(x) q(., ds, dx)$$

is, by definition, a real martingale with finite variation and jumps only at times T_n, with $M_0^\varphi = 0$ and

(31.6.2) $$\Delta M_{T_n}^\varphi = \int_{\mathbb{E}} \varphi(x) q(., \{T_n\}, dx)$$

or

(31.6.3) $$\Delta M_{T_n}^\varphi 1_{\{T_n \leqslant t\}} = \int_{\mathbb{E}} \varphi(x) 1_{\{T_n \leqslant t\}} q(., \{T_n\}, dx) = \iint_{]0,t] \times \mathbb{E}} \varphi(x) 1_{\{T_n = s\}} q(., ds, dx).$$

Since $|\varphi|^2 \in \mathscr{C}^p$ (clearly $\sup_{x \in K} |\varphi(x)|^2/_{p(x)} \leqslant \sup_{x \in K} (|\varphi(x)|^2/_{(p(x))^2}) \sup_{x \in \mathbb{E}} p(x)$, for every compact $K \subset \mathbb{E}$)

$$E|\langle \varphi, F_q(t) \rangle|^2 = E|\int_{\mathbb{E}} \varphi(x) q(.,]0,t], dx)|^2 \leqslant$$
$$\leqslant \{ \iint_{]0,t] \times \mathbb{E}} |\varphi(x)|^2 |q|(., ds, dx)\} < \infty.$$

The real martingale M^φ is therefore an L^2-martingale and, since it has finite variation, $M^\varphi \in \mathscr{M}^{2,d}$. Therefore, in $\mathscr{M}^{2,d}$ (see chapter IV, Th. 17.7)

(31.6.4) $$M^\varphi := \sum_n \Delta M_{\tau_n}^\varphi 1_{[\tau_n, \infty[} + \sum_n (\Delta M_{\sigma_n}^\varphi 1_{[\sigma_n, \infty[} - \tilde{A}_n),$$

where $\tilde{A}_n$ is the (continuous) dual predictable projection of $\Delta M_{\sigma_n} 1_{[\sigma_n, \infty[}$.

Set

(31.6.5) $$\mu^\tau(\omega, ds, dx) := 1_{\{\tau = s\}} q(\omega, ds, dx)$$

(31.6.6) $$\nu^\tau(\omega, ds, dx) := \text{dual predictable projection of } \mu^\tau.$$

One should remember that ν^τ is null if τ is predictable. In this case, for any $\mathscr{P} \otimes \mathscr{B}_{\mathbb{E}}$-measurable Y, the function $1_{\{\tau = s\}} Y$ is predictable and $(\iint_{]0,t] \times \mathbb{E}} Y \mu^\tau(\omega, ds, dx))_{t \geqslant 0}$ is a martingale, which implies $\nu^\tau = 0$.

With these definitions the formula (31.6.4) can be written

(31.6.7) $$M_t^\varphi = \sum_n \iint_{]0,t]\times\mathbb{E}} \varphi(x)(\mu^{T_n} - \nu^{T_n})(\omega, ds, dx).$$

If we notice that

$$|\mu^{T_n}| = |(\Delta F_q)_{T_n}| = |\mu|(\omega, \{T_n\}, \mathbb{E})$$

and

$$\iint_{]0,t]\times\mathbb{E}} p(x)|\nu^{T_n}|(\omega, ds, dx) = \iint_{]0,t]} p(x)|\mu^{T_n}|(\omega, ds, dx).$$

the a.s. summability in L^1 of $\sum \|(\Delta F_q)_{T_n}\|_p$ implies the convergences stated in the theorem. □

31.7 Proposition. *Let q be a white random measure of order p, with jumps at times T_n as in Theorem 31.6, and $\mathbb{H}$ a separable Hilbert space.*

Let Y be a $\mathscr{P}\otimes\mathscr{B}_{\mathbb{E}}$-measurable $\mathbb{H}$-valued function on $\mathbb{R}^+\times\Omega\times\mathbb{E}$ such that

(a) $$E\{\iint_{]0,t]\times\mathbb{E}} \|Y(s,.,x)\|\,|q|(ds,dx)\} < \infty$$

and

(b) $$E\{\iint_{]0,t]\times\mathbb{E}} \|Y(s,.,x)\|^2\,|q|(ds,dx)\} < \infty.$$ Then

(1°) $\int Ydq$ *(as defined in* 31.1*) is an L^2-martingale.*

(2°) *This L^2-martingale is the only $N\in\mathscr{M}^{2,d}$ with no jump outside $\bigcup_n [T_n]$ and such that*

(31.7.1) $$\Delta N_s(\omega) = \int_{\mathbb{E}} Y(s,\omega,x)q(\omega,\{s\},dx)$$

(3°) *The martingales $(N^n)_{n\geqslant 0}$ defined by*

$$N_t^n := \iint_{]0,t]\times\mathbb{E}} Y(s,\omega,x)(\mu^{T_n} - \nu^{T_n})(\omega, ds, dx)$$

are mutually strongly orthogonal (in fact: very strongly orthogonal: see chap. 4, exercise E. 7) *and their sums $\sum_{n\leqslant k} N_t^n$ converge in $\mathscr{M}^{2,d}(\mathbb{H})$ to $\int Ydq$.*

Moreover,

(31.7.2) $$E(\langle N\rangle_t) \leqslant E\{\iint_{]0,t]\times\mathbb{E}} \|Y\|^2|q|(ds,dx)\}.$$

Proof. (1°) Since, by definition, the dual predictable projection of q is zero, $\int Ydq$ is a local martingale according to 31.3 (2°). But the inequality

$$\|\iint_{]0,t]\times\mathbb{E}} Y(\omega,s,x)q(\omega,ds,dx)\|^2 \leqslant \iint_{]0,t]\times\mathbb{E}} \|Y(\omega,s,x)\|^2|q|(\omega,ds,dx)$$

and the hypothesis of the proposition show that $\int Ydq\in\mathscr{M}^2(\mathbb{H})$.

(2°) The inequality a) shows immediately that $\int Ydq$ has finite variation. It is therefore an element of $\mathcal{M}^{2,d}(\mathbb{H})$ with jumps at times T_n only, given by

$$\Delta N_{T_n}(\omega) = \int_{\mathbb{E}} Y(T_n, \omega, x) q(\omega, \{T_n\}, dx).$$

(3°) Since N is the sum in $\mathcal{M}^{2,d}(\mathbb{H})$ of the mutually strongly orthogonal martingales

$$\Delta N_{T_n} 1_{[T_n, \infty[} - \tilde{A}_n \qquad \text{(see chapter IV)},$$

where $\tilde{A}_n$ is the dual predictable projection of $\Delta N_{T_n} 1_{[T_n, \infty[}$, given by $\iint_{]0,t] \times \mathbb{E}} Y(s, \omega, x) \nu^{T_n}(\omega, ds, dx)$, according to (31.7.1) and the definition of ν^{T_n}, the convergence property and the mutual strong orthogonality of the N^n are clear.

The following chain of equalities and inequalities

$$\begin{aligned} E\langle N\rangle_t = \sum_n E\langle N^n\rangle_t &= \sum_{T_n \leqslant t} E|\Delta N_{T_n}|^2 = \\ &= \sum_{T_n \leqslant t} E\{\|\int_{\mathbb{E}} Y(T_n, \omega, x) q(\omega, \{T_n\}, dx)\|^2\} \\ &\leqslant E\{\iint_{]0,t] \times \mathbb{E}} \|Y(s, \omega, x)\|^2 |q|(ds, dx)\} \end{aligned}$$

proves (31.7.2). □

We state now a lemma, which will be used in subsequent proofs.

31.8 Lemma. *Let $\mathbb{B}$ be a separable Banach space, $\mathbb{B}'$ its dual, U a $\mathbb{B}'$-valued function on $\mathbb{R}^+ \times \Omega$ such that, for every $y \in \mathbb{B}$, the real process $\langle y, U\rangle$ is a right-continuous, adapted process with finite variation. Let us assume that Q is an increasing adapted process such that, for every predictable (resp. optional) subset A of $\mathbb{R}^+ \times \Omega$ and every $y \in \mathbb{B}$ with $\|y\| \leqslant 1$, the following inequality holds.*

$$\text{(31.8.1)} \qquad E(\int 1_A(s,.)\langle y, dU_s\rangle) \leqslant E(\int 1_A(s,.) dQ(s)).$$

Then there exists a $\mathbb{B}'$-valued process u such that, for every $y \in \mathbb{B}$, $\langle y, u\rangle$ is predictable (resp. optional) and for every A predictable (resp. optional)

$$E(\int 1_A(s,.)\langle y, dU_s\rangle) = E(\int 1_A(s,.)\langle y, u(s,.)\rangle dQ(s)).$$

Moreover, $\|u\|_{B'} \leqslant 1$.

Proof. The inequality (31.8.1) says that $E(\int 1_A(s,.) dU_s)$ as a function of A is a $\mathbb{B}'$-valued measure, the variation of which is smaller than the positive measure $A \frown E(\int 1_A(s,.) dQ(s))$. The proof of the existence of u and of its properties goes exactly as in the proof of the existence of g and of the properties of g in formula (15.8.5), chapter 3. □

31.9 Theorem. *Let q be a white random measure of order p.*

(1°) *There exists an $\mathfrak{M}^{p\otimes p}(\mathbb{E}\times\mathbb{E})$-valued process $\overset{\circ}{q}$ and an increasing, positive, adapted, right-continuous process b with the following properties.*

(a) $\forall \Phi \in \mathscr{C}^{p\otimes p}(\mathbb{E}\times\mathbb{E})$ *the process* $\left(\int_{\mathbb{E}\times\mathbb{E}} \Phi(x,y)\overset{\circ}{q}(t,.,dx\otimes dy)\right)_{t\geq 0}$ *is predictable.*

(b) *For every real, bounded, predictable process W and $\varphi \in \mathscr{C}^p_{\mathbb{H}}(\mathbb{E})$*

$$(31.9.1)\qquad E\{\sum_{s\leq t} \|\int_{\mathbb{E}} \varphi(x)W(s,.)q(.,\{s\},dx)\|^2\} =$$

$$= E\{\int_{]0,t]} db(s) \int_{\mathbb{E}\times\mathbb{E}} \langle\varphi(x),\varphi(y)\rangle_{\mathbb{H}} W(s,.)\overset{\circ}{q}(s,.,dx\otimes dy)\}$$

(c) *The real function on $\mathscr{C}^p_{\mathbb{H}}(\mathbb{E})\times\mathscr{C}^p_{\mathbb{H}}(\mathbb{E})$, defined by*

$$(f,g) \curvearrowright E\{\int_{]0,t]} db(s) \int_{\mathbb{E}\times\mathbb{E}} \langle f(x),g(y)\rangle_{\mathbb{H}}\overset{\circ}{q}(s,.,dx\otimes dy)$$

is a positive symmetric bilinear form.

(2°) *For every $\mathbb{H}$-valued $\mathscr{P}\otimes\mathscr{B}_{\mathbb{E}}$-measurable function Y let us define*

$$(31.9.2)\qquad \lambda_s(Y) := \int_{\mathbb{E}} \langle Y(s,.,x,Y(s,.,y)\rangle_{\mathbb{H}}\overset{\circ}{q}(s,.,dx\otimes dy).$$

Then

$$(31.9.3)\qquad E\{\int_{]0,t]} db(s)\lambda_s(Y)\} = E\{\sum_{s\leq t} \|\int Y(s,.,x)q(.,\{s\},dx)\|^2\}$$

$$\leq E\{\iint_{]0,t]\times\mathbb{E}} \|Y(s,.,x)\|^2|q|(.,\{s\},dx)\}.$$

Proof. We see immediately that $\sum_{\tau\leq t} 1_{\{\tau=s\}}q(\omega,ds,dx)\otimes q(\omega,ds,dy)$ defines for every t a random measure with values in the dual of the space of continuous functions on $\mathbb{E}\times\mathbb{E}$ weighted by $p\otimes p$ and we call β the dual predictable projection of this measure. We denote by $b(t,\omega)$ the variation on the interval $[0,t]$ of the $\mathfrak{M}^{p\otimes p}$-valued measure $\beta(\omega,ds,.)$ for the norm $\|\ \|_{p\otimes p}$.

Applying lemma 31.8, we define an $\mathfrak{M}^{p\otimes p}(\mathbb{E}\times\mathbb{E})$-valued process $\overset{\circ}{q}$ with the following properties. For every $\Phi\in\mathscr{C}^{p\otimes p}(\mathbb{E}\times\mathbb{E})$ the real process $\int_{\mathbb{E}\times\mathbb{E}} \Phi(x,y)\overset{\circ}{q}(s,\omega,dx\otimes dy)$ is predictable and, for every real, bounded, predictable process Y and any t,

$$(31.9.4)\qquad E\{\int_{]0,t]\times\mathbb{E}\times\mathbb{E}} Y(s,\omega)\Phi(x,y)\beta(\omega,ds,dx\otimes dy)\}$$

$$= E\{\int_{]0,t]} db(s)[\int_{\mathbb{E}\times\mathbb{E}} Y(s,\omega)\Phi(x,y)\overset{\circ}{q}(s,\omega)dx\otimes dy]\}.$$

For W and φ as in the statement of the theorem we may therefore write

$$E\{\sum_{s\leqslant t}\|\int_{\mathbb{E}}\varphi(x)W(s,.)q(.,\{s\},dx)\|^2\}$$

$$=E\{\sum_{s\leqslant t}\int_{\mathbb{E}\times\mathbb{E}}\langle\varphi(x),\varphi(y)\rangle_{\mathbb{H}}W(s,.)q(.,\{s\},dx)\otimes q(.,\{s\},dy)\}$$

$$=E\{\iint_{]0,t]\times\mathbb{E}\times\mathbb{E}}\langle(\varphi(x),\varphi(y)\rangle_{\mathbb{H}}W(s,.)\beta(.,ds,dx\otimes dy)\}$$

$$=E\{\int_{]0,t]}db(s)\int_{\mathbb{E}\times\mathbb{E}}\langle\varphi(x),\varphi(y)\rangle_{\mathbb{H}}\,W(s,.)\overset{\circ}{q}(s,.,dx\otimes dy)\}\,.$$

This proves (1°) (a) and (b) of the theorem.
The (c) is trivial from the definition of q and b.
(2°) Exactly as above we may write

$$E\{\int_{]0,t]}\lambda_s(Y)db(s)=E\{\iint_{]0,t]\times\mathbb{E}\times\mathbb{E}}\langle Y(s,.,x),Y(s,.,y)\rangle_{\mathbb{H}}\beta(.,ds,dx\otimes dy)\}$$

$$=E\{\sum_{s\leqslant t}\int_{\mathbb{E}\times\mathbb{E}}\langle Y(s,.,x),Y(s,.,y)\rangle_{\mathbb{H}}q(.,\{s\},dx)\otimes q(.,\{s\},dy)\}$$

$$=E\{\sum_{s\leqslant t}\|\int Y(s,.,x)q(.,\{s\},dx)\|^2\}$$

$$\leqslant E\{\iint_{]0,t]\times\mathbb{E}}\|Y(s,.,x)\|^2|q|(.,\{s\},dx)\}\,.\quad\square$$

31.10 Theorem *(The isometric L^2-stochastic integral). Let q, $\overset{\circ}{q}$ and b, λ be as in Theorem* 31.9. *We denote by Λ the space of $\mathscr{P}\otimes\mathscr{B}_{\mathbb{E}}$-measurable $\mathbb{H}$-valued functions on $\mathbb{R}^+\times\Omega\times\mathbb{E}$ such that for every t*

$$E\{\int_{]0,t]}\lambda_s(Y)db(s)\}<\infty,\quad\text{with the pre-Hilbert seminorms}$$

(31.10.1) $$\|Y\|_{\Lambda,t}=[E\{\int_{]0,t]}\lambda_s(Y)db(s)\}]^{\frac{1}{2}}\,.$$

Let $(T_n)_{n\geqslant 0}$ be the stopping times as in 31.6. *For every $Y\in\Lambda$ there exists a unique $N\in\mathscr{M}^{2,d}(\mathbb{H})$* with $\langle N\rangle_t=\int_{]0,t]}\lambda_s(y)db_s$ such that

(31.10.2) $$\Delta N_{T_n}=\int_{\mathbb{E}}Y(T_n,.,x)q(\omega,\{T_n\},dx)$$

(Notice that this integral has a meaning according to (31.9.3)*).*

The mapping $Y\curvearrowright N$, denoted by $\int Ydq$, is a continuous extension to Λ of the integral defined in 31.1.b *and is called the stochastic integral with respect to q.*

As usual we shall write

$$(\int Ydq)_t:=\iint_{]0,t]\times\mathbb{E}}Y(s,\omega,x)q(\omega,ds,dx)$$

even though the integral on the right-hand side has not the meaning of an ordinary integral.

Proof. The T_n having disjoint graphs, a martingale $N \in \mathcal{M}^{2,d}(\mathbb{H})$ is uniquely determined by the condition that it has no jump outside $\bigcup_n [T_n]$ and by the knowledge of its jumps (see chapter 4, 17.7), with the condition $E(\sum_n \|\Delta N_{T_n}\|^2 1_{\{T_n \leq t\}}) < \infty$ for all t.

Set

$$h_n(\omega) := \int_{\mathbb{E}} Y(T_n, \omega, x) q(\omega, \{T_n\}, dx).$$

The martingale

$$N_t^n := \iint_{]0,t] \times \mathbb{E}} Y(s, \omega, x)(\mu^{T_n} - \nu^{T_n})(\omega, ds, dx)$$

is an element of $\mathcal{M}^{2,d}(\mathbb{H})$ with one jump h_n at time T_n and

$$E(\langle N^n \rangle_t) = E\{\|h_n\|^2 1_{[T_n, \infty]}(t)\}.$$

The martingales N^n are strongly orthogonal since the T_n have disjoint graphs and $\int_{]0,t]} \lambda_s(Y) db_s$ being by construction the dual predictable projection of $\sum_n \|\Delta N_{T_n}\|^2 1_{\{T_n < t\}}$ the equation

$$\sum_n E\langle N^n \rangle_t = \sum_n E\{1_{\{T_n \leq t\}} \| \int_{\mathbb{E}} Y(T_n, \omega, x) q(\omega, \{T_n\}, dx)\|^2\} = $$
$$= \int_{]0,t]} db(s) \lambda_s(Y)$$

shows that the sequence $(N^n)_{n \geq 0}$ converges in $\mathcal{M}^{2,d}(\mathbb{H})$ to an N with

$$\langle N \rangle_t = \int_{]0,t]} db(s) \lambda_s(Y). \tag{31.10.3}$$

This proves the theorem. □

31.11 Example. *The integral of Skorokhod.* Let $m(\omega, ds, dx)$ be a time-homogeneous Poisson point process of order $r > 0$, that is to say, of order p with $p(x) = \dfrac{\|x\|^r}{1 + \|x\|^r}$, $x \in \mathbb{E} := \mathbb{R}^d - \{0\}$, and Levy measure α.

Consider the random measure

$$q(\omega, ds, dx) := m(\omega, ds, dx) - ds \otimes \alpha(dx)$$

For such a random measure $q(\omega, \{s\}, dx) = m(\omega, \{s\}, dx)$ and for every continuous function Φ with compact support in $\mathbb{E} \times \mathbb{E}$,

$$E(\sum_{s \leq t} \int_{\mathbb{E} \times \mathbb{E}} \Phi(x, y) q(\omega, \{s\}, dx) \otimes q(\omega, \{s\}, dy)) = E \sum_{s \leq t} \int_{\mathbb{E}} \Phi(x, x) q(\omega, \{s\}, dx).$$

Note that, for every $(s, \omega), q(\omega, \{s\}, dx)$ is either zero or a Dirac measure on $\mathbb{E}$. The measure $q(\omega, \{s\}, dx) \otimes q(\omega, \{s\}, dy)$ is therefore either zero or the corresponding Dirac measure on the "diagonal" of $\mathbb{E} \times \mathbb{E}$.

The latter formula shows that the dual predictable projection of the measure β, as defined in 31.1, is the measure $\Phi \curvearrowright \iint_{]0,t]\times\mathbb{E}} \Phi(x,x)ds\,\alpha(dx)$. In this case, $b(t) = \bar{\alpha}t$ with $\bar{\alpha} = \int p(x)\alpha(dx)$ and $\overset{0}{q}(\omega, s, dx \otimes dy)$ is the non-radom measure $\Phi \curvearrowright \int \Phi(x,x)\frac{1}{\bar{\alpha}}\alpha(dx)$ concentrated on the diagonal in $\mathbb{E}\times\mathbb{E}$. The positive number $\bar{\alpha}$ is called the *intensity of the process*. The functional λ is therefore none other than

$$\lambda_t(Y) = \frac{1}{\bar{\alpha}} \int ||Y(.,t,y)||^2_{\mathbb{H}}\,\alpha(dy).$$

The integral of functions Y for which $\lambda_t(Y) < \infty$ was first introduced by Skorokhod (see [Sko]).

Remark that, in this case $||\ ||_{\Lambda,t}$, is, for each t, the Hilbert norm of $L^2_{\mathbb{H}}([0,t]\times\Omega\times\mathbb{E}, \mathscr{P}\otimes\mathscr{B}_{\mathbb{E}}, ds\otimes P\otimes\alpha)$ and, if we call $\mathbb{L}_1$ the Banach space of $\mathbb{H}$-valued Borel functions f on $\mathbb{E}$ such that $\sup_{x\in\mathbb{E}} (||f(x)||/p(x)) < \infty$, the vector space $\mathscr{E}(\mathbb{L}_1)$ of elementary predictable $\mathbb{L}_1$-valued processes is therefore dense in Λ.

31.12 Example. More generally, let μ be a point process with masses only on totally inaccessible stopping times (σ_n) and ν its dual predictable projection. The process F_ν is then continuous and, if we set $q = \mu - \nu$, we have

$$q(\omega, \{s\}, dx) = \mu(\omega, \{s\}, dx).$$

Since, by definition, $\mu(\omega, ds, dx) = \sum_n \varepsilon_{\sigma_n, h_n}$, where the h_n are $\mathscr{F}_{\sigma_n}$-measurable $\mathbb{E}$-valued random variables, for every $\mathbb{H}$-valued $\mathscr{P}\otimes\mathscr{B}_{\mathbb{E}}$-measurable function Y,

$$||Y||^2_\Lambda = E\{\sum_{\sigma_n\leqslant t} ||Y(\sigma_n,.h_n)||^2_{\mathbb{H}}\} =$$

$$= E\{\iint_{]0,t]\times\mathbb{E}} ||Y(s,.,x)||^2_{\mathbb{H}}\,\mu(.,ds,dx)\}.$$

In this case, as in example 31.11, the seminorms $||\ ||_{\Lambda,t}$ are the seminorms of $L^2_{\mathbb{H}}(]0,t]\times\Omega\times\mathbb{E}, \mathscr{P}\otimes\mathscr{B}_{\mathbb{E}}, m)$, where m is the measure on $\mathscr{P}\otimes\mathscr{B}_{\mathbb{E}}$ defined by

$$m(A) = E\{\iint 1_A(s,x)\,\mu(.,ds,dx)\}.$$

The vector space $\mathscr{E}(\mathbb{L}_1)$, as defined in 31.11, is also dense in Λ.

31.13 Example. If μ is a point process with masses on predictable stopping times (τ_n) and totally inaccessible stopping times (σ_n), with dual predictable projection ν, the space Λ associated with $q = \mu - \nu$ still admits $\mathscr{E}(\mathbb{L}_1)$ as a dense subspace.

The proof of this fact is indicated in exercise E.4 at the end of the chapter.

31.14 Stochastic integral with respect to locally white random measures

Let q be a random measure for which there exists an increasing sequence (S_n) of stopping times converging to $+\infty$ such that, for every n, $1_{]0,S_n]}q(\omega, ds, dx)$ is white random measure. Let Y be a process such that the increasing process $\sum_{s \leqslant t} \| \int_{\mathbb{E}} Y(\omega, s, x) q(\omega\{s\} dx)\|^2$ is locally integrable. One may therefore find a localizing sequence (S'_n) of stopping times such that $1_{]0,S'_n]}q(\omega, ds, dx)$ is a white random measure for every n and $E\{\sum_{s \leqslant t} \| \int_{\mathbb{E}} 1_{]0,S'_n]} Y(\omega, s, x) q(\omega, \{s\}, dx)\|^2\} < \infty$ for every n and t. By an argument we have often used, one easily sees that there exists a unique (up to P-equality) locally square integrable martingale N with the property

$$N_{S'_n \wedge t} = (\int 1_{]0,S'_n]} Y dq)_t \qquad \text{for all } n.$$

This locally square integrable martingale can be characterized as the unique $N \in \mathscr{M}^{2,d}_{\text{loc}}(\mathbb{H})$ with jumps only at stopping times T_n for which $P\{|q|(.,|T_n\}, .) > 0\} > 0$ and such that

$$\Delta N_{T_n} = \int_{\mathbb{E}} Y(T_n, ., x) q(., \{T_n\}, dx).$$

It is defined for all Y such that the increasing process $\int_{]0,t]} \lambda_s(Y) db_s$ is locally integrable, with λ and b defined as in 31.9.

Moreover,

$$(31.14.1) \qquad \langle N \rangle_t = \int_{]0,t]} \lambda_s(Y) db_s.$$

We shall write $N = \int Y dq$.

31.15 Extension of the stochastic integral with respect to q

If the white random measure q is such that, for every predictable stopping time τ, the variation $|q|(\omega, \{\tau(\omega)\}, E)$ is zero a.s. (as in example 31.12 above) the process $N := (\int Y dq)$ has no predictable jump and we know, as an immediate consequence of the Doob inequality and the continuity of $\langle N \rangle$, that

$$(31.15.1) \qquad E(\sup_{s<\tau} \|N_s\|^2) \leqslant 4E\langle N \rangle_\tau = 4E\langle N \rangle_{\tau-} \qquad \text{for every stopping time } \tau.$$

From formula (31.14.1) we may write, for every $Y \in \mathscr{E}(\mathbb{L}_1)$,

$$(31.15.2) \qquad E(\sup_{s<\tau} \|N_s\|^2) \leqslant 4E(\int_{[0,\tau[} \lambda_s(Y) db(s)).$$

Assume $\mathscr{E}(\mathbb{L}_1)$ is dense in Λ, as it is the case for a random measure in examples above. The same procedure as the one used for defining the integral with respect to semimartingales shows that the mapping $Y \curvearrowright N$ can be extended to the class of all functions Y on $\mathbb{R}^+ \times \Omega \times \mathbb{E}$ measurable for $\mathscr{P} \otimes \mathscr{B}(\mathbb{E})$ and such that the increasing process $(\int_{]0,t]} \lambda_s(Y)db(s))_{t\in\mathbb{R}^+}$ is finite. This integral, still denoted by $\int Ydq$, is such that for every stopping time τ

(31.15.3) $$E(\sup_{s<\tau} \|\iint_{]0,t]\times\mathbb{E}} Y(s,.,x)q(.,ds,dx)\|^2) \leqslant 4E(\int_{]0,\tau[} \lambda_s(Y)db(s)).$$

Remark: When F_q has jumps at predictable stopping times the integral $\int Ydq$ of $Y\in\mathscr{E}(\mathbb{L}_1)$ may have also jumps at predictable stopping times, so (31.15.3) no longer holds. In this case, to obtain an analogous extension of the stochastic integral, one can use "a control couple other than $\overset{\circ}{q}$ and b". This is done in exercise E.5 at the end of the chapter.

32. Local characteristics of a semimartingale • Diffusions • Martingale problems

32.1 Theorem. *Let $X\in\mathscr{Q}_{\mathrm{loc}}(\mathbb{R}^d)$. There exists an $\mathbb{R}^d$-valued predictable R.R.C. process V with locally integrable variation, an $\mathbb{R}^d\otimes\mathbb{R}^d$-valued continuous process C with finite variation, the values of which are positive symmetric matrices and a predictable random measure ν of order 2 on $\mathbb{E}:=\mathbb{R}^d-\{0\}$ such that, for every twice continuously differentiable function φ on $\mathbb{R}^d$, the process M^φ, defined by*

(32.1.1) $$\begin{aligned} M_t^\varphi := \varphi(X_t) - \varphi(X_0) - &\int_{]0,t]} \sum_{i=1}^d D^i\varphi(X_{s^-})dV_s^i - \\ &-\tfrac{1}{2}\int_{]0,t]} \sum_{i,j=1}^d D^{i,j}\varphi(X_{s^-})dC_s^{i,j} \\ &- \iint_{]0,t]\times\mathbb{R}^d-\{0\}} (\varphi(X_{s^-}+x)-\varphi(X_{s^-}) - \\ &- \sum_{i=1}^d D^i\varphi(X_{s^-})x^i)\nu(ds,dx) \end{aligned}$$

is a local martingale.

The triple (V, C, ν) satisfying the above properties is unique.

Proof. Let us consider the (unique) decomposition of X as $X = M + V$, where M is a local martingale and V is a predictable process with locally integrable variation. We define $C:=[X]^c = [M]^c$ and ν to be the dual predictable projection of the point process of jumps of X (see 31.1). If we apply the transformation formula in the form (31.2.1) we obtain

(32.1.2) $$\begin{aligned} M_t^\varphi = &\int_{]0,t]} \sum_{i=1}^d D^i\varphi(X_{s^-})dM_s^i + \iint_{]0,t]\times\mathbb{R}^d-\{0\}} (\varphi(X_s+x)-\varphi(X_{s^-}) - \\ &- \sum_{i=1}^d D^i\varphi(X_{s^-})x^i)(\mu-\nu)(ds,dx). \end{aligned}$$

We know from Theorem 24.4. 4° that the stochastic integral with respect to M is a local martingale. If we set

$$Y(s,\omega,x) = \varphi(X_{s^-}(\omega)+x) - \varphi(X_{s^-}(\omega)) - \sum_{i=1}^{d} D^i\varphi(X_{s^-}(\omega))x^i$$

and assume the boundedness of the derivatives $D^2\varphi$, the function $\|Y(s,\omega,x)\| \leqslant K\|x\|^2$ is integrable for every ω with respect to the point process of order 2 $\mu - \nu$ (see 31.2). In this case, the second integral in (31.1.2) is, according to 31.7, a locally square integrable martingale. The case of a function φ with unbounded second-order derivatives is easily reduced to the previous case by defining the sequence (T_n) of stopping times by

$$T_n := \inf\{t : D^2\varphi(X_t) > n\}$$

and replacing φ for each n by a function $\tilde{\varphi}_n$ equal to φ in a proper ball.

This proves that M^φ is a local martingale.

Conversely, assuming that M^φ is a local martingale for every twice continuously differentiable φ and successively considering $\varphi(x) = x$ and $\varphi(x) = x^i x^j$, we obtain the following.

(1°) $X_t - X_0 - V_t$ is a martingale M. Since V is assumed to be predictable this defines V and M uniquely.

(2°) $C_t^{ij} = [M_t^i, M_t^j]^c$ and ν is the predictable dual projection of the point process of jumps of X. This defines C and ν uniquely.

32.2 Definition (*Local characteristics*). If $X \in \mathscr{Q}_{\mathrm{loc}}(\mathbb{R}^d)$, the elements of the triple (V, C, ν) defined by Theorem 32.1 are called the *local characteristics of* X.

32.3 Remark. It is possible to define the local characteristics for a general semimartingale (see the book by Jacod (Jac. 2]) but in this case V is unique up to some artificial convention (see [Jac. 2], section 41.3).

32.4 Definition *(Diffusions and diffusions with jumps)*. (1°) A semimartingale $X \in \mathscr{Q}_{\mathrm{loc}}(\mathbb{R}^d)$ is called a *diffusion* if it is continuous (this implies $\nu = 0$) and if the other local characteristics V and C are of the form

a) $V_t = \int_0^t b(s, X_s)\,ds$, where $(t,x) \curvearrowright b(t,x)$ is an $\mathbb{R}^d$-valued Borel function on $\mathbb{R}^+ \times \mathbb{R}^d$,

b) $C_t^{i,j} = \int_0^t a^{i,j}(s, X_s)\,ds$, where $(t,x) \curvearrowright a(t,x)$ is a $d \times d$-matrix-valued Borel measurable function on $\mathbb{R}^+ \times \mathbb{R}^d$

c) $\nu = 0$.

(2°) A semimartingale $X \in Q_{\text{loc}}(\mathbb{R}^d)$ is called a *diffusion with jumps* if its local characteristics V and C are of the form a) and b) and if

c) $$v(\omega, ds, du) = f(s, X_{s^-}(\omega))ds \otimes L(s, X_{s^-}(\omega), du),$$

where $(t, x) \curvearrowright f(t, x)$ is a positive measurable function called the *intensity of jumps* (*at point x and time t*) and for each $(t, x), L(t, x, du)$ is a probability law on $\mathbb{R}^d - \{0\}$ with the property that $\forall B \in \mathscr{B}_{\mathbb{E}}$ $(t,x) \curvearrowright L(t, x, B)$ is a Borel function on $\mathbb{R}^d$. The measure $L(t, x, du)$ is the *distribution law of amplitude of jumps at point x and time t*. The measure $L(t, x, x + du)$ is called the *distribution of jumps* at point x.

32.5 Remarks and examples. (1°) Since X has only denumerably many jumps on any finite interval for each ω, the processes V_t and C_t in conditions a) and b) can just as well be written $V_t = \int_0^t b(s, X_{s^-})ds$ and $C_t = \int_0^t a(s, X_{s^-})ds$. The same applies for replacing $f(X_{s^-})$ by $f(X_s)$ in c) but X_{s^-} cannot be replaced by X_s in $L(X_{s^-}(\omega), du)$ because one would lose the predictability of v.

(2°) Exercise E.4.2 in chapter III and the Dynkin formula show that if X is a pure jump Markov process it is a diffusion with jumps with $C = 0$ and $b(s,x) = \int u L(s,x,du)$

(3°) Theorem 28.5 says that the only *diffusion* with local characteristics $(0, \int_0^t Q ds, 0)$, where Q is a constant matrix, is the Brownian motion with covariance matrix Q.

(4°) The important problem is to know precisely whether the local characteristics of a process $X \in \mathscr{Q}_{\text{loc}}$ uniquely determine the law of X. There are only partial answers: see [Jac. 2], chapter 3.

The simplest partial answer is given by the following theorem.

32.6 Theorem. *An $\mathbb{R}^d$-valued stochastically continuous R.R.C. process X with bounded jumps and such that $t \curvearrowright E(X_t)$ is a function with finite variation is a process with increments independent of the past, if and only if it is a semimartingale in $\mathscr{Q}_{\text{loc}}$ and its local characteristics are not random (they do not depend on ω). That is*

$$V_t(\omega) = B(t),\ C_t(\omega) = Q(t),\ v(\omega, ds, du) = F(ds, du),$$

where B and Q are respectively $\mathbb{R}^d$-valued and matrix-valued Borel functions of t and F is a Borel measure on $\mathbb{R}^+ \times (\mathbb{R}^d - \{0\})$.

Proof. We know that if X is a stochastically continuous R.R.C. process with independent increments and bounded jumps, it can be written

$$X_t = E(X_t) + M_t := B(t) + M_t,$$

where M is a square integrable martingale and $t \curvearrowright E(X_t)$ is continuous. (See exercise E.12, chapter 4).

Actually, M is a process with centered increments which are independent of the past. This shows immediately that

$$C_t^{ij} = E(M_t^i M_t^j - M_0^i M_0^j) = Q^{ij}(t).$$

We saw in exercise E.3.4, chapter 3, that the point process of jumps of X is a Poisson process with Levy measure $F(ds, du)$.

The converse is obtained immediately by applying the transformation formula in the form (31.2.1). □

32.7 Corollary. *An $\mathbb{R}^d$-valued stochastically continuous R.R.C. process X with bounded jumps and such that $t \curvearrowright E(X_t)$ is a function with finite variation is a process with increments which are stationary and independent of the past, if and only if it is a semimartingale in $\mathcal{Q}_{\text{loc}}$ and its local characteristics have the following form.*

$$V_t(\omega) = bt,\ C_t(\omega) = Ct,\ \nu(\omega, ds, du) = \lambda\, du L(dx),$$

where $b \in \mathbb{R}^d$, C is a $d \times d$-valued positive matrix, $\lambda \in \mathbb{R}^+$ and L is a Borel measure on $\mathbb{R}^d - \{0\}$.

In this case, the characteristic function $\Psi_t(u) := E\{\exp iu \cdot (X_t - X_0)\}$, $u \in \mathbb{R}^d$ is given by the Levy-Kintchine Formula

$$\Psi_t(u) = \exp(t\{iu \cdot b - \tfrac{1}{2} u \cdot Q(u) + \lambda \int_{\mathbb{R}^d - \{0\}} (e^{iu \cdot x} - 1 - iu \cdot x) L(dx)\})$$

Proof. The first part of the corollary follows trivially from Theorem 32.6.

The evaluation of the characteristic function Ψ_t follows from the definition of the local characteristics. Let us indeed apply this definition with $\varphi(x) = \exp iu \cdot x$. We get

$$\begin{aligned}\Psi_t(u) &= i \int_0^t \Psi_s(u) u \cdot b\, ds - \tfrac{1}{2} \int_0^t \Psi_s(u) u \cdot Q(u) \mathrm{d}s \\ &\quad + \int_0^t \lambda \Psi_s(u) ds \left[\int_{\mathbb{R}^d - \{0\}} (e^{iu \cdot x} - 1 - iu \cdot x) L(dx) \right] \\ &= \int_0^t \Psi_s(u) ds \left[iu \cdot b - \tfrac{1}{2} u \cdot Q(u) + \lambda \int_{\mathbb{R}^d - \{0\}} + (e^{iu \cdot x} - 1 - iu \cdot x) L(dx) \right]\end{aligned}$$

The solution of this linear differential equation in the function $t \curvearrowright \Psi_t(u)$ immediately yields the Levy-Kintchine formula.

32.8 Remark. The hypothesis on boundedness of jumps in 32.6 is unnecessary. A more general result can be found in [Jac. 2], chapter 3. We made the assumption only to insure that $X \in \mathcal{Q}_{\text{loc}}$.

We end this chapter with a definition which formalizes a special case of the general question set in 32.5. 4°.

32.9 Martingale problems

Let $\tilde{\Omega} = \mathbb{C}_{\mathbb{R}^d}(\mathbb{R}_+)$ (resp. $\mathbb{D}_{\mathbb{R}^d}(\mathbb{R}^+)$) be the set of all continuous (resp. right-continuous with left limits) $\mathbb{R}^d$-valued functions on $\mathbb{R}_+$. On $\tilde{\Omega}$ we consider the right-continuous filtration $(\mathscr{C}_t)$ (resp. $(\mathscr{D}_t)$) defined by $\mathscr{C}_t = \mathscr{C}^0_{t^+}$ (resp. $\mathscr{D}_t = \mathscr{D}^0_{t^+}$), where $\mathscr{C}^0_t$ (resp. $\mathscr{D}^0_t$) is the smallest σ-algebra of subsets of $\tilde{\Omega}$ for which the functions $\tilde{\omega} \curvearrowright \tilde{\omega}(A)$, $s \leqslant t$ are measurable. We shall denote by ξ the canonical process on $\tilde{\Omega}: \xi_t(\tilde{\omega}) := \tilde{\omega}(t)$. Let us assume that for every twice continuously differentiable function φ with bounded derivatives we are given an adapted process $(\mathfrak{A}(\varphi, s))_{s \leqslant t}$ on $\tilde{\Omega}$. A probability law $\tilde{P}$ on $\mathscr{C}_\infty$ (resp. $\mathscr{D}_\infty$) is called *a solution of the martingale problem associated with* $\mathfrak{A}$ *and the initial condition* $x_0 \in \mathbb{R}^d$ if

$$\varphi(\xi_t) - \varphi(x_0) - \int_0^t \mathfrak{A}(\varphi, s)ds$$

is a $\tilde{P}$-martingale with respect to the filtration $(\mathscr{C}_t)$ (resp. $(\mathscr{D}_t)$) for every φ.

Let us remark that, with this definition, X is a diffusion with local characteristics V and C as in 32.4, if and only if its law is the solution of the martingale problem $(\mathfrak{A}, x_0)$ with

$$\mathfrak{A}\,\varphi(x) = \sum_{i=1}^{d} b^i(s, X_s) \cdot D^i \varphi(X_s) + \sum_{i,j=1}^{d} a^{i,j}(s, X_s) D^{ij} \varphi(X_s)$$

In this case the functional $\mathfrak{A}$ possesses the following localization property.

$[\ell]$: If $\tilde{\omega}^1_t = \tilde{\omega}^2_t$ for every $t \leqslant \tau$ (τ: a stopping time) then $\mathfrak{A}(\varphi, t, \tilde{\omega}^1) = \mathfrak{A}(\varphi, t, \tilde{\omega}^2)$ for every $t \leqslant \tau$.

Now if φ is a twice continuously differentiable function, one may consider for each n a function, which is twice differentiable, has compact support and such that $\varphi_n(x) = \varphi(x)$ for every $x \in \mathbb{R}^d$, $\|x\| \leqslant n$. When property $[\ell]$ is valid one may extend $\mathfrak{A}$ to φ by setting

$$\tilde{\mathfrak{A}}(\varphi, t, \tilde{\omega}) = \mathfrak{A}(\varphi_n, t, \tilde{\omega}) \qquad \text{for} \qquad t < \tau_n,$$

where $\tau_n(\tilde{\omega}) := \inf\{t : \xi_t(\tilde{\omega}) > n\}$.

See exercise E.2 for the use of this remark.

Exercises and supplements

E.1. *The martingale problem of a branching process.* At time 0, x_0 particles are present. At random times τ_n^- one of the particles present dies and gives birth to a random number $N_n \geqslant 0$ of particles. Given $\mathscr{F}_{\tau_{\bar{n}}}$ the conditional law of the couple $(\tau_{n+1} - \tau_n$, $N_n)$ is given by $X_{\tau_n^-} e^{-\lambda X_{\tau n} s} ds\,\gamma(k)$ where $\lambda > 0$ (the intensity of death of one particle) and γ is a probability-law on $\mathbb{N}$. We denote by (X_t) the right-continuous process representing the number of particles living at time t.

(1°) What is the martingale problem satisfied by the law of X?

Hint: Use the formula of exercise E.3.1., chapter 3, and the Ito formula applied to $\varphi(X_t)$ in the form

$$\varphi(X_t) = \varphi(X_0) + \iint_{]0,t]\times\mathbb{Z}} \varphi(X_{s^-} + x) - \varphi(X_{s^-}))\nu(\omega, ds, dx) + \text{Martingale}$$

(2°) Define $m := \sum_{k\geq 0} k\gamma(k), \alpha := \sum_{k\geq 0} (k-1)^2\gamma(k)$. Show that $X_t - X_0 - \int_0^t \lambda(m-1)X_s ds$ is a martingale M and $\langle M\rangle_t = \int_0^t \lambda\alpha X_s ds$.

(3°) Using $X_t = X_0 + \int_0^t \gamma(m-1)X_s ds + M_t$, show that

$$X_t = e^{\lambda(m-1)t} X_0 + \int_0^t e^{\lambda(m-1)(t-s)} dM_s$$

(*Hint*: Use the solution of a linear equation as in exercise E.6, chapter 6).

Evaluate $E(X_t)$. (*Answer*: $e^{\lambda(m-1)t}x_0$)

(4°) What is the asymptotic behaviour of X_t when $m<1$ (resp. $m>1$).

Call τ the death-time ($\tau := \inf\{t : X_t = 0\}$). Show that if $m=1$ $E(\tau) = +\infty$. (*Hint*: If $m=1$, use the martingale property to derive a contradiction).

E.2. *Equivalent forms of the martingale problem.* Let $\tilde{\Omega} = \mathbb{C}_{\mathbb{R}^d}(\mathbb{R}^+)$ and let $\mathfrak{A}(\varphi, s)$ be as in the definition 32.9. Show that $\tilde{P}$ is a solution of the martingale problem associated with $\mathfrak{A}$ and x_0 if and only if one of the following conditions (1) to (3) holds.

(1) For every twice continuously differentiable function φ on $\mathbb{R}^d$, with compact support, the process $(\varphi(\xi_t) - \varphi(x_0) - \int_0^t \tilde{\mathfrak{A}}(\varphi, s)ds)_{t\geq 0}$ is a martingale.

(2) Call B^i, $i = 1 \dots d$, the process $(\tilde{\mathfrak{A}}(h^i, t))_{t>0}$, where $h^i(x) = x^i$ and A^{ij} the process $(\tilde{\mathfrak{A}}(h^i h^j, t))_{t\geq 0}$. Then the process $M := (\xi_t - x_0 - \int_0^t B_s ds)_{t\geq 0}$ is a local martingale with

$$\langle\!\langle M\rangle\!\rangle_t = \int_0^t A_s ds.$$

(3) For every $u \in \mathbb{R}^d$ the process

$$(\exp\{u.(\xi_t - x_0 - \int_0^t B_s ds) - \tfrac{1}{2}\int_0^t u.A_s(u)ds\})_{t\geq 0} \quad \text{is a local martingale.}$$

Hint: Use the transformation formula and the exponential formula of chap. 6.

E.3. *Deterministic solutions of a martingale problem.* Let $\tilde{\Omega} = \mathbb{C}_{\mathbb{R}^d}(\mathbb{R}^+)$ and let $b(t,x)$ be a continuous (in x and t) d-dimensional vector field on $\mathbb{R}^d$. We define $\mathfrak{A}(\varphi, s, \tilde{\omega}) := \sum_{i=1} b^i(t,x) D^i \varphi(\xi_s(\tilde{\omega}))$.

Show that if $\tilde{P}$ is a solution of the martingale problem $(\mathfrak{A}, x_0)$ then

$$\xi_t(\tilde{\omega}) = x_0 + \int_0^t b(t, \xi_s(\tilde{\omega}))\, ds. \quad \text{a. s.}$$

In other words, the probability $\tilde{P}$ gives probability one to the set of $\tilde{\omega} \in \tilde{\Omega}$ for which $\xi_t(\tilde{\omega})$ is a solution of the differential equation

$$\begin{cases} \dfrac{d\xi_t}{dt} = b(t, \xi_t)\mathrm{dt} \\ \xi_0 = x_0 \end{cases}$$

Hint: Use Exercise E.2., formulation (2) of the martingale problem and observe that $M := 0$ $\tilde{P}$ – a.s.

E.4. *Proof of the assertion in 31.13.* To prove this statement one may do the following.

Let τ_n be the predictable stopping times and σ_n the totally inaccessible stopping times where q has masses.

Write

$$\|Y\|^2_{\Lambda,t} = E\{\sum_{\tau_n \leq t} \| \iint_{\{\tau_n = s\}} Y(s,.,x) q(., ds, dx)\|^2\}$$

$$+ E \iint_{]0,t] \times \mathbb{E}} \|1_{\{s \notin \bigcup_n [\tau_n]\}} Y(s,.,x)\|^2 q(., ds, dx)$$

The sequences $1_{\{s \notin \bigcup_n [\tau_n]\}} Y + \sum_{k \leq n} 1_{\{\tau_n \neq s\}} Y$ converge to Y in Λ. (The martingales stochastic integrals converge in $\mathcal{M}^{2,d}$).

$1_{\{s \notin \bigcup_n [\tau_n]\}}$ can be approximated in Λ by a sequence of functions in $\mathscr{E}(\mathbb{L}_1)$ as in 31.12.

Every $1_{\{\tau_k = s\}} Y$ can be approximated in Λ by $1_{\{\tau_k = s\}} (Y \wedge np(x))$ and every W such that $E(\iint_{]0,t] \times \mathbb{E}} \|W\|^2_{\mathbb{H}} d|q|) < \infty$ can be approximated in Λ by a sequence in $\mathscr{E}(\mathbb{L}_1)$.

E.5. „*Control couple*“ *of a random measure.* (See remark at the end of 31.15). Let $\{\tau_n\}_{n \geq 0}$ be a family of predictable stopping times such that F_q, the primitive process of the white random measure q, has no jump on any predictable stopping time with graph disjoint from $\bigcup_n [\tau_n]$.

Show the existence of an optional process $\gamma(\omega, s, dx \otimes dy)$ with values in $\mathfrak{M}^{p \otimes p}(\mathbb{E} \times \mathbb{E})$ and an increasing real process Q such that for every $Y \in (\mathbb{L}_1)$

$$E(\|\sup_{t < \tau} \iint_{]0,t] \times \mathbb{E}} Y dq\|^2) \leq E(\int_{]0,\tau[} dQ_s \int \langle Y(.,s,x), Y(.,s,y)\rangle_{\mathbb{H}} \gamma(s, dx \otimes dy))$$

Hints: Apply the inequality of Theorem 19.4 to the martingale $N = \int Y dq$, remark that

$$[\check{N}]_t = \sum_n \int_{\mathbb{E}\times\mathbb{E}} < Y(\,.\,,\tau_n, x)\, Y(\,.\,,\tau_n, y) >_{\mathbb{H}} q(\,.\,,\{\tau_n\}, ds) \otimes q(\,.\,,\{\tau_n\}, dy)$$

and apply Lemma 31.8 to the random measure

$$\beta(\omega, ds, dx \otimes dy) + \sum_n 1_{\{\tau_n = s\}}\, q(\,.\,, ds, dx) \otimes q(\,.\,, ds, dy)\,.$$

Historical and bibliographical notes

The notion of point processes has long been known in connection with the point process of jumps of a process with independent increments and its role in constructing the process from the Poisson process of jumps (P. Levy [Lev. 1] anf K. Ito [Ito. 7]). This latter paper contains the stochastic integral with respect to the Poisson point process. It is, however, in Skorokhod's book [Sko] that we find the stochastic integral with respect to Poisson point processes connected with martingale theory in the modern sense. The introduction of the dual predictable projection of a point process (Jacod [Jac. 3], Grigelionis [Grig. 2]) made possible the extension of the theory. We have essentially given Jacod's construction but added continuity properties of the integral connected with the „control couple“ as introduced in [Met. 8].

The notion of local characteristics of a semimartingale was introduced by Grigelionis [Grig. 4] and generalized by Jacod and Memin [JaM. 1]. The characterization of the law of processes with independent increments has been extended to to processes with conditionally independent increments by Grigelionis [Grig. 5].

The martingale problem was first formulated by Stroock and Varadhan for diffusion processes [StV. 1]. The reference book on this problem is [StV. 2].[1]). General martingale problems in their present state are studied in chapters 12 and 13 of [Jac. 2], referring to numerous works on the subject.

[1]) A short and substantial introduction to the subject may be found in Priouret [Pri]. See also Williams [Wil].

Chapter 8

Stochastic differential equations

We have already met stochastic equations in the course of this book. In chapter 5 (exercise E.1), we solved the Langevin equation. In chapter 6 a linear differential equation led to the exponential formula (section 29). Another type of linear equation was considered in exercise E.6, chapter 6.

The theory of „differential stochastic equations“ is similar in many ways to the theory of ordinary differential equations: „local solutions“ are obtained, under Lipschitz-type hypotheses, as fixed points of a contractive mapping in a proper metric space. Then the existence of maximal solutions is shown and their properties are studied. In particular, are the maximal solutions defined for all $t \in \mathbb{R}_+$? When the maximal solutions are defined only on $[0, \tau[$, $\tau < \infty$ (here τ is a stopping time) what is their behaviour when $t \nearrow \tau$?

There is, however, a serious difference with the case of ordinary differential equations. Several concepts of solutions can be introduced, namely, strong end weak solutions. In the strong problem the probability space is given and the unknown process on this space is a function of the stochastic driving term in the equation, this term being defined on the same probability space.

In the weak problem, the probability space is not imposed. Only the law of the stochastic driving term is given.

This chapter is mainly devoted to the study of strong solutions for a wide class of equations under Lipschitz-type hypotheses. The driving term in the equation involves a semimartingale and a point process. After examples in section 33, we study existence and uniqueness of the strong solutions in section 34, give conditions for non-explosion in section 35 and study some regularity properties of the solutions with respect to a parameter in section 36.

In section 37, we introduce the problem of weak solutions in the particular classical case of Ito- equations, indicating the relations between this problem and the martingale problem formulated at the end of chapter 7.

33. Examples of stochastic equations – Definitions

33.1 The Langevin equation and the Ornstein-Uhlenbeck process

This equation is one of the first introduced to describe physical phenomena. Langevin proposed the following differential equation to describe the movement of a free particle submitted to infinitely many uncorrelated shocks.

(33.3.1) $$\frac{d^2 X}{dt^2} + \alpha \frac{dx}{dt} = \sigma \frac{d\beta}{dt}$$

In this equation X is the coordinate of the free particle, α is a „viscosity coefficient" and $\frac{d\beta}{dt}$ is intended to represent the effect of a small shock at time t. β is thus a process which is the sum of random frequent, independent and small Dirac masses alternatively positive and negative. Such a process is usually approximated by Brownian motion. The term $\frac{d\beta}{dt}$ has then no rigourous mathematical meaning because of the structure of Brownian paths and the equation (31.1.1) can be read

(31.1.2) $$\frac{dX}{dt}(t) = \frac{dX}{dt}(0) - \alpha(X_t - X_0) - \sigma\beta_t$$

We solved such an equation in exercise E.1., chapter 5, as an application of the Ito formula (see also [MeP. 4]).

The process $V(t) = \frac{dX}{dt}(t)$ is the so-called *Ornstein-Uhlenbeck speed process.* This is a Gaussian process with covariance

(33.1.3) $$C(s,t) := E(V(s)\,V(t)) = \int_0^s e^{\alpha(t-u)}\sigma^2 e^{\alpha(s-u)}\,du$$

The process V can itself be expressed as a stochastic integral:

(33.1.4) $$V_t = e^{\alpha t} V_0 + \sigma e^{\alpha t} \int_{]0,t]} e^{-\alpha u} d\beta(u)$$

33.2 Ito-Skorokhod equations

Equations first studied by Ito are a natural generalization of the above equation in as much as the coefficients α and σ may depend on the position X_t of the particle at time t and on time as well. These equations, usually written

(33.2.1) $$dX(t) = b(t, X_t)dt + \sigma(t, X_t)d\beta_t,$$

where β is a Brownian motion, are to be interpreted as

(33.2.2)
$$X_t = X_0 + \int_{]0,t]} b(s, X_s)ds + \int_{]0,t]} \sigma(s, X_s)d\beta_s .$$

A solution is a process X defined on a pre-given stochastic basis, on which β itself is defined, such that both members of (33.2.1) are P-equal processes.

K. Ito showed the existence and uniquencess of these solutions for a given „initial condition“ x_0, and an assumption of the following kind on b and σ: b and σ are measurable functions of (t, x) and there exists a constant K such that for every $t \geqslant 0$, $x \in \mathbb{R}^d$ and $y \in \mathbb{R}^d$,

(33.2.3)
$$\|b(t,x) - b(t,y)\| + \|\sigma(t,x) - \sigma(t,y)\| \leqslant K\|x-y\| .$$

This "uniform Lipschitz condition" was later weakened into a local Lipschitz condition plus an extra condition to avoid explosion. Other weakenings were also proposed. We do not enter into these details because all these results will be included in the rather general result, which will be stated and proved in the next section. A good account of Ito's equations, besides the original papers of K. Ito (see [Ito. 2], [Ito. 4], [Ito. 6]) can be found in P. Priouret [Pri].

The solutions of equations (33.2.2) are continuous processes because of the continuity of β. (This gives meaning to the stochastic integral in (33.2.2), the process $(\sigma(t, X_t))_{t \geqslant 0}$ then being predictable).

Skorokhod considered equations (see Skorokhod [Sko]) in which the stochastic perturbation is not only a continuous one defined with the help of Brownian motion but contains Poisson-type terms. Let μ be a Poisson point process of order α as defined in section 15.11, let L be its Levy measure and let q be the random measure

$$q(\omega; ds, dx) := \mu(\omega; ds, dx) - ds \otimes L(dx) .$$

A Skorokhod equation is of the type

(33.2.4)
$$X_t = X_0 + \int_{]0,t]} b(s, X_s)ds + \int_{]0,t]} \sigma(s, X_s)d\beta_s + \\ + \iint_{]0,t] \times \mathbb{E}} F(s, X_{s-}, u)q(\,.\,, ds, du) ,$$

where b and σ are as in (33.2.2) and $F(s, x, u)$ is a $\mathscr{B}_{\mathbb{R}^+} \otimes \mathscr{B}_{\mathbb{R}^d} \otimes \mathscr{B}_{\mathbb{E}}$-measurable, $\mathbb{R}^d$-valued function.

A solution of (33.2.4) is an R.R.C. process X such that the second member of (33.2.4) is defined and is P-equal to X.

Skorokhod proved the existence and uniqueness (up to P-equivalence) of the solutions of (33.2.4) for a given initial condition X_0 and under the following "uniform Lipschitz condition". There exists a constant K such that, for every t, x and y,

(33.2.5)
$$\|b(t,x) - b(t,y)\|^2 + \|\sigma(t,x) - \sigma(t,y)\|^2 + \\ + \int_{\mathbb{E}} \|F(t,x,u) - F(t,y,u)\|^2 L(du) \leqslant K^2 \|x-y\|^2 .$$

This condition can also be weakened. This will follow in particular from statements below.

33.3 Ito-Skorokhod equations with memory

In general, authors (see in particular Liptzer and Shiryaev [LiS. 2]) have considered Ito equations in which the coefficients b and σ depend not only on t and the state of the process X at time t but on the past history of X before t. Thus,

$$dX_t = b(t, X)dt + \sigma(t, X)d\beta_t, \tag{33.3.1}$$

where $(b(t, X))_{t \geqslant 0}$ and $(\sigma(t, X))_{t \geqslant 0}$ are adapted processes such that $X_s = Y_s$ for every $s \leqslant t \Rightarrow b(t, X) = b(t, Y)$ and $\sigma(t, X) = \sigma(t, Y)$.

In their book Liptzer and Shiryaev show the existence and uniqueness of the solution under a Lipschitz hypothesis of the following type. There exist a constant K and a right-continuous increasing function L such that for every couple X, Y,

$$\begin{aligned} &||b(t, X) - b(t, Y)||^2 + ||\sigma(t, X) - \sigma(t, Y)||^2 \leqslant \\ &\leqslant \int_0^t ||X_s - Y_s||^2 dL_s + K ||X_t - Y_t||^2 \end{aligned} \tag{33.3.2}$$

They call the solution processes Ito processes.

Analogous generalizations have been made for Ito-Skorokhod equations. (See for example [Gal. 2]).

All these cases will be included in the situation considered in section 34.

33.4 Doleans-Dade-Protter equations

C. Doleans-Dade and Ph. Protter considered generalizations of the Ito equation where the processes t and β are respectively replaced by a $\mathbb{R}^q$-valued process A with finite variation and a $\mathbb{R}^p$-valued local martingale. Such an equation reads

$$dX_t = b(t, X)dA_t + \sigma(t, X)dM_t \tag{33.4.1}$$

or, better, in the integral form

$$X_t = X_0 + \int_{]0, t]} b(s, X)dA_s + \int_{]0, t]} \sigma(s, X)dM_s \tag{33.4.2}$$

where b and σ are random functionals of the process X. This means, that for every R.R.C. $\mathbb{R}^d$-valued process X $(b(t, X))_{t \geqslant 0}$ and $(\sigma(t, X))_{t \geqslant 0}$ are predictable processes with values in $\mathscr{L}(\mathbb{R}^q, \mathbb{R}^d)$ and $\mathscr{L}(\mathbb{R}^p, \mathbb{R}^d)$ respectively, with the property $Y_s = X_s$ for every $s < t \Rightarrow b(s, X) = b(t, Y)$ a.s. and $\sigma(t, X) = \sigma(t, Y)$ *a.s.*

A solution X of (33.4.2) is an $\mathbb{R}^d$-valued R.R.C. process such that the second member of (33.4.2) is defined and is a process P-equal to X.

Existence and uniqueness of the solution of (33.4.2) were obtained separately by

C. Doléans-Dade and Ph. Protter under a Lipschitz hypothesis in C. Doléans-Dade [Dol. 4], Ph. Protter [Pro. 1]. See also C. Doléans-Dade and P. A. Meyer [DoM. 2].

An account of these results with different methods extending to Hilbert-valued processes is given in the book by M. Metivier and J. Pellaumail [MeP. 1]. Because of their simplicity we will follow the same path as in this latter book in a context including Skorokhod-type perturbation terms.

To conclude this introduction to various types of stochastic equations, let us remark that in chapter 6 we considered a particular case of the above equations, namely, the equation

$$X_t = X_0 + \int_{]0,t]} X_{s^-} dZ_s ,$$

where Z_s is a semimartingale.

We gave an explicit form of the solution X_t which is the exponential $\mathscr{E}(Z)$ for $X_0 = 1$.

Let us also remark that if Z is the $\mathbb{R}^q \times \mathbb{R}^p$-valued semimartingale (A, M) the equation (33.4.2) has the form

$$X_t = X_0 + \int_{]0,t]} a(s, X) dZ_s . \tag{33.4.3}$$

It is actually under this form that it was considered by C. Doleans-Dade and in [Jac. 2].

33.5 Remark. The above examples clearly show that the so-called stochastic differential equations must actually be interpreted as stochastic integral equations.

We shall therefore speak rather of stochastic integral equations or, in short, of stochastic equations.

34. Strong solutions under Lipschitz hypotheses

34.1 The equations under consideration

We are given a stochastic basis $(\Omega, \mathfrak{A}, (\mathscr{F}_t)_{t \geq 0}, P)$ satisfying the usual hypothesis, on this basis a $\mathbb{G}$-valued semimartingale S, $\mathbb{G}$ being a separable Hilbert space and a locally white random measure $q\{\omega, ds, dx\}$ of order p on the locally compact space $\mathbb{E}$ (see section 31).

Let $\mathbb{H}$ be a separable Hilbert space. With every R.R.C. $\mathbb{H}$-valued process X is associated an $\mathscr{L}(\mathbb{G}; \mathbb{H})$-valued predictable process $(a^1(t, X))_{t \geq 0}$ and a $\mathscr{P} \otimes \mathscr{B}_{\mathbb{E}}$-measurable, $\mathbb{H}$-valued function $(t, \omega, u) \curvearrowright a^2(t, \omega, X, u)$ having the following property.

(34.1.1) If X and Y are two $\mathbb{H}$-valued R.R.C. processes and τ is a stopping time such that $X 1_{[0,\tau[} = Y 1_{[0,\tau[}$, then

$$1_{[0,\tau]}a^1(.\,,X) = 1_{[0,\tau]}a^1(.\,,Y)$$

and

$$1_{[0,\tau]}a^2(.\,,.\,,X,.) = 1_{[0,\tau]}a^2(.\,,.\,,Y,.).$$

V being a given $\mathbb{H}$-valued R.R.C. process, the stochastic integral equation which will be considered in this section is the following.

$$X_t = V_t + \int_{]0,t]} a^1(s,X)dS_s + \iint_{]0,t]\times\mathbb{E}} a^2(s,.\,,X,u)q(.\,;ds,du) \tag{34.1.2}$$

Notation. The process $(a^1(t,X))_{t\geqslant 0}$ will often be abbreviated to $a^1(X)$ and the function $(t,\omega,u)\frown a^2(t,\omega,X,u)$ denoted by $a^2(X)$.

34.2 Definition. *Strong solution on a closed stochastic interval.* An $\mathbb{H}$-valued R.R.C. continuous process defined on the stochastic interval $[0,\sigma]$ is said to be a strong solution of (34.1.2) on $[0,\sigma]$ if the integral on the right of (34.1.2) is defined on $[0,\sigma]$ and, on this interval, is a process P-equal to X.

34.3 Remarks. (1°) This definition makes perfect sense since the existence of the stochastic integral processes on $[0,\sigma]$ involves the values of $a^1(.\,,X)$ and $a^2(.\,,.\,,X,.)$ only on $[0,\sigma]$ and because these values depend only on the values of X on $[0,\sigma[$.

(2°) As to condition (34.1.1), it is clear that it is satisfied by the coefficients F of equations (33.2.4) since, in this case, $a^2(s,X,u)$ depends only on X_{s-}. As to the coefficients b an σ, it should be remarked that the continuity in x of the functions $b(t,x)$ and $\sigma(t,x)$ and the particular properties of integration with respect to ds and to Brownian motion allow one to write (33.2.4) in the equivalent form

$$X_t = X_0 + \int_{]0,t]} b(s,X_{s-})ds + \int_{]0,t]} \sigma(s,X_{s-})d\beta_s + \\ + \iint_{]0,t]\times\mathbb{E}} F(s,X_{s-},u)\,q(.\,;ds,du) \tag{34.3.1}$$

34.4 Definition. *Strong solutions on an open stochastic interval.* An $\mathbb{H}$-valued R.R.C. process defined on the open stochastic interval $[0,\sigma[$ is called a strong solution of (34.1.2) on $[0,\sigma[$ if there exists a family $\mathcal{T}$ of stopping times such that $[0,\sigma[\subset \bigcup_{\tau\in\mathcal{T}} [0,\tau]$ and X is a strong solution of (34.1.2) on every $[0,\tau]$, $\tau\in\mathcal{T}$.

34.5 Lipschitz conditions. We will consider the following extended Lipschitz conditions.

$[L_1]$ *(Global Lipschitz condition)*

For every control process A of S (see section 23) there exists an increasing, positive, right-continuous, adapted process L such that, for every couple X and Y of $\mathbb{H}$-valued R.R.C. process and every $t\in\mathbb{R}^+$,

(34.5.1)
$$\int_{]0,t]} \|a^1(s,Y)-a^1(s,X)\|^2_{\mathscr{L}(\mathbb{G},\mathbb{H})}\,dA_s + \int_{]0,t]} \lambda_s(a^2(X)-a^2(Y))\,db(s) \leqslant \leqslant \int_{]0,t]} \sup_{r<s} \|X_r - Y_r\|^2_{\mathbb{H}}\,dL_s\,,$$

where λ and b are respectively the functional and the increasing processes associated with q as in section 31. (See 31.9).

Remark. We recall that, when μ is a Poisson point process with Levy measure L and $q = \mu - dt \otimes L$, $db(s) = \bar{\alpha}ds$, where $\bar{\alpha}$ is the intensity of the Poisson process and $\lambda_s(a^2(X)-a^2(Y)) = \frac{1}{\bar{\alpha}} \int_{\mathbb{E}} \|a^2(s,.,X,u) - a^2(s,.,Y,u)\|^2_{\mathbb{H}} L(du)$

[L_2] *(Local Lipschitz condition)*

For every $\gamma > 0$ and every $A \in \mathscr{A}(S)$ there exists an increasing, right-continuous, positive adapted process L^γ such that, for every couple X and Y of $\mathbb{H}$-valued R.R.C. processes and every $t \in \mathbb{R}^+$,

(34.5.2)
$$\int_{]0,t]} \|a^1(s,Y)-a^1(s,X)\|^2\,dA_s + \int_{]0,t]} \lambda_s(a^2(X)-a^2(Y))\,db(s) \leqslant \leqslant \int_{]0,t]} \sup_{r<s} \|X_u - Y_u\|^2_{\mathbb{H}}\,dL^\gamma_s$$

for every $\omega \in \{\omega : \sup_{s<t} (\|Y_s(\omega)\| \vee \|X_s(\omega)\| \leqslant \gamma)$.

34.6 Classical forms of the Lipschitz conditions

To make clear the meaning of conditions [L_1] and [L_2] we remind the reader of classical Lipschitz conditions which trivially imply either [L_1] or [L_2].

[$\tilde{L}_1$] *(Classical global Lipschitz condition)*

There exists a constant K such that, for every couple X and Y of $\mathbb{H}$-valued R.R.C. process and every $t \in \mathbb{R}^+$,

(34.6.1)
$$\|a^1(t,Y)-a^1(t,X)\|^2 + \lambda_t(a^2(X)-a^2(Y)) \leqslant K \sup_{s<t} \|X_s - Y_s\|^2$$

[$\tilde{L}_2$] *(Classical local Lipschitz condition)*

For every $\gamma > 0$ there exists a constant K_γ such that, for every couple X and Y of $\mathbb{H}$-valued R.R.C. processes and every t,

(34.6.2)
$$\|a^1(t,Y)-a^1(t,X)\|^2 + \lambda_t(a^2(X)-a^2(Y)) \leqslant K_\gamma \sup_{s<t} \|X_s - Y_s\|^2$$

on the event $\{\omega : \sup_{s<t} \|Y_s(\omega)\| \vee \|X_s(\omega)\| \leqslant \gamma\}$.

Conditions of the type (33.3.2) as considered by Liptzer and Shiryaev can clearly be written in the form [L_1].

The most classical Lipschitz case is the case of Ito equations where the coefficients $\sigma(t,x)$ and $b(t,x)$ are continuously differentiable functions of x. If the first derivatives are bounded, we have condition $[\tilde{L}_1]$, and if they are continuous unbounded, we have only condition $[\tilde{L}_2]$. (Remember that the Ito equation (33.2.2) can be equivalently written

$$X_t = X_0 + \int_{]0,t]} b(s, X_{s^-})ds + \int_{]0,t]} \sigma(s, X_{s^-})d\beta_s .$$

See remark 34.3, (2°).

34.7 Theorem *(Existence and uniqueness of strong solutions). Besides the general hypothesis of section 34.1 we assume the local Lipschitz condition* $[L_2]$. *Then there exists a P.a.s. uniquely defined stopping time* τ *and a unique (up to P-equality) process* X *on* $[0, \tau[$ *such that*

(i) *X is a strong solution of (34.1.2) on $[0,\tau[$;*

(ii) *for every strong solution X' of (34.1.2) on a closed stochastic interval $[0,\sigma]$, one has $[0,\sigma] \subset [0,\tau[$ and $X' = X$ on $[0,\sigma]$;*

(iii) *τ is predictable and* $\limsup_{t\to\tau} \|X_t\| = +\infty$ *on* $\{\tau < \infty\}$.

Proof. The proof (similar to the one in [MeP. 4]) will be carried out in several steps in a sequence of lemmas.

Before stating and proving these lemmas let us introduce the following abbreviated notation.

If X is an $\mathbb{H}$-valued R.R.C. process such that the stochastic integrals in the second member make sense on a stochastic interval $[0,\sigma]$, we call $T(X)$ the stochastic process defined on $[0,\sigma]$ by this expression. In other words,

(34.7.1) $$T_t^\sigma(X) := V_t + \int_{]0,t]} a^1(s, X)dS_s + \iint_{]0,t]\times\mathbb{E}} a^2(s,.,X,u)q(.\,; ds, du)$$

for $t \leqslant \sigma$.

We also define, for every $\mathscr{L}(\mathbb{G};\mathbb{H})$-valued predictable process α^1 and every $\mathscr{P}\otimes\mathscr{B}$ -measurable $\mathbb{H}$-valued function α^2, A and b being the increasing processes entering $[L_2]$,

(34.7.2) $$\tilde{b}_s := A_s + b_s ,$$

(34.7.3) $$\tilde{\lambda}_s(\alpha^1, \alpha^2) := \|\alpha_s^1\|^2 \frac{dA_s}{d\tilde{b}_s} + \lambda_s(\alpha^2)\frac{db_s}{d\tilde{b}_s},$$

(34.7.4) $$\tilde{A}_t = 2(A_t \vee 1).$$

With this notation one may concisely write the inequality in $[L_2]$ as

(34.7.5) $$\int_{]0,t]} \tilde{\lambda}_s(a^1(X)-a^1(Y), a^2(X)-a^2(Y))\, d\tilde{b}_s \leqslant \int_{]0,t]} \sup_{r\leqslant s} \|X_r - Y_r\|_{\mathbb{H}}^2\, dL_s^\gamma$$

and the properties of stochastic integrals show that, for every stopping time τ and every X such that the right-hand side of (34.1.2) is defined, the following inequality holds.

$$(34.7.6) \qquad E(\sup_{s<\tau}[\| \int_{]0,t]} a^1(s,X)\,dS_s\|^2_{\mathbb{H}} + \| \iint_{]0,t]\times\mathbb{E}} a^2(s,.,X,u)\,q(.\,;ds,du)\|^2_{\mathbb{H}}]) \leqslant$$
$$\leqslant E(\tilde{A}_{\tau-} \int_{[0,\tau[} \tilde{\lambda}_s(a^1(X),a^2(X))\,d\tilde{b}_s).$$

The main steps of the proof are as follows.

(1°) Proof of uniqueness (Lemma 34.8).

(2°) Proof of local existence and a principle of extension (Lemmas 34.9 and 34.10). The set $\mathfrak{S}$ of couples (Y,τ), where Y is a strong solution on $[0,\tau]$, is not empty and if $P\{\tau<\infty\}>0$, it is always possible to extend Y to a strong solution on a larger stochastic interval $[0,\sigma]$ with $P\{\tau<\sigma<\infty\}$ arbitrarily close to $P\{\tau<\infty\}$.

(3°) A maximal solution (X,τ) is defined with τ the essential supremum of $\{\sigma: (Y,\sigma)\in\mathfrak{S}\}$. The existence of a finite $\limsup_{t\uparrow\tau}\|X_t\|$ is shown to contradict the extension principle.

34.8 Lemma *(Uniqueness of solutions). Let X and Y be two strong solutions of (34.1.2) on a stochastic interval $[0,\sigma]$. Then $X=Y$ (P.a.s.) on $[0,\sigma]$.*

Proof. Let A be a control process of S such that $a^1(X)$ and $a^1(Y)$ are both A-integrable on $[0,\tau]$ with respect to S. If X and Y are both solutions on $[0,\sigma]$ and using the inequality (34.7.6), then, for every $\tau\leqslant\sigma$,

$$(34.8.1) \qquad E(\sup_{t<\tau}\|X_t-Y_t\|^2)\leqslant E(\tilde{A}_{\tau-}\int_{]0,\tau[}\tilde{\lambda}_s(a^1(X)-a^1(Y),a^2(X)-a^2(Y))\,d\tilde{b}_s)$$

If we define the stopping time

$$\tau_n := \inf\{t:\|X_t\|+\|Y_t\|>n\}\wedge\inf\{t:\tilde{A}_t+L_t>n\}\wedge\sigma,$$

we see that $\lim_{n\to\infty} P\{\tau_n<\sigma\}=0$ and that it is therefore enough to prove the equality of X and Y on $[0,\tau_n]$ for each n.

But the inequality (34.8.1) and the local Lipschitz condition $[L_2]$ give, for every $\tau\leqslant\tau_n$,

$$(34.8.2) \qquad E(\sup_{t<\tau}\|X_t-Y_t\|^2)\leqslant n\,E\int_{]0,\tau[}(\sup_{s\leqslant t}\|X_s-Y_s\|^2)\,dL_r.$$

We may therefore apply the "Gronwall Lemma" proved in section 29.1 which now gives

$$E(\sup_{t<\tau_n}\|X_t-Y_t\|^2)=0.$$

But the equality of X and Y on $[0,\tau_n[$ implies the equality of X and Y on $[0,\tau_n]$ because of

$$\Delta X_{\tau_n} = \Delta V_{\tau_n} + a^1_{\tau_n}(X)\Delta S_{\tau_n} + \int_{\mathbb{E}} a^2(\tau_n, X, u)q(.,\{\tau_n\}, du)$$

and of property (34.1.1) of the functionals a^1 and a^2.

This proves the lemma. □

34.9 Lemma. *Let X be a strong solution of (34.7.2) on $[0, \tau]$ and X' be a strong solution on $[0, \tau']$. Then there exists a strong solution Y on $[0, \tau' \vee \tau]$.*

Proof. Because of the uniqueness Lemma, 34.8 X and X' coincide on $[0, \tau \wedge \tau']$. It is therefore clear that the process Y defined on $[0, \tau \vee \tau']$ by

$$Y := \begin{cases} X & \text{on } [0, \tau] \\ X' & \text{on } [0, \tau'] \end{cases}$$

is a strong solution of (34.7.2). □

34.10 Lemma *(Extension principle, local existence under $[L_1]$). Let us assume $[L_1]$ instead of $[L_2]$ and that (34.1.2) has a strong solution ξ on a closed stochastic interval $[0, \tau]$ (τ may be identically 0. In this case, $\xi_0 = V_0$).*

Let us assume $\alpha := P\{\tau < \infty\} > 0$ and let $\varepsilon > 0$ with $0 < \varepsilon < \alpha$. Then there exists a stopping time σ and a strong solution X on $[0, \sigma]$ such that

(a) $P\{\sigma > \tau\} > \alpha - \varepsilon$

(b) *X is a strong solution on $[0, \sigma]$ with $1_{[0,\tau]}X = 1_{[0,\tau]}\xi$.*

Proof. Let σ be a stopping time with $\sigma \geqslant \tau$. We define the following complete metric space $\mathbb{M}^\sigma$ of $\mathbb{H}$-valued R.R.C. processes on $[0, \sigma[$.

$$\mathbb{M}^\sigma := \{X : X\ \mathbb{H}\text{-valued R.R.C.}, X 1_{[0,\tau]} = \xi 1_{[0,\tau]}, E(\sup_{0<t<\sigma} \|X_t\|^2 1_{\{\sigma>\tau\}} < \infty)\}$$

with the distance

$$\delta(X, Y) := [E(\sup_{\tau \leqslant t < \sigma} \|X_t - Y_t\|^2)]^{\frac{1}{2}}$$

We intend to define σ in such a way that $P\{\sigma > \tau\} > \alpha - \varepsilon$ and the mapping $X \curvearrowright T^\sigma(X)$ (see (34.7.1)) is a contraction from $\mathbb{M}^\sigma$ into $\mathbb{M}^\sigma$ for the distance δ. The existence of X such that $X_t = T^\sigma_t(X)$ for $t < \sigma$ will follow. Since the definition of $T^\sigma_\sigma(X)$ depends only on the values of X on $[0, \sigma[$, we find the solution X on $[0, \sigma]$ by setting $X_\sigma = T^\sigma_\sigma(X)$.

The lemma will then be proved.

(1°) *Let us prove now that T^σ maps $\mathbb{M}^\sigma$ into $\mathbb{M}^\sigma$.*

Let us also call ξ a right-continuous process on $\mathbb{R}^+$ equal to the given ξ on $[0, \tau]$. Let l be a positive number. We define

$$F_\ell := \{\sup_{s\leqslant\tau} \|\xi_s\|^2 + \|V\|_\tau^2 + \tilde{A}_\tau \int_{]0,\tau]} \tilde{\lambda}_s(a^1(\xi), a^2(\xi))d\tilde{b}_s \leqslant \ell\} \cap \{\tau < \infty\}.$$

Since the random variable inside the parentheses is finite by definition of the stochastic integrability, we can choose ℓ in such a way that $P(F_\ell) > \alpha - \varepsilon$ with $0 < \varepsilon < \alpha$.

We set

(34.10.1) $$\sigma' := \inf\{t : t > \tau, \tilde{A}_t(L_\tau - L_\tau) > \tfrac{1}{2}\} \wedge \inf\{t : t > \tau, \|\xi_t\|^2 + \|V_t\|^2 + $$
$$+ \tilde{A}_t \int_{]0,t]} \tilde{\lambda}_s(a^1(\xi), a^2(\xi))d\tilde{b}_s > 2\ell\} \wedge \inf\{t : t > \tau, \|V_t - V_\tau\|^2 > \ell\}$$

Since F_ℓ is $\mathscr{F}_\tau$-measurable we obtain a stopping time σ by setting

(34.10.2) $$\sigma := \begin{cases} \sigma' \text{ on } F_\ell \\ \tau \text{ on } F_\ell^c \end{cases}$$

The right-continuity of A and L implies $\{\sigma > \tau\} = F_\ell$ and therefore $P\{\sigma > \tau\} > \alpha - \varepsilon$.

With these definitions, $\xi \in \mathbb{M}^\sigma$. The set $\mathbb{M}^\sigma$ is therefore not empty.

Let X be an element of $\mathbb{M}^\sigma$. Using the inequality (37.7.6) and the equality $X = \xi$ on $[0, \tau]$ we obtain

$$E(\sup_{0\leqslant t<\sigma} \|T_t^\sigma(X)\|^2 1_{\{\sigma>\tau\}}) \leqslant 4E(\sup_{0\leqslant t<\tau} \|\xi_t\|^2 1_{\{\sigma>\tau\}}) + 4E(\sup_{\tau\leqslant t<\sigma} \|V_t - V_\tau\|^2)$$
$$+ 4E(\tilde{A}_{\sigma-} \int_{[\tau,\sigma[} \tilde{\lambda}_s(a^1(X) - a^1(\xi), a^2(X) - a^2(\xi))d\tilde{b}_s +$$
$$+ 4E(A_{\sigma-} \int_{[\tau,\sigma[} \tilde{\lambda}_s(a^1(\xi), a^2(\xi))d\tilde{b}_s)$$

Using then $\{\sigma > \tau\} = F_l$, the definition of F_l and σ and condition $[L_1]$ in the form (34.7.5) we obtain

$$E(\sup_{0\leqslant t<\sigma} \|T_t^\sigma(X)\|^2 1_{\{\sigma>\tau\}}) \leqslant 16l + 4E(\tilde{A}_{\sigma-} \int_{[\tau,\sigma[} \sup_{r\leqslant s} \|X_r - \xi_r\|^2 dL_s)$$

However the stopping time σ has been defined in such a way that $A_{\sigma-}(L_{\sigma-} - L_\tau) \leqslant \frac{1}{2}$. We may therefore write

$$E(\sup_{0<t<\sigma} \|T_t^\sigma(X)\|^2 1_{\{\sigma>\tau\}} \leqslant 16l + \tfrac{4}{2} E(\sup_{s<\sigma} \|X_s - \xi_s\|^2 1_{\{\tau<\sigma\}}).$$

This inequality shows that $T^\sigma(X) \in \mathbb{M}^\sigma$.

(2°) *Let us prove that T^σ is a contraction on $\mathbb{M}^\sigma$.*

Let us consider X and X' in $\mathbb{M}^\sigma$.

$$[\delta(T^\sigma(X), T^\sigma(X'))]^2 = E(\sup_{\tau \leqslant t < \sigma} \|T^\sigma(X) - T^\sigma(X')\|^2)$$

$$= E(\sup_{\tau \leqslant t < \sigma} \| \int_{]0,t]} (a^1(X) - a^1(X')dS + \int_{]0,t]} (a^2(X) -$$

$$- a^2(X'))dq\|^2)$$

Using the equality of X and X' on $]0, \tau]$ and the inequality for stochastic integrals, expressed in the form (34.7.6) we get

$$[\delta(T^\sigma(X), T^\sigma(X'))]^2 \leqslant E(\tilde{A}_{\sigma^-} \int_{[\tau,\sigma[} \tilde{\lambda}_s(a^1(X) - a^1(X'), a^2(X) - a^2(X'))d\tilde{b}_s).$$

The Lipschitz condition $[L_1]$ and the definition of σ then show immediately that

$$[\delta(T^\sigma(X), T^\sigma(X')]^2 \leqslant \tfrac{1}{2}\delta(X, X')$$

and thus yield the contraction property for T^σ.

The lemma is thus proved. □

34.11 Lemma *(Extension principle and local existence under $[L_2]$. The conclusions of Lemma 34.10 are still valid if the uniform condition $[L_1]$ is replaced by the local condition $[L_2]$.*

Proof. Let τ, ξ and l be as in the proof of Lemma 34.10.

For every $\beta > 0$ and every R.R.C. process Y we define a new process Y^β by setting

(34.11.1) $$Y_s^\beta(\omega) := \left(1 \wedge \frac{\beta}{\|Y_s(\omega)\|}\right) Y_s(\omega)$$

and we define the functionals ($i = 1, 2$)

(34.11.2) $$a_s^{i,\beta}(Y) := a_s^i(Y^\beta).$$

Since Y^β is bounded by β and holds the inequality $\|\left(1 \wedge \frac{\beta}{\|y\|}\right) y - \left(1 \wedge \frac{\beta}{\|y'\|}\right) y'\| \leqslant$ $\leqslant \|y - y'\|$ holds for every $y, y' \in \mathbb{H}$ the stochastic equation

(34.11.3) $$X_t = V_t + \int_{]0,t]} a^{1,\beta}(X)dS + \int_{]0,t]} a^{2,\beta}(X)dq$$

satisfies the hypothesis of Lemma 34.8 with uniform Lipschitz condition $[L_1]$, namely, that for every X and Y,

(34.11.4) $$\tilde{\lambda}_s(a^{1,\beta}(X) - a^{1,\beta}(Y), a^{2,\beta}(X) - a^{2,\beta}(Y)) \leqslant \int_{]0,t]} \sup_{r \leqslant s} \|X_r - Y_r\|^2 dL_s^\beta.$$

Let us consider

(34.11.5) $$\tau' := \inf\left\{t : \|\xi_t\|^2 > \frac{\beta}{2}\right\} \wedge \tau .$$

We now choose β in such a way that $\beta > 4l$.

Since, by definition, the processes $a^i(\xi)$ and $a^{i,\beta}(\xi)$ are equal on $[0, \tau']$, ξ is a solution of (34.11.3) on $[0, \tau']$. Since $\beta > 2l$, $F_l \subset \{\tau' = \tau\}$.

Let us now define σ' and σ as in (34.10.1) and (24.10.2), replacing only τ by τ', L by L^β and a^i by $a^{i,\beta}$. One has $\{\sigma > \tau'\} = F_l$, which implies $\{\sigma > \tau'\} = \{\sigma > \tau\} = F_l$ and therefore

(34.11.6) $$P\{\sigma > \tau\} = P(F_l) > \alpha - \varepsilon .$$

The proof of Lemma 34.10 shows the existence of a solution X of (34.11.3) on $[0, \sigma]$, which is an extension of ξ on $[0, \tau']$. Define now

$$\sigma_1 := \inf\{t : t > \tau', \|X_t\| > \beta\} \wedge \sigma .$$

The right-continuity of X and the property $\|X\|_\tau = \|X\|_{\tau'} \leqslant l < \beta$ on $F_l = \{\sigma > \tau\}$ imply $\{\sigma_1 > \tau'\} = \{\sigma_1 > \tau\} = F_l$. Therefore, $P\{\sigma_1 > \tau'\} > \alpha - \varepsilon$. Since, moreover, $\|X\| \leqslant \beta$ on $[0, \sigma_1[$ the process X is also a solution of (34.1.2) on $[0, \sigma_1]$. Applying Lemma 34.9 we obtain the existence of a solution Y of (34.1.2) on $[0, \sigma_1 \vee \tau]$ which coincides with ξ on $[0, \tau]$. This proves the lemma. □

34.12 End of proof of Theorem 34.7. We consider the set $\mathfrak{S}$ of all pairs (Y, σ), where Y is an R.R.C. $\mathbb{H}$-valued process on $[0, \sigma]$ such that Y is a strong solution of (34.1.2) on $[0, \sigma]$. This set is non-empty in view of 34.11. Lemma 34.9 shows that if $(Y_1, \sigma_1) \in \mathfrak{S}$ and $(Y_2, \sigma_2) \in \mathfrak{S}$, there exists $(Y, \sigma_1 \vee \sigma_2) \in \mathfrak{S}$. This shows that the family $\{\sigma; \sigma$ stopping time, $\exists (Y, \sigma) \in \mathfrak{S}\}$ is directed for the order relation $\leqslant$. If we set

(34.12.1) $$\tau := ess.\ \sup.\ \{\sigma : (Y, \sigma) \in \mathfrak{S}\}$$

there exists an increasing sequence (σ_n) of stopping times and for each n a strong solution Y_n of (34.1.2) on $[0, \sigma_n]$ such that

(34.12.2) $$\tau = \lim_{n \to \infty} \sigma_n \ a.s.$$

with $Y_m = Y_n$ on $[0, \sigma_m]$ for $m \leqslant n$.

According to definition 34.3 this shows the existence of a strong solution Y on $[0, \tau[$, which is necessarily unique, as a consequence of 34.8.

We have now only to prove (iii). Let us therefore consider the following stopping times.

$$\tau_{n,k} := \sigma_n \wedge \inf\{t : \|X'_t\| > k\} \quad n, k \in \mathbb{N},$$

$$\tau_k := \sup_n \tau_{n,k} .$$

We have $\tau_k \leqslant \tau$, $\lim \tau_k = \tau$. We shall prove

(34.12.3) $P\{\tau_k = \tau < \infty\} = 0.$

It will follow that τ is predictable.

Since $\tau_k < \tau$ implies $\tau_k = \tau_{n,k}$ for n big enough, $\|X_{\tau_k}\| \geqslant k$ on $\{\tau_k < \tau\}$. This will immediately imply (iii).

If (34.12.3) were not true, there would exist $k \in \mathbb{N}$ such that

$$P\{\tau_k = \tau < \infty\} = \alpha > 0.$$

Applying Lemma 34.11 we may find a stopping time τ' with $P(\{\tau' > \tau_k\} \cap \{\tau_k = \tau < \infty\}) > 0$ and a strong solution X on $[0, \tau']$. However, since $[0, \tau']$ would not be contained in $[0, \tau]$, this would contradict the definition of τ. The inequality (34.12.3) is therefore established and this completes the proof of the theorem. □

35. Conditions for non-explosion

We give a condition for "non-explosion".

35.1 Explosion times

Let X be a strong solution of equation (34.1.2) on $[0, \tau[$. If $\limsup_{t \to \tau(\omega)} \|X_t(\omega)\| = +\infty$ for each ω such that $\tau(\omega) < \infty$ and if $P\{\tau < \infty\} > 0$, the stopping time τ is called an *explosion time.*

Property (iii) of Theorem 34.7 states that, if (X, τ) is a maximal strong solution of (34.1.2), and if $P\{\tau < \infty\} > 0$, τ is an explosion time.

We say that the strong solutions have no explosion if for any maximal solution (X, τ), $P\{\tau = \infty\} = 1$.

35.2 Theorem *(Sufficient condition for non explosion). We consider the equation (34.1.2) with the hypothesis of Theorem 34.7. We assume, moreover, the following condition.*

[K] *For every control process A of S, there exists an increasing, positive, right-continuous, adapted process L such that, for every R.R.C. $\mathbb{H}$-valued process X and every $t \in \mathbb{R}^+$,*

(35.2.1)
$$\int_{]0,t]} \|a^1(s, X)\|^2_{\mathscr{L}(\mathbb{G};\,\mathbb{H})} dA_s + \int_{]0,t]} \lambda_s(a^2(X))\, db(s) \leqslant \int_{]0,t]} (1 + \sup_{u<s} \|X_u\|^2) dL_s$$

Then any strong solution of (34.1.2) has no explosion. In other words the equation (34.1.2) has a unique strong solution defined on $\mathbb{R}^+ \times \Omega$.

Proof. Let us write $X_t^* := \sup_{s<t} \|X_s\|$ and $V_t^* := \sup_{s<t} \|V_s\|$.

We show that, under the hypothesis of the theorem, $P\{\tau \leqslant \alpha\} = 0$ for every $x \in \mathbb{B}$ This will prove the theorem.

Let A be a control process of Z. We set $\tilde{A} := 2(A_t \wedge 1)$ as in section 34 and define the following sequence $(\tau_n)_{n \geqslant 1}$ of stopping times.

$$\tau_n := \inf\{t : L_t \vee V_t^* \vee A_t \vee \tilde{b}_t > n\} \wedge \tau \wedge \alpha,$$

where $\tilde{b}$ is defined by (34.7.2). Clearly,

$$\lim_{n\to\infty} P\{\tau_n = \tau \leqslant \alpha\} = P\{\tau \leqslant \alpha\} \tag{35.2.2}$$

By definition,

$$X_t = V_t + \int_{]0,t]} a^1(s, X)dS_s + \iint_{]0,t]\times\mathbb{E}} a^2(s, X, u)q(.\,, ds, du). \tag{35.2.3}$$

Using the notation of section 34 and, in particular, the inequality (34.7.6), we may write for every $\sigma < \tau_n$,

$$E(\sup_{t<\sigma} \|X_t\|^2) \leqslant 2E(|V_\sigma^*|^2) + 2E(\tilde{A}_{\sigma-} \int_{[0,\sigma[} \tilde{\lambda}_s(a^1(X), a^2(X))d\tilde{b}_s) \tag{35.2.4}$$

and, as a consequence of the definition of τ_n and (35.2.1),

$$E((X_{\sigma-}^*)^2) \leqslant 2n^2 + 2nE \int_{[0,\sigma[} (1 + (X_s^*)^2)dL_s$$

$$\leqslant 4n^2 + 2n \int_{[0,\sigma[} (X_s^*)^2\, dL_s$$

We may how apply Lemma 29.1, which gives

$$E((X_{\tau_n}^*)^2) \leqslant 8n^2 \sum_{j=0}^{4n^2} (4n^2)^j.$$

Therefore,

$$\limsup_{t\to\tau_n} \|X_t\|^2 < \infty$$

and this implies

$$P\{\tau_n = \tau \leqslant \alpha\} = 0.$$

In view of (35.2.2), the theorem is thus proved. □

We give now a sufficient condition for non-explosion in a more traditional form.

35.3 Corollary. *The following condition is sufficient for the non-explosion of the strong solution as in Theorem 34.7.*

$[K_1]$ *There exists a constant K such that for every $t \geqslant 0$ and every regular process X,*

$$\|a^1(t, X)\|^2_{\mathscr{L}(\mathbb{G};\mathbb{H})} + \lambda_t(a^2(X) \leqslant K(1 + \sup_{s<t} \|X_s\|^2).$$

Proof. This is clearly a trivial consequence of Theorem 35.2. □

When the functionals $a^1(t, .)$ and $a^2(t, . , u)$ depend only on the values of X at time t^-, thus,

$$a^1(t, X) := \alpha^1(t, X_{t^-}),$$
$$a^2(t, X, u) := \alpha^2(t, X_{t^-}, u)$$

and q is a centered Poisson process with Levy measure L, the condition $[K_1]$ reads

(35.3.2) $$\|\alpha^1(t, X_{t^-})\|^2 + \int_{\mathbb{E}} \|\alpha^2(t, X_{t^-}, u)\|^2 L(du) \leqslant K(1 + \sup_{s<t} \|X_s\|^2).$$

This condition is usually referred to as a condition of "at most linear increase at infinity".

35.4 Remarks. Let us assume that for one process Y (for example for $Y = 0$) the condition (35.2.1) holds and that the stochastic equation verifies the global Lipschitz condition $[L_1]$ of section 34.5. Then condition $[K]$ is true for every $\mathbb{H}$-valued R.R.C. process.

The most classical case is when

$$a^1(t, X) = \alpha^1(t, X_{t^-}),$$
$$a^2(t, X, u) = \alpha^2(t, X_{t^-}, u),$$

where α_1 (resp. α_2) is a measurable mapping from $(\mathbb{R}_+ \times \mathbb{H})$ (resp. $(\mathbb{R}_+ \times \mathbb{H} \times \mathbb{E})$) into $\mathscr{L}(\mathbb{G}; \mathbb{H})$ (resp. into $\mathbb{H}$).

If we assume the global Lipschitz conditions in the form

(35.4.1) $$\|a^1(t, x) - a^2(t, y)\|^2 \leqslant K\|x - y\|^2 \quad \text{for all} \quad t \geqslant 0, \quad x, y \in \mathbb{H},$$

(35.4.2) $$\int_{\mathbb{E}} < \alpha_2(t, x, u) - \alpha_2(t, y, u), \alpha_2(t, x, v) - \alpha_2(t, y, v) > \overset{\circ}{q}(s, . , du \otimes dv) \leqslant$$
$$\leqslant K\|x - y\|^2,$$

then condition (35.3.2) is automatically satisfied.

36. Pathwise regularity of solutions of equations depending on a parameter

36.1. The goal of this section is to prove that, under sufficient regularity conditions on the coefficients, stochastic differential equations of the type

$$(36.1.1)\qquad X^u(t) = V^u(t) + \int_{]0,t]} \sigma(u, s, X^u_{s-})dS_s + \iint_{]0,t]\times\mathbb{E}} f(u, s, X^u_{s-}, x)q(ds, dx)$$

where S is a $\mathbb{G}$-valued semimartingale, q a random measure with zero dual predictable projection and u a parameter taking its values in a bouned open subset G of $\mathbb{R}^d$, admit, for each u, a solution which can be determined in such a way that P.a.s. the functions $u \curvearrowright X^u(t, \omega)$ are continuous for every t and even continuously differentiable if more regularity is imposed on the coefficients.

36.2 Notation. To make the notation more concise we shall write Z for the couple (S, q) and g for the couple (σ, f). The equation (36.1.1) will therefore be written

$$(36.2.1)\qquad X^u(t) = V^u(t) + \int_{]0,t]} g(u, s, X^u_{s-})dZ_s.$$

Let $\tilde{\lambda}$ and $\tilde{b}$ be defined as in equalities (34.7.2) and (34.7.3). For every locally bounded predictable $\mathscr{L}(\mathbb{G}; \mathbb{H})$ valued process α_1 and every $\mathscr{P}\otimes\mathscr{B}_{\mathbb{E}}$ -measurable $\mathbb{H}$-valued function α_2, we write $\tilde{\lambda}(\alpha)$ or $\tilde{\lambda}\circ\alpha$ for $\tilde{\lambda}(\alpha_1, \alpha_2)$.

We shall make use of a slight extension of the notion of control process for a semimartingale. This extension is proposed as exercise E.5 at the end of this chapter.

This extension asserts the existence, for every $p \geqslant 2$, of an increasing process $\tilde{A}^p$ with the following property for every stopping time τ.

$$(36.2.2)\qquad E(\sup_{t<\tau} \| \int_{]0,t]} \alpha_s dZ_s \|^p) \leqslant E(\tilde{A}^p_{\tau-} \int_{]0,\tau[} |\tilde{\lambda}_s \circ \alpha|^{p/2} d\tilde{b}_s).$$

36.3 Hypothesis. We introduce the following hypotheses, which will ensure the continuity of X^u as a function of u. L is an increasing, positive, adapted process and p a positive real number with $p \geqslant d + \varepsilon$ for some $\varepsilon > 0$.

With this notation we formulate the following hypotheses.

(H_1)
$$\sup_{s\leqslant t} \|V^u_s - V^v_s\| \leqslant L_t \|u - v\| \quad \text{for all} \quad t \geqslant 0, \quad u \quad \text{and} \quad v \in G$$

and

$$\sup_{u\in\mathbb{G}\, s<t} \|V^u_t\| < \infty.$$

(H_2) (Lipschitz hypotheses)

$$\forall t\in\mathbb{R}^+ \qquad \int_{]0,t]} [\lambda_s \circ (g(u, \xi) - g(u, \xi'))]^{p/2} d\tilde{b}_s \leqslant \int_{]0,t]} \sup_{r\leqslant s} \|\xi_r - \xi'_r\|^p dL_s$$

for every couple (ξ, ξ') of $\mathbb{H}$-valued R.R.C. processes, P.a.s.

(H_3) $$\int_{]0,t]} [\lambda_s \circ g(u,\xi)]^{p/2}\, d\tilde{b}_s \leqslant \int_{]0,t]} (1 + \sup_{r \leqslant s} \|\xi_s\|^p)\, dL_s$$

for every $u \in G$ every $\mathbb{H}$-valued R.R.C. ξ, P.a.s.
(Note that (H_3) is implied by (H_2) in most classical cases; see 35.4)

(H_4) ψ being a given positive increasing (possibly constant) function on $\mathbb{R}^+$, for every stopping time τ the following inequality holds for every $\mathbb{H}$-valued R.R.C. ξ, every u and v in G.

$$E(\sup_{t<\tau} [\lambda_t \circ [g(u,\xi) - g(v,\xi)]]^{p/2}) \leqslant \|u - v\|^{d+\epsilon} \psi(E(\sup_{t<\tau} \|\xi_t\|^p))$$

36.4 Theorem (Continuity with respect to u). (1°) *Unter the above hypotheses (H_1) to (H_4), the equation (36.2.1) has, for each u, a unique strong solution X^u on $\mathbb{R}^+$ and the random function $(t, \omega, u) \curvearrowright X_t^u(\omega)$ can be determined in such a way that $u \curvearrowright X_t^u(\omega)$ is continuous on G for every t and ω, while the mapping $t \curvearrowright X_t(.)(\omega)$ is, for each ω, right-continuous from $\mathbb{R}^+$ into the set $C_b^{\mathbb{H}}(G)$ of bounded, continuous, $\mathbb{H}$-valued functions on G endowed with the uniform topology.*

(2°) *There exists an increasing sequence (σ_n) of stopping times and constants $K(\psi, n, p, Z)$ such that*

$$\text{a) } \lim_n P\{\sigma_n < T\} = 0 \quad \textit{for every} \quad T > 0,$$

$$\text{b) } E(\sup_{t<\sigma_n} \|X^v(t) - X^u(t)\|^p) \leqslant K(Y, n, p, Z) \|u - v\|^p.$$

A simple particular form of this theorem is the following

Corollary: *Let $(u, h) \rightsquigarrow \sigma(u, h)$ be a continuous mapping from $(\mathbb{G} \times \mathbb{H})$ into $\mathscr{L}(\mathbb{G}; \mathbb{H})$ and $(u, h, x) \curvearrowright F(u, h, x)$ a continuous mapping from $(\mathbb{G} \times \mathbb{H} \times \mathbb{E})$ into $\mathbb{H}$. We take for q a centered Poisson point process with Levy measure L. Assume the existence of a constant K such that, for every $u, v \in \mathbb{G}$ $h, h' \in \mathbb{H}$, the following inequality holds.*

$$\|\sigma(u,h) - \sigma(u,h')\|^2 + \int_{\mathbb{E}} \|F(u,h,x) - F(v,h',x)\|^2 L(dx) \leqslant K(\|u - v\|^2 + \|h - h'\|^2).$$

Assume also (H_1).

Then there exists a function $(u, t, \omega) \curvearrowright X(u, t, \omega)$ which is continuous in u for every (t, ω) and such that for every u the process $(X(u, t, .))_{t>0}$ is a regular solution of (36.1.1).

Proof. The stopping times σ_n are defined as follows.

$$\sigma_n := \inf\{t\colon \tilde{A}_t^p \vee L_t \vee \sup_{\substack{u \in G \\ s \leqslant t}} \|V_t^u\|^p \vee \tilde{b}_t > n\}$$

Next we use the following Lemmas.

36.5 Lemma 1.

$$E(\sup_{t<\sigma_n} \|X_t^u\|^p) \leqslant 2^p(n+n^2) \sum_{j=0}^{2^p n^2} (2^p n^2)^j$$

Proof of Lemma 1. We remark that $\tilde{b}^p_{\sigma_n^-} \leqslant n$, $L_{\sigma_n^-} \leqslant n$, $\sup_{t<\sigma_n} \sup_u \|V_t^u\|^p \leqslant n$.

We then apply inequality (36.2.2) to the second member of (36.2.1) and get

$$E(\sup_{t<\sigma_n} \|X_t^u\|^p) \leqslant 2^{(p-1)}n + 2^{(p-1)} E(\tilde{A}^p_{\sigma_n^-} \int_{]0,\sigma_n[} [\tilde{\lambda}_s \circ g(u, X^u)]^{p/2} d\tilde{b}_s)$$

and property (H_3) gives, for every stopping time $\tau \leqslant \sigma_n$,

$$E(\sup_{t<\sigma_n} \|X_t^u\|^p) \leqslant 2^{(p-1)}(n+n^2) + 2^{(p-1)} n E(\int_{]0,\tau[} (\sup_{s<t} \|X_s^u\|^p) dL_s)$$

Applying the "Gronwall Stochastic Lemma" 29.1 we obtain the inequality of the Lemma.

36.6 Lemma 2. *There exist constants* $K(\Psi, n, p, \tilde{b}, \tilde{A}^p)$ *such that*

$$\forall u, v \quad E(\sup_{t<\sigma_n} \|X_t^u - X_t^v\|^p) \leqslant K(\Psi, n, p, \tilde{b}, \tilde{A}^p)\|u-v\|^p.$$

Proof of Lemma 2. Applying again inequality (36.2.2) to the stochastic integrals

$$\int_{]0,t]} (g_s(u, X_{s-}^u) - g_s(v, X_{s-}^u)) dZ_s \qquad \text{and}$$

$$\int_{]0,t]} [g_s(v, X_{s-}^u) - g_s(v, X_{s-}^v)] dZ_s$$

and using properties (H_1), (H_2) and (H_4), we can write, for every stopping time $\tau \leqslant \sigma_n$,

$$E(\sup_{s<\tau} \|X^u(s) - X^v(s)\|^p) \leqslant 3^{p-1} n^p \|u-v\|^p + 3^{(p-1)} n \Psi(E(\sup_{s<\tau} \|X_s^u\|^p))$$
$$+ 3^{(p-1)} n E(\int_{]0,\tau[} (\sup_{t<s} \|X^u(s) - X^v(s)\|^p) dL_s)$$

Applying as above the same "Gronwall Inequality" we obtain the Lemma.

Theorem 36.4 is now a direct consequence of the following Lemma which is a straightforward extension of a Lemma as stated by Neveu in [Nev. 6] (see also P. Priouret [Pri], chap. 3, Lemma 13). The proof of this Lemma is proposed as exercise E.4 at the end of the chapter.

36.7 Lemma 3. *Let* $\{Y(t,\omega,u): t \in \mathbb{R}^+, \omega \in \Omega, u \in G\}$ *be an* $\mathbb{H}$*-valued random function such that, for every* u, $(Y(t,.,u))_{t \geqslant 0}$ *is a regular process and for every* t,

$$E(\sup_{s \leqslant t} \|Y_{s,u} - Y_{s,v}\|^p) \leqslant a_{t,p} \|u-v\|^{d+\varepsilon}.$$

Then there exists a mapping $Y^*: (t, \omega, u) \curvearrowright Y^*(t, \omega, u) \in \mathbb{H}$ *such that*

(a) $u \curvearrowright Y^*(t, \omega, u)$ *is continuous,*

(b) $\forall u \in G,\ Y(t, u, .) = Y^*(t, u, .)$ *for all* t *a.s.,*

(c) $t \curvearrowright Y^*(t, ., \omega)$ *is a right-continuous mapping from* $\mathbb{R}^+$ *into* $C_b^{\mathbb{H}}(G)$ *endowed with the topology of uniform convergence on compact subsets of* G.

This completes the proof of Theorem 36.4. □

36.8 Hypotheses for differentiability

We consider the same equation (36.1.1) or, in abreviated notation, (36.2.1).

For a couple $g := (\sigma, f)$ of "coefficients", as in (36.1.1), we write for simplicity

$$\|g(u, s, \omega, h, .)\|_\Lambda := [\|\sigma(u, s, \omega, h)\|^2_{\mathscr{L}(\mathbb{K};\, \mathbb{H})} + \int_{\mathbb{E}\times\mathbb{E}} \langle f(u, s, \omega, h, x), f(u, s, \omega, h, y)\rangle_{\mathbb{H}} \ \mathring{q}(\omega, s, dx \otimes dy)]^{\frac{1}{2}}.$$

We set $v_t^* := \sup_{u \in G} \sup_{s < t} \|D_u V_s^u\| + \|V_s^u\| + \|D_{u^2}^2 V_s^u\|$,

where $D_u \Phi$ denotes the first-order derivative and $D_{u^2}^2 \Phi$ the second-order derivative with respect to u of a function Φ of u.

In the hypotheses below C is a constant.

[D_1] For all t and ω the derivatives $D_u V^u(t, \omega)$ and $D_{u^2}^2 V^u(t, \omega)$ exist and $v_t^* < \infty$

(D_2] The derivatives $D_u g(s, u, x)$ $D_u^2 g(s, u, x)$ $D_{u,x} g(s, u, x)$ and $D_x g(s, u, x)$ exist and

$$\sup_{u, s, x} (\|D_u g(s, u, x)\|_\Lambda + \|D_{u^2}^2 g(s, u, x)\|_\Lambda + \|D_{ux}^2 g(s, u, x)\|_\Lambda + \|D_x g(s, u, x)\|_\Lambda) \leqslant 0$$

[D_3] For all x, y u and v

$$\|D_x g(s, u, x) - D_x g(s, v, y)\|_\Lambda \leqslant K_t(\|y - x\| + \|u - v\|)$$

36.9 Theorem. *Under the above hypotheses* [D_1] *to* [D_3] *equation* (36.2.1) *has a unique (up to P-equality) solution* X^u *on* $\mathbb{R}^+$ *and there exists a version* $(\omega, t, u) \curvearrowright X_t^u(\omega)$ *of this random function such that, for P-almost all* ω, *the following hold.*

(a) $u \curvearrowright X_t^u(\omega)$ *is continuously differentiable for every* t.

(b) $t \curvearrowright X_t^{(.)}(\omega)$ *and* $t \curvearrowright D_u X_t^{(.)}(\omega)$ *are right-continuous when the spaces* $C_b(G; \mathbb{H})$ *and* $C_b(G; \mathscr{L}(G; \mathbb{H}))$ *are endowed with the topology of uniform convergence on compact subsets of* G.

(c) *For every* u *the stochastic process* $(D_u X_t^u)_{t \geqslant 0}$ *is a strong solution of the following stochastic equation (where* X^u *is the process solution of* (36.2.1) *as in Theorem* 36.4

(36.9.1) $$Y^u(t) = D_u V_t^u + \int_{]0,t]} (D_u g(s, u, X_{s-}^u) + D_x g(s, u, X_{s-}^u) \circ Y_s^u) dZ_s$$

Proof. The proof is in several steps corresponding to Lemmas 4 and 5 and section 36.12 below.

36.10 Lemma 4. *Under the hypotheses* $[D_1]$, $[D_2]$, $[D_3]$, *equations* (36.2.1) *and* (36.9.1) *satisfy the conditions* $[H_1]$ *to* $[H_4]$ *of section* 36.3 *for every* $p \geqslant 2$ *on any interval* $]0, \sigma_n]$ *as defined in Theorem 1.*

Proof. Let us first consider equation (36.2.1). (H_1) is trivially implied by $[D_1]$ to $[D_3]$ also implies the Lipschitz property (H_2) and conditions (H_3) and (H_4), which is expressed here in the much stronger form $||g(s, u, x) - g(s, v, x)||_\Lambda \leqslant C||u - v||$. We turn now to equation (36.9.1), which is linear in Y^u. The only condition (H_i) which is not immediately implied by the hypotheses of the Lemma is condition (H_4). We write

$$\begin{aligned}
&||D_u g(s, v, X_{t-}^v) - D_u g(s, u, X_{t-}^u) + D_x g(s, v, X_{t-}^v) \circ \xi_{t-} - D_x g(s, u, X_{t-}^u) \circ \xi_{t-}||_\Lambda^p \\
&\leqslant 4^{p-1}\{||D_u g(s, v, X_{t-}^v) - D_u g(s, u, X_{t-}^v)||_\Lambda^p\} + \\
&\quad + 4^{p-1}\{||D_u g(s, u, X_{t-}^v) - D_u g(s, u, X_{t-}^u)||_\Lambda^p\} \\
&\quad + 4^{p-1}\{||[D_x g(s, v, X_{t-}^v) - D_x g(s, u, X_{t-}^v)] \circ \xi_{t-}||_\Lambda^p\} \\
&\quad + 4^{p-1}\{||[D_x g(s, u, X_{t-}^v) - D_x g(s, u, X_{t-}^u)] \circ \xi_{t-}||_\Lambda^p\} \\
&\leqslant 4^{p-1} C^p(||u - v||^p + ||X_{t-}^u - X_{t-}^u||^p + \\
&\quad + 4^{p-1} C^p ||u - v||^p ||\xi_{t-}||^p + 4^{p-1} C^p ||(X_{t-}^v - X_{t-}^u) \circ \xi_{t-}||^p
\end{aligned}$$

One knows from Theorem 36.4 that there exists an increasing sequence (σ_n) of stopping times and constants C_n such that

$$E \sup_{s < \sigma_n} ||Y^u(s) - Y^v(s)||^{2p} \leqslant C_n ||u - v||^{2p}.$$

Let us write, for every stopping time τ,

$$E(\sup_{t < \tau \wedge \sigma_n} ||(X_t^v - X_t^u) \circ \xi_{t-}||^p) \leqslant$$

$$[E(\sup_{t < \tau \wedge \sigma_n} ||X_t^u - X_t^u||^{2p})]^{\frac{1}{2}} [E(\sup_{t < \tau \wedge \sigma_n} ||\xi_t||^{\frac{2p}{2p-1}})]^{\frac{2p-1}{2}}$$

$$\leqslant C_n^{\frac{1}{2}} ||u - v||^p E(\sup_{t < \tau \wedge \sigma_n} ||\xi_t||^\alpha)^{p/\alpha},$$

with $\alpha = \dfrac{2p}{2p-1}$.

Therefore,

$$E(\sup_{s<\tau\wedge\sigma_n} ||g(s,u,\xi_{s^-}) - g(s,v,\xi_{s^-})||_\Lambda^p) \leqslant 4^{(p-1)} C^p ||u-v||^p [1 + C_n + $$
$$+ E(\sup_{t<\tau\wedge\sigma_n} ||\xi_{t^-}||^p)] + $$
$$+ C_n^{\frac{1}{2}} [E(\sup_{t<\tau\wedge\sigma_n} ||\xi_{t^-}||^\alpha)]^{p/\alpha}$$

If we remark that $E(\sup_{t<\tau\wedge\sigma_n} ||\xi_{t^-}||^p) \geqslant [E(\sup_{t<\tau\wedge\sigma_n} ||\xi_{t^-}||^\alpha)]^{p/\alpha}$, we see that property ($H_4$) holds with

$$\Psi(\varrho) = 1 + C_n + (1 + C_n^{\frac{1}{2}})\,\varrho\,.$$

36.11 Lemma 2. *If we define*

$$\Phi_t(e,u,\lambda) = \frac{1}{\lambda}[X_t^{u+\lambda e} - X_t^u - \lambda Y_t^u \circ e]\,,$$

there exists an increasing sequence (τ_n) *of stopping times such that* $\lim_n P\{\tau_n < T\} = 0$ *and a sequence* C_n *of constants such that*

$$E\{\sup_{t<\tau_n} ||\Phi_t(e,.,\lambda)||^2_{L^2(G)}\} \leqslant C_n \lambda^2\,.$$

Proof. For each u the process $(\Phi_t(e,u,\lambda))_{t<T}$ is solution of

(36.11.1)
$$\Phi_t(e,u,\lambda) = \frac{1}{\lambda}(V_t^{u+\lambda e} - V_t^u - \lambda D_e V_t^u) + $$
$$+ \int_{]0,t]} \frac{1}{\lambda}[g(s,u+\lambda e, X_{s^-}^{u+\lambda e}) - g(s,u,X_{s^-}^u) - $$
$$- \lambda D_e g(s,u,X_{s^-}^u) - \lambda D_x g(s,u,X_{s^-}^u) \circ Y_{s^-}^u \circ e]\,dZ_s$$

We may write for $x, y \in \mathbb{H}$ and $\eta \in \mathscr{L}(\mathbb{H};\mathbb{H})$

(36.11.2)
$$g(s,u+\lambda e,y) - g(s,u,x) - \lambda D_e g(s,u,x) - \lambda D_x g(s,u,x)\circ\eta\circ e =$$
$$\lambda D_e g(s,u,y) + D_x g(s,u,x)\circ(y-x) - \lambda D_e g(s,u,x) -$$
$$- \lambda D_x g(s,u,x)\circ\eta\circ e + h(s,u,x,y,\eta,\lambda,e)$$
$$= D_x g(s,u,x)\circ(y - x - \lambda\eta\circ e) + \tilde{h}(s,u,x,y,\eta,\lambda)\,,$$

with

(36.11.3)
$$||\tilde{h}(s,u,x,y,\eta,\lambda)||_\Lambda \leqslant |\lambda| K(||y-x|| + |\lambda|)$$

for some constant K.

The equation (36.11.1) can therefore be written

(36.11.4)
$$\Phi_t(e,u,\lambda) = H_t(u,\lambda,e) + \int_{]0,t]} D_x g(s,u,X_{s^-}^u)\circ\Phi_{s^-}(e,u,\lambda)\,dZ_s\,,$$

where the process $H(u, \lambda, e)$ satisfies

(36.11.5) $$\|H_t(u, \lambda, e)\|_{\mathbb{H}} \leqslant |\lambda| v_t^* + \| \int_{]0,t]} \frac{1}{\lambda} \tilde{h}(s, u, X_{s-}^{u+\lambda e}, X_{s-}^u, Y_{s-}^u \circ e) dZ_s \| .$$

Using (36.11.4) we obtain from (36.11.5), for every stopping time σ,

$$E(\sup_{t<\sigma} \|H_t(u, \lambda, e)\|^2 \leqslant 2\lambda^2 v_{\sigma-}^* + E(\tilde{A}_{\tau-} \cdot \int_{]0,\tau[} [\lambda^2 + C^2 \|X_{s-}^{u+\lambda e} - X_s^u\|^2] d\tilde{b}_s).$$

Using then Theorem 36.4, we see that there exists a sequence (σ_n) of stopping times and a sequence of constants (K_n) such that

(36.11.6) $$\sup_{s<\sigma_n} (\tilde{A}_s \vee \tilde{b}_s) \leqslant n \quad \text{and}$$

(36.11.7) $$E(\sup_{t<\sigma_n} \|H_t(u, \lambda, e)\|^2) \leqslant K_n \lambda^2$$

(use a standard stopping procedure for processes v^*, $\tilde{A}$ and $\tilde{b}$).

This implies

(36.11.8) $$E(\sup_{t<\sigma_n} \int_G \|H_t(u, \lambda, e)\|^2 du) \leqslant \int_G K_n \lambda^2 du \leqslant \tilde{K}_n \lambda^2 .$$

We next consider the $L^2(G)$ valued process $(\Phi_t(e, ., \lambda))_{t \leqslant T}$.

Since $D_x g$ is bounded by some constant C, inequality (36.11.5) shows that the $L^2(G)$-valued process Φ_t satisfies an inequality of the following type for every stopping time $\tau \leqslant \sigma_n$.

$$E\{\sup_{t<\tau} \|\Phi_t(e, ., \lambda)\|_{L^2(G)}\} \leqslant 2\tilde{K}_n \lambda^2 + 2E(\tilde{A}_{\tau-} \int_{[0,\tau[} C^2 \sup_{s<t} \|\Phi_s(e, ., \lambda)\|^2_{L^2(G)} d\tilde{b}_s)$$

$$\leqslant 2\tilde{K}_n \lambda^2 + 2nC^2 \int_{[0,\tau[} \sup_{s<t} \|\Phi_s(e, ., \lambda)\|_{L^2(G)} d\tilde{b}_s$$

The "Gronwall Inequality" 29.1 immediately shows the existence of a constant C_n as in the Lemma.

36.12 *End of the proof of the theorem.* We make use of the following property, which is easily established. Let $f \in L^2(G; \mathbb{H}) \cap C(G; \mathbb{H})$ and $\bar{f} \in L^2(G; \mathscr{L}(\mathbb{H}; \mathbb{H}) \cap C(G; \mathscr{L}(\mathbb{H}; \mathbb{H}))$ such that for all $e \in \mathbb{R}^d$, all $u \in \mathbb{R}^d$ and some decreasing sequence $\lambda_k \downarrow 0$,

$$\lim_{k\to\infty} \|f(u + \lambda_k e) - f(u) - \lambda_k \bar{f}(u) \circ e\|_{L^2(G, \mathbb{H})} = 0 ,$$

Then $\bar{f}$ is the derivative of f in the sense of distributions and therefore in the ordinary sense in every point $u \in G$. Let us consider for each ω and n a P-negligable set Ω_n and a sequence λ_k such that $\lambda_k \downarrow 0$ and $\lim_{k\to\infty} \sup_{t<\tau_n(\omega)} \|\Phi_t(e, ., \omega, \lambda_k)\|_{L^2(G)} = 0$ for every $\omega \notin \Omega_n$.

The above property shows that for every $\omega \notin \Omega_n$ and $t < \tau_n(\omega)$, $Y_t^u(\omega)$ is the derivative of $u \curvearrowright X_t^u(\omega)$ at point the u. Therefore, $Y_t^u(\omega)$ is the derivative of $u \curvearrowright X_t^u(\omega)$ for all $t < \tau_n(\omega)$ and $\omega \notin (\bigcup_n \Omega_n)$.

This proves the theorem. □

37. Weak solutions of some stochastic differential equations

This concluding section of the book touches on the important subject of weak solutions of stochastic differential equations.

37.1 The martingale problem associated with an Ito equation

We come back to the following Ito-Skorokhod equation.

(37.1.1) $$X_t = x + \int_{]0,t]} b(X_s)ds + \int_{]0,t]} \sigma(X_s)dW_s,$$

where
- $x \in \mathbb{R}^d$,
- $x \curvearrowright b(x)$ is a measurable mapping from $\mathbb{R}^d$ into $\mathbb{R}^d$,
- $x \curvearrowright \sigma(x)$ is a measurable mapping from $\mathbb{R}^d$ into the set of $d \times d$-matrices,
- W is a d-dimensional Brownian motion with identity covariance matrix, given on a stochastic basis $(\Omega, (\mathscr{F}_t)_{t \geq 0}, P)$.

Let us define the following operator A on C^2-functions φ on $\mathbb{R}^d$, with bounded derivatives.

(37.1.2) $$A\varphi(x) := \sum_{i=1}^{d} b^i(x)D^i\varphi(x) + \sum_{i,j=1}^{d} \tfrac{1}{2}a^{ij}(x)D^{ij}\varphi(x), \quad \text{where}$$

(37.1.3) $$a(x) = \sigma \circ \sigma^*(x) \quad (\sigma^* \text{ being the transpose of } \sigma)$$

If ξ is the canonical process on $\mathbb{C}_{\mathbb{R}^d}(\mathbb{R}_+)$, set

(37.1.4) $$\mathfrak{A}(\varphi, s) = A\varphi(\xi_s).$$

Let X be a strong solution of (37.1.1). A straightforward application of the Ito formula shows that, for every bounded φ having bounded first- and second-order derivatives, the process

(37.1.5) $$(\varphi(X_t) - \varphi(x) - \int_{]0,t]} A\varphi(X_s)\,ds)_{t \geq 0}$$

is a martingale on $(\Omega, (\mathscr{F}_t)_{t \geq 0}, P)$, with $P\{X_0 = x\} = 1$.

Let $\tilde{P}_x$ be the law of X on $(\mathbb{C}_{\mathbb{R}^d}(\mathbb{R}_+), \mathscr{C}_\infty)$: this is the image of P by $\omega \curvearrowright X(\omega) \in \mathbb{C}_{\mathbb{R}^d}$.

Formula (37.1.5) leads immediately to the conclusion that P_x is a solution of the martingale problem on $(\mathbb{C}_{\mathbb{R}^d}, (\mathscr{C}_t)_{t \geq 0})$ associated with $\mathfrak{A}$, defined by (37.1.4) and the initial point x.

There is a converse to this property, expressed by the following theorem.

37.2 Theorem. *For every $x \in \mathbb{R}^d$ we are given a symmetric positive $d \times d$ matrix $a(x)$ and a vector $b(x) \in \mathbb{R}^d$.*

Assume the matrix field $x \curvearrowright a(x)$ is continuous and $x \curvearrowright b(x)$ is measurable.

Let $\tilde{P}_x$ be a solution of the martingale problem $(\mathfrak{A}, x)$, where $\mathfrak{A}$ is given by (37.1.4). *Then there exists a stochastic basic $(\Omega, (\mathscr{F}_t)_{t \geq 0}, P)$ and, on this basis, a d-dimensional Brownian motion W and a process X such that*

(a) *X is a strong solution of the Ito equation*

$$X_t = x + \int_{]0,t]} b(X_s)ds + \int_{]0,t]} \sigma(X_s)dW_s,$$

(b) $\sigma \circ \sigma^* = a$,

(c) *the law P^X of X on $\mathbb{C}_{\mathbb{R}^d}(\mathbb{R}_+)$ is equal to $\tilde{P}_x$.*

Proof. We give the proof under the extra assumption that rank $(a(x)) = d$ for all x. When this is not true, the theorem still holds. We refer the reader to Priouret ([Pri], chap. 3) for the complete proof in this case.

Since $a(x)$ is symmetric and positive there exists an orthogonal matrix $u(x)$ such that

$$u(x) \circ a(x) \circ u^*(x) = \begin{bmatrix} \lambda_1(x) & 0 \cdots 0 \\ 0 & \ddots & \vdots \\ \vdots & & \\ 0 & & \lambda_d(x) \end{bmatrix} \quad \text{with } \lambda_i(x) > 0$$

for all x, the functions u and λ_i being continuous.

We set

$$\Phi(x) = u(x) \begin{bmatrix} (\lambda_1(x))^{-\frac{1}{2}} & 0 & & 0 \\ 0 & \ddots & & \vdots \\ \vdots & & & \\ 0 & & & (\lambda_d(x))^{-\frac{1}{2}} \end{bmatrix}$$

and $\sigma(x) = \Phi^{-1}(x)$.

The equality $\sigma \circ \sigma^*(x) = a(x)$ is evident from the definitions.

By assumption, for every twice continuously differentiable function φ, bounded and with bounded derivatives, the process

$$(\Phi(\xi_t) - x - \int_{]0,t]} A\varphi(\xi_s)ds)_{t \geq 0}$$

is a martingale. If we set

(37.2.1) $$M_t := \xi_t - x - \int_{]0,t]} b(\xi_s)ds,$$

we saw in chapter 7, exercise (E_2), that this latter property is equivalent to M being a locally square integrable martingale with

$$\langle\!\langle M \rangle\!\rangle_t = \int_{]0,t]} a(\xi_s)\,ds\,.$$

If, on the canonical space $(\tilde{\Omega}, (\tilde{\mathscr{F}}_t)_{t \geq 0}, \tilde{P}) = (\mathbb{C}_{\mathbb{R}^d}(\mathbb{R}^+), (\mathscr{C}_t)_{t \geq 0}, \tilde{P}_x)$, we define

$$W := \int \Phi(\xi)\,dM\,, \tag{37.2.2}$$

then W is a continuous local martingale such that

$$\langle\!\langle W \rangle\!\rangle_t = \int_{]0,t]} \Phi(\xi) \otimes \Phi(\xi)\,d\langle\!\langle M \rangle\!\rangle$$
$$= t\,.$$

This shows that W is a Brownian motion and, as a consequence of (37.2.1) and (37.2.2),

$$\xi_t = x + \int_{]0,t]} b(\xi_s)\,ds + \int_{]0,t]} [\Phi(\xi_s)]^{-1}\,dW_s\,.$$

In other words, the canonical process ξ is a strong solution of the equation

$$X_t = x + \int_{]0,t]} b(X_s)\,ds + \int_{]0,t]} \sigma(\xi_s)\,dW_s$$

on the stochastic basis $(\mathbb{C}_{\mathbb{R}^d}(\mathbb{R}_+), (\mathscr{C}_t)_{t \geq 0}, \tilde{P}_x)$.

If rank $\sigma(x)$ is not always d, the canonical space has to be extended to provide a stochastic basis $(\Omega, (\mathscr{F}_t)_{t \geq 0}, P)$ on which a Brownian motion and a process X together can be defined with the properties of the theorem (see Priouret [Pri], chap. 3, Prop. 5). □

This theorem leads to the following notion of a weak solution of a stochastic differential equation.

37.3 The concept of weak solutions

Suppose we have

(1°) a measurable mapping $\alpha\colon x \curvearrowright \alpha(x)$ from $\mathbb{R}^d$ into $\mathscr{L}(\mathbb{R}^k; \mathbb{R}^d)$,

(2°) $x \in \mathbb{R}^d$.

Definition (*$\tilde{P}$-weak solutions and weak uniqueness property*). (1°) Let $\tilde{P}$ be a probability law on $(\mathbb{D}_{\mathbb{R}^k}(\mathbb{R}'_+), \mathscr{D}_\infty)$ such that the canonical process ξ is a semimartingale for $\tilde{P}$. We say that $(\Omega, (\mathscr{F}_t)_{t \geq 0}, P, Z, X)$, where $(\Omega, (\mathscr{F}_t)_{t \geq 0}, P)$ is a stochastic basis and Z and X two regular processes on this basis, is a weak $\tilde{P}$-solution of the equation

$$X_t = x + \int_{]0,t]} \alpha(X_{s^-})\,dZ_s \tag{37.3.1}$$

if
a) the law P^Z of Z is $\tilde{P}$,
b) X is a strong solution of (37.3.1) on $(\Omega, (\mathscr{F}_t)_{t \geq 0}, P)$.

(2°) We say that the equation (37.3.1) has the weak uniqueness property with respect to $\tilde{P}$ when, for any two solutions $(\Omega.(\mathscr{F}_t)_{t>0}, P, X, Z)$ and $(\Omega', (\mathscr{F}_t')_{t>0}, P', X', Z')$ and every $x \in \mathbb{R}^d$, the processes X and X' have same probability law on $\mathbb{D}_{\mathbb{R}^d}(\mathbb{R}_+)$, i.e. $P^X = P^{X'}$.

Examples and remarks. (1°) If Z is given on a basis $(\Omega, (\mathscr{F}_t), P)$ and is a semimartingale with law $\tilde{P}$, every strong solution X in the sense of section 34 gives a weak solution $(\Omega, \mathscr{F}, P, Z, X)$.

(2°) In section 37.2 we gave an example of a weak $\tilde{P}$-solution, where $\tilde{P}$ is the law of $(W_t, t)_{t \geqslant 0}$ on $\mathbb{D}_{\mathbb{R}^{d+1}}(\mathbb{R}_+)$ and $\alpha(x) = \begin{pmatrix} a(x) \\ b(x) \end{pmatrix}$.

(3°) There exist equations with weak solutions and no strong solution. (See [Cir] and [IkW]).

37.4 Definition *(Pathwise uniqueness property)*. We say that the equation (37.3.1) has the pathwise uniqueness property with respect to $\tilde{P}$ if, for every $x \in \mathbb{R}^d$ and any two $\tilde{P}$-weak solutions $(\Omega, (\mathscr{F}_t)_{t \geqslant 0}, P, X, Z)$ and $(\Omega, (\mathscr{F}_t)_{t \geqslant 0}, P', X', Z')$ on the same stochastic basis, the P-equality of Z and Z, implies the P-equality of X and X'.

37.5 Theorem. *Let us assume the existence of a constant K such that, for every x and y in $\mathbb{R}^d$,*

$$\|\alpha(x) - \alpha(y)\|_{\mathscr{L}(\mathbb{R}^k, \mathbb{R}^d)} \leqslant K \|x - y\|.$$

Then the equation (37.3.1) *has the pathwise uniqueness property with respect to every probability law $\tilde{P}$ on $\mathbb{D}_{\mathbb{R}^k}(\mathbb{R}_+)$ for which the canonical process ξ is a $\tilde{P}$-semimartingale.*

Proof. This is a straightforward consequence of the uniqueness of strong solutions under Lipschitz conditions on the coefficients, as proved in section 34. □

37.6 Theorem *(Yanada-Watanabe). If the equation* (37.3.1) *has the pathwise uniqueness property with respect to $\tilde{P}$, it has the weak uniqueness property with respect to $\tilde{P}$.*

Proof. Let $(\Omega^i, (\mathscr{F}_t^i)_{t \geqslant 0}, P^i, X^i, Z^i)$, $i = 1, 2$, be two weak solutions. We consider the measure space

$$(\bar{\Omega}, (\mathfrak{A}_t)_{t \in \mathbb{R}^+}) := (\mathscr{X}_1 \times \mathscr{X}_2 \times \mathscr{X}_3, (\mathfrak{A}_t^1 \otimes \mathfrak{A}_t^2 \otimes \mathfrak{A}_t^3)_{t \geqslant 0}),$$

where

$$\begin{aligned} &\mathscr{X}_1 = \mathscr{X}_2 := \mathbb{D}_{\mathbb{R}^d}(\mathbb{R}_+), \\ &\mathscr{X}_3 := \mathbb{D}_{\mathbb{R}^k}(\mathbb{R}_+), \\ &(\mathfrak{A}_t^1) = (\mathfrak{A}_t^2) := \text{canonical filtration on } \mathbb{D}_{\mathbb{R}^d}(\mathbb{R}_+), \\ &(\mathfrak{A}_t^3) := \text{canonical filtration on } \mathbb{D}_{\mathbb{R}^k}(\mathbb{R}_+). \end{aligned}$$

We intend to prove that if we call $\tilde{P}^i$ $(i=1,2)$ the probability law on $\mathfrak{A}^i \otimes \mathfrak{A}^3$, which is the image measure of P^1 (resp. P^2) by the mapping $\omega \curvearrowright (X^i(\omega) \times Z^i(\omega))$, there exists a probability law $\bar{P}$ on $(\bar{\Omega}, \mathfrak{A}_\infty)$ with the following properties.
a) $\tilde{P}^i$ $(i=1,2)$ is the projection of $\bar{P}$ on $(\mathscr{X}_i \times \mathscr{X}_3, \mathfrak{A}^i_\infty \otimes \mathfrak{A}^3_\infty)$.
b) If ξ^1, ξ^2, ξ^3 are the canonical projections of $\mathscr{X}_1 \times \mathscr{X}_2 \times \mathscr{X}_3$ on $\mathscr{X}_1, \mathscr{X}_2, \mathscr{X}_3$ respectively, the σ-algebras generated respectively by $\{\xi^1_s : s \leqslant t\}$, $\{\xi^2_s : s \leqslant t\}$, are $\bar{P}$-conditionally independent given the σ-algebra $\{\xi^3_s : s \leqslant t\}$.

Let us therefore consider regular conditional probabilities $Q^1_t(\xi^3; d\xi^1)$ (resp. $Q^2_t(\xi^3; d\xi^2)$ of ξ^1 (resp. ξ^2) with respect to ξ^3. These regular conditional probabilities exist since $\mathscr{X}_1, \mathscr{X}_2$ and $\mathscr{X}_3$ are complete metric spaces. Since by definition, $P^{Z^1} = P^{Z^3} = \tilde{P}$, properties a) and b) for $\bar{P}$ will be satisfied if we define $\bar{P}$ by

$$\bar{P}(A_1 \times A_2 \times A_3) = \int_{A_3} P^Z(d\xi^3) Q^1_t(\xi^3; A_1) Q^2_t(\xi^3; A_2) \tag{37.6.1}$$

for every $A_1 \times A_2 \times A_3 \in \mathfrak{A}^1_t \otimes \mathfrak{A}^2_t \otimes \mathfrak{A}^3_t$.

Let us prove now that ξ^3 is a semimartingale on $(\bar{\Omega}, (\mathfrak{A}_t)_{t \geqslant 0}, \bar{P})$. Since the canonical process ξ on $(\mathbb{D}_{\mathbb{R}^k}(\mathbb{R}_+), (\mathscr{D}_t))$ is a semimartingale by definition, there exist for $P^Z = \tilde{P}$ a local square integrable martingale $M(t, \xi)$ and a process with finite variation $V(t, \xi)$ on $(\mathscr{X}_3, \mathfrak{A}^3, P^Z)$ with the property $\xi_t = M(t, \xi) + V(t, \xi)$ for every ξ and t. We therefore only have to prove that $M(t, \xi^3)$ is a martingale on $(\mathscr{X}_1 \otimes \mathscr{X}_2 \times \mathscr{X}_3, (\mathfrak{A}^1_t \otimes \mathfrak{A}^2_t, \mathfrak{A}^3_t)_{t \geqslant 0}, \bar{P})$.

Let $F_1 \times F_2 \times F_3 \in \mathfrak{A}^1_s \otimes \mathfrak{A}^2_s \otimes \mathfrak{A}^3_s$. We immediately obtain

$$\int_{F_1 \times F_2 \times F_3} (M(t, \xi^3) - M(s, \xi^3)) \bar{P}(d\xi) =$$

$$\int_{F_3} (M(t, \xi^3) - M(s, \xi^3)) Q^1_s(\xi^3; F_1) Q^2_s(\xi^3; F_2) \; P^Z(d\xi^3) = 0 .$$

This proves the semimartingale property for ξ^3 on the basis $(\bar{\Omega}, \mathfrak{A}, \bar{P})$.
Since, by definition,

$$X^i_t = x + \int_{]0,t]} \alpha(X^i_{s-}) dZ^i_s, \qquad i = 1, 2,$$

we may write

$$\xi^i_t = x + \int_{]0,t]} \alpha(\xi^i_{s-}) d\xi^3_s, \qquad i = 1, 2,$$

the pathwise uniqueness property implies the $\bar{P}$-equality of ξ^1 and ξ^2 on $(\bar{\Omega}, \mathfrak{A}, \bar{P})$.
Therefore, $P^{X^1} = P^{X^2}$. □

37.7 Theorem. *Let a and b be as in Theorem 37.2 and $\tilde{P}$ be the probability law on $\mathscr{C}_{\mathbb{R}^{d+1}}(\mathbb{R}_+)$, which is the law of $(W_t, t)_{t \geqslant 0}$ for any d-dimensional Brownian motion W with covariance I_d.*

Let us assume that a has the global Lipschitz property and b is bounded and *measurable. Then the equation*

$$X_t = x + \int_{]0,t]} b(X_s)ds + \int_{]0,t]} a(X_s)dW_s$$

has a $\tilde{P}$-weak solution and the $\tilde{P}$-weak uniqueness property.

The martingale problem $(\mathfrak{A}, x)$, where $\mathfrak{A}$ is given by (37.1.4) also has a unique solution P_x.

Proof. It is clear from Theorem 37.2 that the existence and uniqueness of the solution of the martingale problem is equivalent to the existence of a $\tilde{P}$-weak solution and the $\tilde{P}$-weak uniqueness property.

Using the Cameron-Martin Theorem proved as exercise E.7, chapter 6, this reduces to having to prove the existence and uniqueness of the solution of the martingale problem associated with

(37.7.1) $$\mathfrak{A}(\varphi, s) = \sum_{i,j=1}^{d} a^{i,j}(\xi_s) D^{ij} \varphi(\xi_s).$$

However, since the stochastic equation

(37.7.2) $$X_t = x + \int_{]0,t]} a(X_s)dW_s$$

has a unique strong solution for every given Wiener process W on a stochastic basis $(\Omega, (\mathscr{F}_t), P)$, equation (37.7.2) has a weak solution with respect to the Wiener measure $\tilde{P}_d$ on $\mathbb{C}_{\mathbb{R}^d}(\mathbb{R}_+)$ and also the weak pathwise uniqueness with respect to $\tilde{P}_d$. This shows the existence of a unique solution to the martingale problem $(\mathfrak{A}, x)$ and therefore establishes the theorem. □

Exercises and supplements

E.1. *Non uniqueness of solutions.* (1°) The theory of ordinary differential equations shows that one may loose uniqueness of solutions if Lipschitz properties are not present. Take for example the equation

(1) $$x(t) = \int_0^t x^{1/3}(s)ds \quad (x \text{ real}).$$

This shows, a fortiori, the non-uniqueness of strong solutions for stochastic differential equations without some Lipschitz hypothesis.

(2°) If w is a Brownian motion, the stochastic equation

(1) $$x(t) = w(t) + \int_{]0,t]} (x - w_s)^{1/3} ds$$

reduces to (1) by a change of variable $y := x - w$ and has, therefore, strong solutions but there is no uniqueness.

This shows that one can have strong solutions without Lipschitz properties on the coefficients.

E.2. *Non existence of strong solutions.* Let $f(x) = -1_{[0,\infty[}(x)$.

The deterministic equation

$$X(t) = \int_0^t f(X_s)\,ds$$

has no solution.

E.3. *Use of Cameron-Martin formula.* We recall as in E.1 that the differential equation

$$dx(t) = x^{1/3}(t)\,dt$$

has no unique solution for the initial condition $x_0 = 0$.

Prove that the Ito equation

$$dX(t) = dW_t + X^{1/3}(t)\,dt \qquad \text{(where } W \text{ is a Brownian motion)}$$

has a unique strong solution for every initial condition.

(*Hint*: Apply the existence of solutions of differential equations with continuous coefficients to the equation $Y(t) = (Y(t) + W(t))^{1/3}\,dt$ and use the Cameron-Martin Theorem to prove uniqueness).

E.4. "*Kolmogorov-Neveu Lemma*". Let G be an open subset of $\mathbb{R}^d$ and $\bar{G}$ its closure. For every $t \in \mathbb{R}_+$ and $u \in \bar{G}$, $Y(t, u, .)$ is a random variable taking its values in the separable Hilbert space $\mathbb{H}$. We make the following assumptions on Y.

(a) For every $u \in \bar{G}$ the stochastic process $(Y(t, u))_{t \geqslant 0}$ has a.s. its paths in $\mathbb{D}_{\mathbb{H}}(\mathbb{R}_+)$.

(b) For every t there exists a constant a_t such, that for some $p \geqslant 1$ and $\varepsilon > 0$,

$$E(\sup_{s \leqslant t} \|Y(s, u) - Y(s, v)\|_{\mathbb{H}}^p) \leqslant a_t \|u - v\|^{d+\varepsilon}$$

for all u and v in $\bar{G}$ with $\|u - v\| \leqslant 1$.

Then there exists a family $\{Y^*(t, u, .) : t \in \mathbb{R}_+, u \in \bar{G}\}$ of $\mathbb{H}$-valued random variables such that the following hold.

(i) For every u, the processes $(Y^*(t, u))_{t \geqslant 0}$ and $(Y(t, u))_{t \geqslant 0}$ are P-equal.

(ii) For every $\omega \in \Omega$ and t the function $u \curvearrowright Y^*(t, u, \omega)$ is continuous on $\bar{G}$.

(iii) For every $\omega \in \Omega$ the mapping $t \curvearrowright Y^*(t, ., \omega)$ is right-continuous as a mapping from $\mathbb{R}_+$ into the space of continuous $\mathbb{H}$-valued functions on $\bar{G}$, viewed with the topology of uniform convergence on compact subsets.

Indication of the proof. The proof follows a method due to Neveu as given in Priouret [Pri], chap. 3, Lemma 3.

Define $D_n := \left\{u : u \in \frac{k}{2^n}\,\mathbb{Z}^d\right\}$. Two points u and u' will be said to be neighbours if

$$\sup_i |u^i - u'^i| \leqslant \frac{1}{2^n}.$$

Set $U_n^t := \sup_{s \leqslant t} \{\|Y(s, v) - Y(s, v')\| : v, v' \in D_n,\ v \text{ and } v' \text{ neighbours},$

$$|v^i| \leqslant n, |v'^i| \leqslant n\}$$

(1°) Show $E((U_n^t)^p) \leqslant a_t' \frac{n^d}{2^{n\varepsilon}}$ for some constant a_t' and conclude that

$$\sum_n U_n^t < \infty \text{ a.s.}$$

(2°) Let $\underset{\sim}{a}$ be a mapping from $\bigcup_n D_n$ into $\mathbb{H}$ and define

$$\Phi_n(\underset{\sim}{a}, u) := \sum_{v \in D_n(u)} \frac{2^{-n} - \|u - v\|}{\sum_{v \in D_n(u)} (2^{-n} - \|u - v\|)}\, \underset{\sim}{a}(v),$$

where $D_n(u)$ denotes the set of neighbours of order n of u. Assume that the following holds for $\underset{\sim}{a}$.

$$\sup\{\|\underset{\sim}{a}(v) - \underset{\sim}{a}(v')\| : v, v' \in D_n, v' \in D_n(v), |v^i| \leqslant n, |v'^i| \leqslant n\} \leqslant u_n$$

with $\sum_n u_n < \infty$.

Then the functions $u \curvearrowright \Phi_n(\underset{\sim}{a}, u)$ converge uniformly on every compact subset K of $\bar{G}$ to a continuous function $\Phi(a, .)$.

(3°) Apply the above to the functions

$$\underset{\sim}{a}(u) := Y(t, u, \omega), \quad u \in \bigcup_n D_n$$

and set

$$Y^*(t, u, \omega) = \Phi(Y(t, . , \omega), u).$$

Show that Y^* has the properties of the Kolmogorov-Neveu Lemma.

E.5. *Majorization in L^p.* Let S be a semimartingale and q a white random measure as in equation (34.1.2). For every couple $\alpha = (\alpha^1, \alpha^2)$, where α^1 is an $\mathscr{L}(\mathbb{G}; \mathbb{H})$-valued, predictable process and α^2 is a $\mathscr{P} \otimes \mathscr{B}_{\mathbb{E}}$-measurable, $\mathbb{H}$-valued function, we define $\tilde{\lambda}_s \circ \alpha$ and $\tilde{b}$ by formulas (34.7.2) and (34.7.3).

Show the existence for every $p \geqslant 2$ of a positive increasing process $\tilde{A}^p$ such that, for every stopping time τ,

$$E(\sup_{t<\tau} \| \int_{]0,t]} \alpha^1, dS + \int_{]0,t]\times \mathbb{E}} \alpha^2 dp \|^p) \leqslant E(\tilde{A}^p_{\tau-} \int_{]0,\tau[} |\tilde{\lambda}_s \circ \alpha|^{p/2} d\tilde{b}_s).$$

Hint: Use the majoration in exercise E.5, chapter 6, and apply E.2, chap. 6, to the local martingale $\int \alpha^2 dq$ (se chapter 7, formula (31.10.3)).

E.6. *Markov solutions of stochastic equations.* Let α be a mapping from $\mathbb{R}^d$ into $\mathscr{L}(\mathbb{R}^d; \mathbb{R}^k)$ with the following Lipschitz properties.

(i) For every $\beta > 0$ there exists a constant L^β such that, for every $x, y \in \mathbb{R}^d$ with $\|x\| \; \exists \|y\| \leqslant \beta$,

$$\|\alpha(x) - \alpha(y)\| \leqslant L^\beta \|x - y\|.$$

(ii) There exists $K > 0$ such that, for every $x \in \mathbb{R}^d$,

$$\|\alpha(x)\|^2 \leqslant K(1 + \|x\|^2).$$

We consider the stochastic equation

$$X_t = x + \int_{]0,t]} \alpha(X_{s^-}) dZ_s,$$

where Z is a semimartingale.

Let $\tilde{\Omega} := \mathbb{D}^d$ be the space of right-continuous functions from $\mathbb{R}^+$ into $\mathbb{R}^d$ which have left limits. On $\tilde{\Omega}$ we consider the filtration $\tilde{\mathscr{D}}^Z$ defined as follows. $\mathscr{D}^0_t$ is the σ-algebra generated by $\{\xi_s : s \leqslant t\}$, where ξ is the canonical process, $\tilde{P}^Z$ is the law of Z, that is, the image of P on $\mathscr{D}^0_\infty$ by the mapping $\omega \curvearrowright Z(.,\omega)$, $\mathscr{N}_\infty$ is the set of subsets of $\mathbb{D}^d$ of exterior $\tilde{P}^Z$-measure zero, and $\tilde{\mathscr{D}}^Z_t := \mathscr{P}_t \bigcup \mathscr{N}_\infty$. The space $(\mathbb{D}^d, \tilde{\mathscr{D}}^Z, \tilde{P}^Z)$ satisfies the usual hypothesis.

We assume that ξ is a semimartingale on $(\mathbb{D}^d, \tilde{\mathscr{D}}^Z, \tilde{P}^Z)$.

(1°) There exists a mapping $(t, x, \tilde{\omega}) \curvearrowright \Phi(t, x, \tilde{\omega})$ from $\mathbb{R}^+ \times \mathbb{R}^d \times \mathbb{D}^d$ into $\mathbb{R}^d$ such that for every x, $\Phi(.,x,.)$ is an R.R.C. process, for every $(t, \tilde{\omega})$, $x \curvearrowright \Phi(t, x, \tilde{\omega})$ is continuous and

$$\Phi(t, x, .) = x + \int_{]0,t]} \alpha(\Phi(s^-, x, .)) d\xi_s$$

Hint: Trivial consequence of the Kolmogorov-Neveu Lemma E.4.

(2°) Let Y_0 be an $\mathbb{R}^d$-valued, $\tilde{\mathscr{D}}^Z_0$-measurable random variable. Then

$$\Phi(t, Y_0, .) = Y + \int_{]0,t]} \alpha(\Phi(s^-, Y_0, .)) d\xi_s$$

Hint: Prove it first for $Y_0 = \sum_{i=1}^n x_i 1_{F_i}$, $F_i \in \tilde{\mathscr{D}}^Z_0$.

(3°) Lor every real bounded $\tilde{\mathscr{D}}^Z_\infty$-measurable h we set $\mu(x, h) = \tilde{E}^Z(h(\Phi(\,.\,;x)))$. Then (Y_0, h) is a version of the conditional expectation $\tilde{E}[h(\Phi(\,.\,;x))|\tilde{\mathscr{D}}^Z_0]$ for every x.

Hint: Use (2°) and the uniqueness of the solution of $\tilde{X}_t = x + \int_{]0,t]} \alpha(\tilde{X}_{s-})d\xi_s$.

(4°) Assume the following property for $\tilde{P}^Z$. Let Θ_τ be the transformation $\xi \frown\!\!\searrow (\xi_{\tau+t} - \xi_\tau)_{t \geqslant 0}$ on $\tilde{\Omega}$ and assume $\tilde{P}^Z$ invariant by Θ_τ for all stopping times τ. Show that the process X, the solution of $X_t = x + \int_{]0,t]} \alpha(X_{s-})dZ_s$, has the strong Markov property with respect to its own filtration.

Hint: The strong Markov property for $\Phi(\,.\,,x,\,.\,)$ with respect to $(\tilde{\mathscr{D}}^Z_t)$ follows from (3°). (See chap. 1, E.18, for the definition of the strong Markov property).

Historical and bibliographical notes

1. Strong solutions

The main historical step concerning stochastic differential equations are mentioned in section 33. The most classical accounts which include the strong Markov property of solutions are [Ito. 2], (Sko], [GiS. 2]. In [Gis. 2] (chap. 5) the reader will find an introduction to stochastic differential Ito equations in a domain of $\mathbb{R}^d$ when boundary conditions (in particular reflexion) are imposed. For the theory of strong solutions under Lipschitz conditions for equations driven by semimartingales, see [DoM. 2], [Pro. 1, 2], [MeP. 3] and a synthetical account in [MeP. 1], chapter 3. It is this presentation which is extended here to the case where the driving term includes a random measure (this also follows [Met. 8]).

Strong solutions may also exist when Lipschitz-type conditions are not present. Such results were first proved for "Ito-type" equations involving infinite dimensional processes. See E. Pardoux [Par. 1, 2], Metivier-Pistone [MPi. 1], Krylov and Rozovski [KrR]. In all these examples the Lipschitz condition is replaced by a monotonicity-condition. The same type of hypothesis was applied by Jacod to equations driven by semimartingales in the finite-dimensional case [Jac. 5]..

The problem of studying the solutions of a stochastic equation depending on a parameter goes back many years. An almost sure regularity result is due to Blagovesčenskii and Freidlin in 1961 [BlF]. Gikhman and Skorokhod have proved L^p-differentiability results. That is, if u is the parameter, $u \frown\!\!\searrow X^u_t(\,.\,) \in L^p(\Omega, \mathscr{F}_t, P)$ is differentiable. (See their book [GiS. 2]). These results have been extended recently by K. Bichteler to equations driven by semimartingales.

The almost sure differentiability of the type studied in the book has been studied by several authors in connection with flows of solutions of Ito equations (the parameter here is only the initial condition). See [Kun. 2], [Mal], [Bis. 2, 3]. The result given here seems rather new.

Properties of stabilities refer to the following problem. When the driving term converges for some topology, does the solution converge for a suitable topology in processes? Considering topologies on semimartingales Emery studied this question in [Eme. 1, 2]. See also [Pro. 3]. Very similar results are given in [MeP. 1], chap. 3.

2. Weak solutions and martingale problems

The concept of weak solution is explicit in [Gri]. The distinction between strong and weak solutions is only implicit in [McK]. Until very recently the weak solutions have essentially been studied for Ito equations in connection with the laws of diffusions. The classical reference book is the book of Stroock and Varadhan [StV. 2]. A very efficient, self-contained account is provided by Priouret's lecture notes [Pri]. Another recent account is in Williams [Wil]. A big new reference book on the subject is the one by Ikeda and Watanabe.

The famous example of an Ito equation having weak but no strong solutions is due to Čirelson [Cir].

The concept of weak solution was extended to a class of equations with discontinuous driving terms by Lepeltier and Marchal [LeM]. Recently, equations driven by semimartingales have been considered independently by Jacod and Memin [JaM. 1] and Pellaumail [Pel. 3]. These are the most general existence results, which are presently known.

3. Stochastic equations in domains with a boundary

We did not consider this problem at all. The most elementary situation for strong solutions is treated in [GiS. 2], as indicated above. The fundamental results go back to Stroock-Varadhan. We should also mention *N.* El Karoui's thesis [Elk] and the account of this at the beginning of [BeL]. Connections with the problem of constructing local times on the boundary is very fundamental. A basic reference for this latter problem is the collective semimar edited by Azema and Yor [AzY].

Bibliography

[Ald] Aldous, D., Weak convergence and the general theory of processes. Preprint.

[All] Allain, M.F., Sur quelques types d'approximation des solutions d'équations différentielles stochastiques. Thèse de 3ème cycle, Université de Rennes, 1974.

[Aus] Austin, D.G., A sample function property of martingales. Ann. Math. Stat., *37*, 1966, 1396–1397.

[Aze] Azéma, J.

(1) Quelques applications de la théorie générale des processus, Invent. Math., *18*, 1972, 293–336.

(2) Théorie générale des processus et retournement du temps. Ann. Sci. ENS *6*, 1973, 459–519.

[AzY] Azema, J., M. Yor, Temps locaux. Asterisque, 52–53, S.M.F., 1978.

[Bac] Bachelier, L., Théorie de la spéculation. Thèse, Paris, 1900. Ann. Ec. Norm. Sup., *13*, Vol. 17, 1900.

[Bae] Baez-Duarte, L., An a.e. divergent martingale that converges in probability. J. Math. Appl., *36*, 1971, 149–150.

[Bar] Barlow, M.T., Study of a filtration expanded to include an honest time. Z.W.*, Vol. 44, 1978, 307–324.

[Bau] Bauer, H., Probability theory and elements of measure theory. New York, 1972.

[BeL] Bensoussan, A., J.L. Lions, Application des inéquations variationelles en contrôle stochastique. Paris 1978.

[BeJ] Benveniste, A., . Jacod, Systèmes de Levy des processus de Markov. Inv. Math., Vol 21, 1978, 183–198.

[Bic] Bichteler, K., Stochastic integration and L^p-theory of semimartingales. Preprint.

[Bil] Billingsley, P., Convergence of probability measures. New York 1968

[Bis] Bismut, J.M.

(1) Intégrales convexes et probabilités. J. of Math. Anal. and Appl., Vol. 42, *3*, June 1973, 639–673.

(2) A generalized formula of Ito and some other properties of stochastic flows. Z.W., *55*, 1981, 331–350.

(3) Mécanique aléatoire. Lecture Notes in Math. *866*, Berlin-Heidelberg-New York 1981.

[BlF] Blagovescenskii, Y.M., M.I. Freidlin, Some properties of diffusion processes depending on a parameter. DAN, *138*, 1961. Sov. Math., *2*, 633–636.

[Bla] Blake, L.H., A generalization of martingales and two consequent convergence theorems. Pacific. J. Math., *25*, 1970, 279–284.

[BlG] Blumenthal, R.M., R.K. Getoor, Markov processes and potential theory. New York 1968.

[Boc] Bochner, S., Harmonic analysis and the theory of probability. Berkeley Univ. of California Press 1955.

[BVW] Boel, R., P. Varaiya, E. Wong, Martingales on jump processes; 1. Representation results; 2. Applications., SIAM J. of Control., *13*, 1975, 999–1021 and 1022–1061.

[Bou] Bourbaki, N.
(1) Integration (chap. 1–4). Paris 1965.
(2) Espaces vectoriels topologiques. Paris 1967.

[Boy] Boylan, E.S., Equiconvergence of martingales. Ann. Math. Stat., *42*, 1971, 552–559.

[Bre] Breiman, L., Probability. Addison-Wesley, 1968.

[Brm] Brémaud, P.
(1) A martingale approach to point processes. PhD-Thesis, Electronic Research Laboratory. Berkeley M 345, 1972.
(2) Point processes and queues: martingale dynamics. Series: "Statistics", Berlin-Heidelberg-New York 1981.

[BrJ] Brémaud, P., J. Jacod, Processus ponctuels et martingales – résultats récents sur la modélisation et le filtrage. Adv. Appl. Proba., Vol. 9, 1977, 362–416.

[BrY] Brémaud, P., M. Yor, Change of filtrations and of probability measures. Z. W., *45*, 1978, 269–296.

[Buc] Bucy, R.S., Stability and positive supermartingales. J. Differential equations, *1*, 1965, 151–155.

[BuJ] Bucy, R.S., P.D. Joseph, Filtering for stochastic processes. Intersciences, 1968.

[Bur] Burkholder, D.L.
(1) Independent sequences with the Stein property. Ann. Math. Stat., *39*, 1968, 1282–1288.
(2) Martingale inequalities. Lecture Notes 190, Berlin-Heidelberg-New York 1970.
(3) Integral inequalities for convex functions of operators on martingales. Sixth Berkeley Symposium.

[Cha] Chacon, R.V., A stopped proof of convergence. Advances in Math., *14*, 1974, 365–368.

[Cht] Chatterji, S.D.,
(1) Martingales of Banach-valued random variables. Bull. A.M.S., *66*, 1960, 395–398.
(2) A note on the convergence of Banach-space-valued martingales. Math. Ann., *153*, 1964, 142–149.
(3) Comments on the martingale convergence theorem. Symposium Loutraki, Berlin-Heidelberg-New York
(4) Martingale convergence and the Radon-Nikodym theorem in Banach spaces. Math. Scand., *22*, 1968, 21–41.

[Chk] Choksi, J.R., Inverse limits of measure spaces. "Proc. of London Math. Soc.", Vol. 8, 1958, 321–342.

[Chw] Chow, Y.S.
(1) Martingale in a σ-finite measure space, indexed by directed sets. Trans. A.M.S., *97*, 1960, 260–285.
(2) Convergence theorems of martingales. Z.W., *1*, 1962, 340–346.
(3) A martingale convergence theorem of Wald's type. Illinois J. Math., 9, 1965, 569–576.

[Chu] Chung, K.L.,
(1) On a stochastic approximation method. Ann. Math. Stat., *25*, 1954, 463–483.
(2) Elementary probability theory with stochastic processes. Berlin-Heidelberg-New York 1979.
(3) Markov chains with stationary transition probabilities. Berlin-Heidelberg-New York 1967.

[ChD] Chung, K.L., J.L. Doob, Fields optionality and measurability. Amer. J. Math., *87*, 1965, 397–424.
[Cin] Činlar, E., Introduction to stochastic processes. Prentice Hall 1975.
[CJP] Činlar, E., J. Jacod, Ph. Protter, M. Sharpe, Semimartingales and Markov processes. Z.W., *54*, 1980, 161–219.
[Cir] Čirelson, B.S. An example of stochastic differential equation having no strong solution. Theo. Proba. and Appl., *20*, 1975, 416–418.
[Cou] Courrèges, Ph., Intégrales stochastiques et martingales de carré intégrable. Séminaire Brelot-Choquet-Deny, 7e année, 1962–1963, exposé n° 7.
[CoP] Courrèges, P., P. Priouret, Temps d'arrêt d'une fonction aléatoire. Théorèmes de décomposition., Publ., Inst. Stat. Paris, *14*, 1965, 245–274; 275–377.
[Cur] Curtain, R.F., Stochastic evolution equations with general white noise disturbance. J. Math. Anal. Appl., *60*, 1977, 570–595.
[CuF] Curtain, R.F., P.L. Falb, Stochastic differential equations in Hilbert spaces. J. Differential Equations, *10*, 1971, 412–430.
[Dar] Darling, D.A., H. Robbins, Some further remarks on inequalities for sample sums. Proc. Nat. Acad. Sci., *60*, 1968, 1175–1182.
[Dav] Davis, B.
(1) Comparison tests for martingale convergence. Ann. Math. Stat., *39*, 1968, 2141–2144.
(2) A comparison test for martingale inequalities. Ann. Math. Stat., *40*, 1969, 505–508.
(3) On the integrability of the martingale square function. Israël J. Math., *8*, 1970, 187–190.
[Daw] Dawson, D.A.
(1) Stochastic evolution equations. Math. Biosci,m *15*, 1972, 287–316.
(2) Stochastic evolution equations and related measure processes. J. Multivariate Anal., *5*, 1975, 1–52.
(2) Stochastic evolution equations and related measure processes. J. Multivariate Anal., *5*, 1975, 1–52.
[Del] Dellacherie, C.
(1) Capacités et processus stochastiques. Ergebn. der Math., Vol. 67, Berlin-Heidelberg-New York
(2) Une représentation intégrale des surmartingales à temps discret. Publ. Inst. de Stat., *17*, 1967, 1–18.
(3) Intégrales stochastiques par rapport aux processus de Wiener et de Poisson. Séminaire de Proba. VIII. Lecture Notes in Math. *381*, Berlin-Heidelberg-New York
(4) Contribution to the International Meeting of Mathematicians, Helsinki 1978.
(5) A survey of stochastic integration. J. Stoch. Proc. Appl., 1960.
[DeM] Dellacherie, C., P.A. Meyer; Probabilités et potentiel; 1: Chap. 1–4; 2: Chap. 5–8. Paris 1975 (éd. refondue).
[Din] Dinculeanu, N., Vector measures. Berlin (Ost) 1966.
[Dol] Doléans-Dade, C.
(1) Existence du processus croissant naturel associé à un potentiel de classe (D). Z.W., *9*, 1968, 309–314.
(2) Une martingale uniformément intégrable mais non localement de carré intégrable. Séminaire de Proba. V, Lecture Notes in Math. *191*, Berlin-Heidelberg-New York 1971.
(3) Intégrales stochastiques dépendant d'un paramètre. Bull. Inst. Stat. Univ. Paris, 16, 1967, 23–34.
(4) On the existence and unicity of solutions of stochastic integral equations. Z.W., *36*, 1976, 93–101.

[DoM] Doléans-Dade, C., P. A. Meyer.
(1) Intégrales stochastiques par rapport aux martingales locales. Séminaire de Proba. IV, Lecture Notes in Math. *124*, Berlin-Heidelberg-New York 1970.
(2) Equations différentielles stochastiques. Séminaire de Proba. XI, Lecture Notes in Math. *581*, Berlin-Heidelberg-New York 1977.
[Doo] Doob, J.L.,
(1) Stochastic processes. New York 1953.
(2) Discrete potential theory and boundaries. J. Math. Mech., *8*, 1959, 433–458.
(3) Notes on martingale theory. Fourth Berkeley Symposium, Vol., II, 1960, 95–102.
[Dos] Doss, H., Liens entre équations différentielles stochastiques et ordinaires. Ann. Inst. Henri Poincaré, *13*, 1977, 99–125.
[Dub] Dubins, L.E.
(1) Rises and upcrossings of semimartingales. Illionois J. Math., *6*, 1962, 226–241.
(2) A note on uprcrossings of semimartingales. Ann. Math. Stat., *37*, 1966, 728.
(3) Rises of nonnegative semimartingales. Illinois J. Math., *12*, 1968, 649–653.
[DuS] Dunford, N., J.T. Schwarz, Linear Operators. Part. 1. New York 1957.
[Dvo] Dvoretsky, A., On stochastic approximation. Proceed. Third Berkeley Symposium. Math. Stat. and Proba., Vol. 1, 1954–1955. University of California Press. Berkeley, 1956, 39–55.
[Dyn] Dynkin, E.B., Markov processes. Vol. I, Berlin-Heidelberg-New York 1965.
[DyY] Dynkin, E.B., A. Yuskevič, Markov processes: theoremes and problems. Plenum Press, 1969.
[Elk] El Karoui, N., Processus de diffusion associés à un opérateur elliptique dégénéré et à une condition frontière. Thèse, Univ. Paris VI, 1971.
[ElL] El Karoui, N., J.P. Lepeltier, Représentation des processus ponctuels multivariés à l'aide d'un processus de Poisson. Z.W.
[Ell] Elliott, R., Stochastic integrals for martingales of a jump process with partially accessible jump times. Z.W., *36*, 1976, 213–226.
[Eme] Emery, M.
(1) Perturbation d'équations différentielles stochastiques; intégrales multiplicatives. C.R. Acad. Sci. Paris Ser., A, *285*, 1977.
(2) Stabilité des solutions des équations différentielles stochastiques. Application aux intégrales multiplicatives stochastiques. Z.W., *41*, 1978, 241–262.
(3) Une topologie sur l'espace des semimartingales. Séminaire de Proba. XIII, Lecture Notes in Math. *721*, Berlin-Heidelberg-New York 1979, 180–260.
[ESh] Engelbert, H.J., A.N. Shiryaev, On absolute continuity and singularity of probability measure. Banach Center Publ. Warsaw, Vol. 6: Mathematical Statistics, 1976.
[Ers] Ershov, M.P., On the absolute continuity of measures corresponding to diffusion processes. Theory of Proba. Appl., Vol. 17, 1972, 173–178.
[Fab] Fabian, V.
(1) Stochastic approximation methods. Czechoslovak Math. J., *10* (85), 1960, 123–159.
(2) On the choice of design in stochastic approximation methods. Ann. of Math. Stat., *39*, 1968, 457–465.
[Fel] Feller, W., An introduction to probability theory and its applications. Vol. II, New York, 1966.
[Fis] Fisk, D.L., Quasimartingales. Trans. A.M.S., *120*, 1965, 369–389.
[Föl] Föllmer, H., The exit measure of a super-martingale. Z.W., *21*, 1972, 154–166.
[Gal] Galtchouk, L.I.
(1) Structure de certaines martingales. Proceed. Séminaire Processus Aléatoires, Druskininkai, *1*, 1974, 7–32 (in Russian).

(2) Existence et unicité pour des équations différentielles stochastiques par rapport à des martingales et des mesures aléatoires. 2d Vilnius Conf. Proba. Math. Stat. Vol. 1, 1977, 88–91.

[Gar] Garsia, A.M., Martingales inequalities. Seminar notes on recent progress. Mathematics Lecture Notes Ser. Benjamin. Reading 1973.

[GeW] Gelfand, I.M., N.Y. Vilenkin, Generalized functions. Applications of harmonic analysis. Vol. 4, New York, 1964.

[GiS] Gikhman, I.I., A.V. Skorokhod

(1) The theory of stochastic processes, *1.* + *2.* Berlin-Heidelberg-New York 1974/1975.

(2) Stochastic differential equations. Berlin-Heidelberg-NewYork 1972.

[Gir] Girsanov, I.V.

(1) On transforming a certain class of stochastic processes by absolutely continuous substitution of measures. Theory Proba. Appl., *5*, 1960, 314–330.

(2) An example of non-uniqueness of the solution of K. Ito's stochastic integral equations. Theory Proba. Appl., *7*, 1962, 336–341.

[Gla] Gladyšev, E.G., On stochastic approximation. Teor. Verojatnost. i. Primenen, *10*, 1965, 297–300 and Theory of Proba. Appl., *10*, 1965, 275–278.

[GrP] Gravereaux, B., J. Pellaumail, Formule de Ito pour des processus à valeurs dans des espaces de Banach. Ann. Inst. H. Poincaré, *10*, 4, 1974, 339–422.

[Gri] Grigelionis, B.

(1) On the stochastic integral representation of square integrable martingales. Lit. Math. J., *14*, 1975, 53–69.

(2) Stochastic point processes and martingales. Lit. Math. J., *15*, 1975, 101–114.

(3) On the martingale characterization of stochastic processes with independent increments. Lit. Math. J., *17*, 1977, 75–86.

(4) On the absolute continuity of measures corresponding to stochastic processes. Lit. Math. J., *11*, 1971, 783–794.

(5) The characterization of stochastic processes with conditionally independent increments. Lit. Math. J., *15*, 1975, 53–60.

[Has] Has'Minskii, R.Z., M.B. Nevel'Son, Stochastic approximation and recursive estimation. Transl. American Math. Soc., Vol. 47, Providence, 1973.

[Hel] Helms, L.L., Mean convergence of martingales. Trans. Amer. Math. Soc., *87*, 1958, 366–433.

[HeJ] Helms, L.L., G. Johnson, Class D super-martingales. Bull. Amer. Math. Soc. *69*, 1963, 59–62.

[Hun] Hunt, G.A. Martingales et processus de Markov. Paris 1966, 145 p.

[IkW] Ikeda, N., S. Watanabe, Stochastic differential equations and diffusion processes. North Holland, 1981

[IoT] Ionescu-Tulcea A., C. Ionescu-Tulcea

(1) Abstract ergodic theorems. Trans. Amer. Math. Soc., *107*, 1963, 105–125.

(2) Topics in the theory of lifting. Berlin-Heidelberg-New York

[Ito] Ito, K.

(1) Extension of stochastic integral. Proceed. of Intern. Symp. on S.D.E., Kyoto, 1976, 95–109.

(2) Lectures on stochastic processes. Tata Institute, Bombay, 1961.

(3) Stochastic integral. Proc. Imp. Acad. Tokyo, *20*, 1944, 519–524.

(4) On stochastic integral equation. Proc. Japan Acad., *22*, 1946, 32–35.

(5) On a formula concerning stochastic differentials. Nagoya Math. J., *3*, 1951, 55–65.

(6) On stochastic differential equations. Mem. Amer. Math. Soc., *4*, 1961.

(7) On stochastic processes. Japan J. Math., *18*, 1942, 261–301.

[ItM] Ito, K., H.P. McKean, Diffusion processes and their sample paths. Berlin-Heidelberg-New York

[ItW] Ito, K., S. Watanabe, Transformation of Markov processes by additive functionals. Ann. Inst. Fourier, Grenoble, *15*, 1965, 13–30.

[Jac] Jacod, J.

(1) Sous-espaces stables de martingales. Z.W., *44*, 1978, 103–115.

(2) Calcul stochastique et problèmes de martingales. Lecture Notes in Math. *714*, Berlin-Heidelberg-New York 1979.

(3) Multivariate point processes: Predictable projection, Radon-Nikodym derivative, representation of martingales. Z.W., *31*, 1975, 235–253.

(4) Projection previsible et décomposition multiplicative d'une semimartingale positive. Séminaire de Proba. Lecture Notes in Math. *469*, Berlin-Heidelberg-New York, 1978, 22–34.

(5) Une condition d'existence et d'unicité pour les solutions fortes d'équations différentielles stochastiques. Stochastics, 1980.

[JaM] Jacod, J., J. Mémin

(1) Existence of weak solutions for stochastic differential equations with driving semimartingales. Preprint.

(2) Caractéristiques locales et conditions de continuité absolue pour les semimartingales. Z.W., *35*, 1976, 1–37.

[Jeu] Jeulin, T., Grossissement d'une filtration et applications. Séminaire Proba. XIII, Lecture Notes in Math. *721*, Berlin-Heidelberg-New York 1979, 574–609.

[Kal] Kallenberg

(1) Random measures. Berlin (Ost) 1975.

(2) On the existence and path properties of stochastic integrals. Ann. Proba., Vol. 3, 1975, 262–280.

[KLS] Kabanov, Y., R.S. Liptzer, A.N. Shiryaev. Martingale methods in the theory of point processes. Proc. School Seminar Vilnius. Ac. Sc. Lit. SSR, Part. II, 1975, 269–354.

[Kai] Kailath, A general likelihood ratio formula for random signals in Gaussian noise. IEEE Trans. I.T., USA, *15*, 1969, 350–361.

[Kaz] Kazamaki, N.

(1) Note on a stochastic integral equation. Lecture Notes Séminaire de Proba. VI, Lecture Notes in Math. *258*, Berlin-Heidelberg-New York

(2) Change of time, stochastic integrals and weak martingales. Z.W., *22*, 1972, 25–32.

[KiW] Kiefer, J., J. Wolfowitz, Stochastic estimation of the maximum of a regression function. Ann. Math. Stat., *23*, 1952, 462–466.

[Kin] Kingman, J., Completely Random measures. Pac. J. Math., *21*, 1967, 59–78.

[Kol] Kolmogorov, A.N.

(1) Grundbegriffe der Wahrscheinlichkeitsrechnung. Berlin-Heidelberg-New York 1953.

(2) Über die analytischen Methoden in der Wahrscheinlichkeitsrechnung. Math. Ann., *104*, 1931, 415–458.

[Kri] Krickeberg, K.

(1) Convergence of martingales with a directed index set. Trans. A.M.S., *83*, 1956, 313–337.

(2) Seminar on martingales. Aarhus, Matematisk Institut, 1959, 58p.

(3) Convergence of conditional expectation operators. Teor. Veroj. Prim., *9*, 1964 and Th. Proba. Appl., *9* 1964, 538–549.

[KrP] Krickeberg, K., C. Pauc, Martingales et dérivations. Bull. Soc. Math., *91*, 1963, 455–544.

[Kry] Krylov, N.V., On Ito's stochastic integral equations. Th. Proba. Appl., *14*, 1969, 330–336.

[KrR] Krylov, N.V., B.L. Rozovski, Equations d'Ito fortement paraboliques dans les espaces de Banach. To appear.

[Kun] Kunita, H.
(1) Stochastic integrals based on martingales taking values in Hilbert spaces. Nagoya Math. J., *38*, 1970, 41–52.
(2) On the decomposition of solutions of stochastic differential equations. In: Stochastic Integrals. Proc. of LMS Durham symposium. Lecture Notes in Math. *851*, Berlin-Heidelberg-New York

[KuW] Kunita, H., S. Watanabe, On square integrable martingales. Nagoya Math. J., *30*, 1967, 209–245.

[KuC] Kushner, H., D.S. Clark, Stochastic approximation for constrained and unconstrained systems. Appl. Math. Sc., Vol. 26, Berlin-Heidelberg-New York 1978.

[Kus] Kussmaul, A.V., Stochastic integration and generalized martingales. Pitman. London 1977.

[Lan] Landau, I.D. (ed.), Outils et modèles mathématiques pour l'automatique, l'analyse des systèmes et le traitement du signal. Ed. du CNRS, Paris, 1981.

[Len] Lenglart, E. Transformation des martingales locales par changement absolument continu de probabilité. Z.W., *39*, 1977, 65–70.

[LeM] Lepeltier, J.P., B. Marchal, Problèmes de martingales et equations différentielles stochastiques associées à un opérateur intégrodifférentiel. Ann. Inst. H. Poincaré (B), *12*, 1976, 43–103.

[Lep] Lépingle, D. La variation d'ordre p des semimartingales. Z.W., *36*, 1976, 295–316.

[LeM] Lépingle, D., J. Ménin, Sur l'intégrabilité uniforme des martingales exponentielles. Z.W., *42*, 1978, 175–204.

[LeO] Lépingle, D., J.Y. Ouvrard, Martingales browniennes hilbertiennes. C.R. Acad. Sci. Paris, *276*, 1225, 1973.

[Lev] Lévy, P.
(1) Théorie de l'addition des variables aléatoires. Paris 1937.
(2) Processus stochastique et mouvement brownien. Paris 1948.

[Lic] Licea, G., On supermartingales with partially ordered parameter set. Theory Proba. Appl., *14*, 1969, 135–137.

[LiS] Liptzer, R.S., A.N. Shiryaev.
(1) Statistics of stochastic processes. Berlin-Heidelberg-New York 1977.
(2) Sur l'absolue continuité des mesures correspondant aux processus de type diffusion, par rapport à la mesure brownienne. Izv. Akad. Nauk. SSSR. Ser. Math., *36*, 1972, 847–889.

[Mal] Malliavin, P. Stochastic calculus of variations and hypoelliptic operators. Proc. Intern. Symp. Stoch. Dif. Equations. Kyoto (1976), New York 1978, 195–263.

[MaS] Mallory, J., M. Sion. Limits of inverse systems of measures. Ann. Inst. Fourier, Grenoble, t.XXI, fasc. 1, 1971, 25–52.

[MSa] Mandrekar, V., H. Salehi. The square integrability of operator-valued functions with respect to a non-negative operator-valued measure and Kolmogorov-isomorphism theorem. Indiana Univ. Math. J., Vol. 20, 6, 1970.

[Mai] Maisonneuve, B., Quelques martingales remarquables associées à une martingale continue. Publ. Inst. Stat. Univ. Paris, *3*, 1968, 13–27.

[McK] McKean, H.P., Stochastic integrals. New York 1969.

[McS] McShane, E.J., Stochastic calculs and stochastic models. New York 1974.

[Mem] Mémin, J., Decompositions multiplicatives de semimartingales exponentielles et applications. Séminaire de Probabilités XII, Lecture Notes in Math. *649*, Berlin-Heidelberg-New York 1978.

[MSh] Mémin, J., A.N. Shiryaev, Un critère prévisible pour l'uniforme intégrabilité des semimartingales exponentielles. Séminaire de Proba. XIII, Lecture Notes in Math. *647*, Berlin-Heidelberg-New York.

[Met] Métivier, M.

(1) Limites projectives de mesures, martingales, applications. Ann. Math. Pura Appl., *63*, 1963, 225–352.

(2) Convergences de martingales à valeurs vectorielles. Bull. Soc. Math. Grèce, *5*, 1964, 54–74.

(3) Martingales à valeurs vectorielles: application à la dérivation des mesures vectorielles. Ann. Inst. Fourier, *17*, 1967, 175–208.

(4) Reelle und vektorwertige Quasimartingale und die Theorie der stochastichen Integration. Lecture Notes in Math. *607*, 1977.

(5) Intégrale stochastique par rapport à des processus à valeurs dans un espace de Banach réflexif. Teor. Veroj. Prim., *19*, 1974, 577–606.

(6) Notions fondamentales de la théorie des Probabilités. 4e éd., Paris 1982.

(7) Un lemme de Gronwall stochastique. C. R. Acad. Sci Paris, *289*, Sér. A, 1979, 287–290.

(8) Stability theorems for stochastic integral equations driven by random measures and semimartingales. J. Integr. Equations, *3*, 1981, 109–135.

(9) Stochastic integrals and vector-valued measures. In: Vector and operator valued measures and applications. New York, 1973.

(10) Stochastic integration with resprect to Hilbert Valued Martingales, Representation theorems and infinite dimensional filtering. in: Measure Theory Applications to Stochastic analysis, edited by G. Kallianpur and D. Kölzow. Lecture Notes in Math. 695, Berlin-Heidelberg-New York, 1978, 13–25.

[MeP] Métivier, M., J. Pellaumail.

(1) Stochastic intégration. New York, 1980.

(2) On Doléans-Föllmer's measure for quasimartingales. J. Math. *77*, 1975, 491–504.

(3) On a stopped Doob's inequality and general stochastic equations. Ann. Proba., *7*, 1979.

(4) Mesures stochastiques dans les espaces L^0. Z.W., *40*, 1977, 101–114.

[MPi] Métivier, M., G. Pistone.

(1) Sur une équation d'évolution stochastique. Bull. Soc. Math. Fr., *104*, 1976, 65–85.

(2) Une formule d'isométrie pour l'intégrale stochastique hilbertienne et équations d'évolution linéaires stochastiques. Z.W., *33*, 1975, 1–18.

[Mey] Meyer, P.A., (See also Dellacherie and Doleans-Dade).

(1) Probabilité et potentiel. Paris 1966.

(2) Une majoration du processus croissant naturel associé à une surmartingale. Sém. Proba. Strasbourg II, 166–170 (Lecture Notes *51* Berlin-Heidelberg-New York 1968).

(3) Un cours sur les intégrales stochastiques. Séminaire de Proba. X. Lecture Notes in Math. *511*, Berlin-Heidelberg-New York 1976.

(4) A decomposition theorem for supermartingales. Illionois J. Math., *6*, 1962, 193–205.

(5) Decomposition for supermartingales: the uniqueness theorem. Illionois J. Math., *7*,, 1963, 1–17.

(6) Sur l'inégalité de Doob de M. Métivier et J. Pellaumail. Sémin. Proba. XIII Lecture Notes in Math. *721*, Berlin-Heidelberg-New York 1979, 611–613.

(7) Intégrales stochastiques.I. Sémin. Proba. I, Lecture Notes in Math. *39*, Berlin-Heidelberg-New York 1967.
(8) Intégrales stochastiques-II. Sémin. Proba. I, Lecture Notes in Math. *39*, Berlin-Heidelberg-New York, 1967.
(9) Note sur les intégrales stochastiques. Intégrales hilbertiennes. Sémin. Proba. XI, Lecture Notes in Math. *581*, Berlin-Heidelberg-New York 1977
[MiL] Millar, P. W., Martingale integrals. Trans. Amer. Math. Soc., Vol. 133, *1*, 1968, 145–146.
[Nel] Nelson, E., Dynamical theory of Brownian motion. Math. Notes, Princeton Univ. Press, Pronceton, New Jersey, 1967.
[Nev] Neveu, J.
(1) Bases mathématiques des probabilités. Paris 1964, 2e éd. 1970.
(2) Martingales multivoques. Ann. Inst. H. Poincaré B-8, 1972, 1–7.
(3) Deux remarques sur la théorie des martingales. Z. W., *3*, 1964, 122–127.
(4) Relations entre la théorie des martingales et la théorie ergodique. Ann. Inst. Fourier, *15*, 1965, 31–42.
(5) Convergence presque sûre de martingales multivoques. Ann. Inst. H. Poincare.
(6) Discrete-parameter martingales. Amsterdam 1975.
(7) Intégrales stochastiques et applications. Cours de 3ème cycle. Univ. Paris VI, 1971–1972.
[Nis] Nisio, M., On the existence of solutions of stochastic differential equations. Osaka J. Math., *10*, 1973, 185–218.
[Ore] Orey, S.
(1) F-processes. Fifth Berkeley Symp., Vol. Ii, *1*, 1965, 301–314.
(2) Limit theorems for Markov chain transition probabilités. New York 1971.
(3) Conditions for the absolute continuity of two diffusions. Trans. Amer. Math. Soc., *193*, 1974, 413–426.
[Ouv] Ouvrard, J. Y.
(1) Représentation de martingales vectorielles de carré intégrable. Z. W., *33*, 1975, 195–208.
(2) Martingales locales et théorèmes de Girsanov dans les espaces de Hilbert réels séparables. Ann. Inst. H. Poincaré, B-9, *4*, 1973, 351–368.
[Par] Pardoux, E.
(1) Sur des équations aux dérivées partielles stochastiques monotones. C. R. cad. Sci. Paris, Ser. A, *275*, 1972, 101–103.
(2) Equations aux dérivées partielles stochastiques non linéaires monotones. Etude de solutions fortes de type Ito. Thèse d'Etat. Univ. Paris-Dud, Orsay, 1975.
[Pel] Pellaumail, J. (see also Métivier).
(1) Sur la décomposition de Doob-Meyer d'une quasimartingale. C. r. Acad. Sci Paris, Ser. A, *274*, 1972, 1563–1565.
(2) Sur l'intégrale stochastique et la décomposition de Doob-Meyer. Asterisque, *9*, S. M. F., 1973.
(3) Solutions faibles et semimartingales. Sém. Proba. XV, Lect. Notes *850*, p. 561.
[Pop] Pop-Stojanovic, Z. R.
(1) Decomposition of Banach-valued quasimartingales. Mat. Systems Theory, Vol. 5, *4*, 1971, 344–348.
(2) On MacShane's belated stochastic integral. SIAM J. Appl. Math., *22*, 87–92.
[Pri] Priouret, P., Processus de diffusion et équations différentielles stochastiques. Ecole d'été de probabilités de Saint-Fluour. Lecture Notes in Math. *390*, Berlin-Heidelberg-New York.
[Pro] Protter, Ph. E.

(1) On the existence, uniqueness, convergence and explosions of solutions of systems of stochastic integral equations. Ann. Prob., Vol. 5, *2*, 1977, 243–261.
(2) Right-continuous solutions of systems of stochastic integral equations. J. Multivariate Anal., *7*, 1977, 204–214.
(3) H^p-stability of solutions of stochastic differential equations. Preprint.
(4) A comparison of stochastic integrals. Ann. Prob., Vol 7, *2*, 1979, 276–289.
[Rao] Rao, K.M.
(1) On decomposition theorems of Meyer. Math. Scand., *24*, 1969, 66–78.
(2) Quasimartingales. Math. Scand., *24*, 1969, 79–92.
[Rau] Raoult, J.P., Sur une généralisation d'un théorème d'Ionescu-Tulcea. C.R. Acad. Sci. Paris, *259*, 1964, 2769–2777.
[RoM] Robbins, L., S. Monro, A stochastic approximation method. Ann. Math. Stat., *22*, 1951, 400–407.
[Roz] Rozanov., Processus aléatoires. Ed. MIR, Moscou 1975.
[Rud] Rùdin, W., Functional analysis. New York 1973.
[Sca] Scalora, F., Abstract martingale convergence theorem. Pacific J. Math., Vol. 11, *1*, 1961, 347–374.
[Sch] Schaefer, A. H., Topological vector spaces. Berlin-Heidelberg-New York 1971.
[Schm] Schmetterer, L., Multidimensional stochastic approximation. Multivariate analysis. Proc. Second Intern. Symp. Dayton, Ohio, 1968. P.R. Krishnaian ed., London-New York 1969.
[Schw] Schwartz, L.
(1) Les semimartingales formelles. Séminaire Proba. XV, Lecture Notes in Math. *850*, Berlin-Heidelberg-New York 1981, 412–487.
(2) Semimartingales sur des variétes et martingales conformes sur des variétés analytiques complexes. lecture Notes in Math. *780*, Berlin-Heidelberg-New York 1980
[Sko] Skorokhod, A.V., (see also Gikhman) Studies in the theory of random processes. Reading, Mass., 1965.
[Str] Stricker, C.
(1) Mesure de Föllmer en théorie de quasimartingales. Sém. Proba. IX, Lecture Notes in Math. *465*, Berlin-Heidelberg-New York, 408–419.
(2) Une caractérisation des quasimartingales. Sém. Proba. IX, Lecture Notes in Math. *465*, Berlin-Heidelberg-New York 420–429.
(3) Quasimartingales, semimartingales et filtrations naturelles. Z.W., *39*, 1977, 55–65.
[StV] Strook, K., S. Varadhan.
(1) Diffusion processes with continuous coefficients-I. Comm. Pure Appl. Math., *22*, 1969, 345–400.
(2) Multidimensional diffusion processes. Berlin-Heidelberg-New York 1959.
(3) Diffusion processes with boundary condition. Comm. Pure Appl. Math., *24*, 1971, 147–225.
[Tre] Trèves, F., Topological vector spaces. Distributions and Kernels. New York 1967.
[Van] Van Cutsem, B., Martingales de multiapplications à valeurs convexes compactes. C.R. Acad. Sci., 1969, 429–431.
[VSW] Van Shuppen, J.H., N.E. Wong, Transformation of local martingales under a a change of law. Ann. Proba.,*2*, 1974, 879–888.
[Var] Varadhan, S., Lectures on diffusion problems and partial differential equations. Lecture Notes. Tata Inst. Bombay, 1980.
[Vil] Ville, J., Etude critique de la notion de collectif. Thèse Paris, 1939.
[Vio] Viot, M.
(1) Solution en loi d'une équation aux dérivées partielles stochastiques non linéaire:

méthode de compacité. C.R. Acad. Sci. Paris, Ser.A, *278*, 1974, 1185–1188.
(2) Equations aux dérivées partielles stochastiques: formulation faibles. Thèse d'Etat, Univ. Paris VI, 1976.
[Wil] Williams, D., Diffusions-Markov processes and martingales. New York 1979.
[Wie] Wiener, N., Differential space. J. Math. Phys. Math. Inst. Tech., *2*, 1923, 131–174.
[WoZ] Wong, E., M. Zakai, (see also van Shuppen).
(1) On the convergence of ordinary integrals to stochastic integrals. Ann. Math. Statist., *36*, 1965.
(2) Martingales and stochastic integrls for processes with a multidimensional parameter. Z.W., *29*, 1974, 109–122.
[Wol] Wolfowitz, J., On the stochastic approximation method of Robbins and Monroe. Ann. Math. Stat., *23*, 1952, 457–461.
[Yam] Yamada, T., Sur une construction des solutions d'équations différentielles stochastiques dans le cas non lipschitzien. Sém. Proba. XII, Lecture Notes in Math. *649*, Berlin-Heidelberg-New York, 1978.
[Yao] Yamada, T., Y. Ogura, On the strong comparison theorems for solutions of stochastic differential equations. Z.W., *56*, 1981, 3–19.
[YaW] Yamada, T., S. Watanabe, On the uniqueness of solutions of stochastic differential equations. J. Math. Kyoto Univ. Vol. 2, *1*, 1971.
[Yor] Yor, M.
(1) Sur les intégrales stochastiques à valeurs dans un Banach. C.R. Acad. Sci, Paris, Ser. A, *277*, 1973, 467–469.
(2) Sous-espaces denses dans L ou H et représentation des martingales. Sém. Proba. XII, Lecture Notes in Math. *649*, Berlin-Heidelberg-New York 1978.
(3) Grossissement d'une filtration et semimartingales: théorèmes généraux. Sém. Proba. XII, Lectures Notes in Math. *649*, Berlin-Heidelberg-New York 1978, 61–69.
[ZvK] Zvonkin, A.K., N.V. Krylov, On strong solutions of stochastic differential equations. School Seminar Vilnius, Ac. Sci. Lit. SSSR, part. II, 1975, 9–88.

Index of notation

General mathematical notations

$\mathbb{R}$	set of real numbers
$\mathbb{R}_+$	set of positive or null real numbers
$\bar{\mathbb{R}}_+$	$\mathbb{R}_+ \cup \{+\infty\}$
$\mathbb{N}$	set of integers
$\bar{\mathbb{N}}$	$\mathbb{N} \cup \{+\infty\}$
$\mathbb{Q}$	set of rational numbers
x^+ (resp. x^-)	$\sup(x, 0)$ (resp. $\sup(-x, 0)$)
ε_x	Dirac mass at point x
$x \wedge y$, $x \vee y$	infimum and supremum of two real numbers x and y
$\mathscr{L}^0_{\mathbb{H}}(\Omega, \mathscr{F}, P)$	set of $\mathbb{H}$-valued $\mathscr{F}$-measurable functions on Ω, with the convergence in probability
$L^0_{\mathbb{H}}(\Omega, \mathscr{F}, P)$	set of P-equivalence classes in $\mathscr{L}^0_{\mathbb{H}}(\Omega, \mathscr{F}, P)$
$\mathscr{L}^p_{\mathbb{H}}(\Omega, \mathscr{F}, P)$	vector space of p-integrable $\mathbb{H}$-valued $\mathscr{F}$-measurable functions with the semi-norm $\\|f\\|_p = [\int \\|f\\|^p dP]^{1/p}$ $(p \geqslant 1)$
$L^p_{\mathbb{H}}(\Omega, \mathscr{F}, P)$ $(p \geqslant 1)$	Banach space of P-equivalence classes in $\mathscr{L}^p_{\mathbb{H}}(\Omega, \mathscr{F}, P)$
$\mathfrak{P}(\Omega)$	set of subsets of Ω
1_A	indicator function of the set A
$\complement A$ or A^c	complementary set of A
$A \backslash B$	difference of two sets A and B
$\mathscr{B}_t$ (resp. $\mathscr{B}_T$)	σ-algebra of Borel subsets of $[0, t]$ (resp. T)
I_d	$d \times d$ identity matrix

Notations for vector spaces

All vector spaces considered are real vector spaces. When $\mathbb{B}$ is a Banach space, its dual (the set of continuous linear forms on $\mathbb{B}$) with the uniform norm topology is denoted by $\mathbb{B}'$. The norm in $\mathbb{B}$ is written $\|.\|_{\mathbb{B}}$ or simply $\|.\|$ when no confusion is possible. The duality bilinear form between $\mathbb{B}$ and $\mathbb{B}'$ is denoted by $\langle x, x' \rangle$.

For a Hilbert space $\mathbb{H}$ and two elements h, h' of $\mathbb{H}$, we usually write $h \cdot h'$ for the scalar product of h and h'. When necessary, in order to avoid confusion, we write this as $\langle h, h' \rangle_{\mathbb{H}}$ or $(h|h')_{\mathbb{H}}$.

$\mathscr{L}(\mathbb{B}; \mathbb{K})$ is the set of bounded linear operators from $\mathbb{B}$ into $\mathbb{K}$

$\mathscr{L}_1(\mathbb{B}; \mathbb{K})$ is the set of nuclear operators from $\mathbb{B}$ into $\mathbb{K}$

$\mathscr{L}_2(\mathbb{B}; \mathbb{G})$ is the set of Hilbert-Schmidt operators from the Hilbert space $\mathbb{H}$ into the Hilbert space $\mathbb{G}$.

The norm in $\mathscr{L}_1(\mathbb{B}; \mathbb{K})$ is written $\|.\|_1$ and in $\mathscr{L}_2(\mathbb{H}; \mathbb{G})$, $\|.\|_2$.

The definitions of $\mathscr{L}_1(\mathbb{B}; \mathbb{K})$ and $\mathscr{L}_2(\mathbb{B}; \mathbb{K})$ are given in the text, p. 139–140

We also redefine the tensor product spaces: $\mathbb{H} \hat{\otimes}_2 \mathbb{H}$ and $\mathbb{B} \hat{\otimes}_1 \mathbb{B}$ p. 137

$\mathfrak{M}^p(E)$ see p. 100

Notations on filtrations and sets

Notations on processes

Spaces of processes

*: theorems and sections marked with a * deal only with infinite dimensional processes and can be omitted by readers interested only in processes with a finite dimensional state-space.

Index

Journal für die reine und angewandte Mathematik

Multilingual Journal

founded in 1826 by

August Leopold Crelle

continued by

C. W. Borchardt, K. Weierstrass, L. Kronecker, L. Fuchs,
K. Hensel, L. Schlesinger, H. Hasse, H. Rohrbach

at present edited by

Otto Forster · Willi Jäger · Martin Kneser
Horst Leptin · Samuel J. Patterson · Peter Roquette

with the cooperation of

M. Deuring, P. R. Halmos, O. Haupt,
F. Hirzebruch, G. Köthe, K. Krickeberg, K. Prachar,
H. Reichardt, L. Schmetterer, B. Volkmann

W DE G

Walter de Gruyter · Berlin · New York 1982

Frequency of publication: yearly approx. 8 volumes
(1983: volume 338–345)
Price per volume: DM 154,–; $70.00

Back volumes: volume 1–250 bound complete DM 35 000,–; $15,910.00
Single volume each DM 168,–; $76.00
